Methods in Molecular Biology

Series Editor
John M. Walker
School of Life and Medical Sciences
University of Hertfordshire
Hatfield, Hertfordshire, UK

For further volumes:
http://www.springer.com/series/7651

For over 35 years, biological scientists have come to rely on the research protocols and methodologies in the critically acclaimed *Methods in Molecular Biology* series. The series was the first to introduce the step-by-step protocols approach that has become the standard in all biomedical protocol publishing. Each protocol is provided in readily-reproducible step-by step fashion, opening with an introductory overview, a list of the materials and reagents needed to complete the experiment, and followed by a detailed procedure that is supported with a helpful notes section offering tips and tricks of the trade as well as troubleshooting advice. These hallmark features were introduced by series editor Dr. John Walker and constitute the key ingredient in each and every volume of the *Methods in Molecular Biology* series. Tested and trusted, comprehensive and reliable, all protocols from the series are indexed in PubMed.

Plant-Nematode Interactions

Edited by

Sergio Molinari

IPSP, Consiglio Nazionale delle Ricerche CNR, Bari, Italy

Humana Press

Editor
Sergio Molinari
IPSP
Consiglio Nazionale delle Ricerche CNR
Bari, Italy

ISSN 1064-3745 ISSN 1940-6029 (electronic)
Methods in Molecular Biology
ISBN 978-1-0716-3637-4 ISBN 978-1-0716-3638-1 (eBook)
https://doi.org/10.1007/978-1-0716-3638-1

This Humana imprint is published by the registered company Springer Science+Business Media, LLC, part of Springer Nature.
The registered company address is: 1 New York Plaza, New York, NY 10004, U.S.A.

Paper in this product is recyclable.

Preface

The selection of chapters for this book has been done to allow researchers from all over the world to know all the techniques, from the classical to the most up-to-date ones, useful to work with plant parasitic nematodes (PPNs) and study their interactions with crop plants. PPNs are microscopic soil-borne animals that attack, parasitize, and damage almost all crops. However, the type of interaction with roots is extremely variable due to the very high numbers of families belonging to this parasitic group. Nematode families have different morphology, lifestyles, parasitism, and host ranges. However, the mouth apparatus of all known PPNs shows a peculiar common characteristic, i.e., the presence of a protrusible stylet by which they pierce and suck the nutrients from root cells. All the invasive forms of nematodes are vermiform; however, nematode families differ in their degree of penetration of the vermiform body inside the root, and can be distinguished into ecto and endoparasites. Nematodes that enter entirely and migrate into the roots are called endoparasites, whilst ectoparasites insert only their stylet inside the root from the soil rhizosphere, where they live and reproduce. If endoparasites, after egg hatching, maintain their vermiform shape and the ability to move inside or outside the roots, they are called migratory; conversely, if the invasive vermiform individuals establish fixed feeding sites and develop into immobile stages, they are called sedentary. All ectoparasites belong to the Order Dorylaimida (*Xiphinema, Longidorus, Trichodorus,* and *Paratrichodorus* spp.). They are not so much important as pests for the damages due to their own parasitism as for their ability to transmit dangerous plant viruses. On the contrary, the Order Tylenchida contains both migratory and sedentary endoparasites. Most genera of migratory endoparasites (*Pratylenchus, Radophulus, Anguina,* and *Ditylenchus*) are economically important because their movement and feeding inside the roots provoke tissue necroses, root system degeneration, and crop yield losses. The most intimate interaction between these parasites and roots occurs in the parasitism of sedentary endoparasites belonging to the family Heteroderidae that includes the most diffused and studied *Globodera*, *Heterodera*, and *Meloidogyne* spp. The peculiarity of the parasitism of this family is the formation of permanent and complex feeding sites through which digestive ferments are injected into the feeding site by invasive juveniles (J2s) and digested nutrients continuously sucked from roots; motile J2s become sedentary, have 2 molts (J3-J4), develop to egg-laying females, and reproduce. Nematodes inject several effectors into root cells that markedly alter gene expression leading to strong modifications of the pierced cells that are transformed into nutrient transfer cells. J2s belonging to *Globodera* and *Heterodera* spp. induce the formation of syncitia, in which few cells merge by dissolving their cell walls; adult females protrude from the roots with the majority of their body, lay their eggs inside the body, and protect the eggs, at the end of their life-cycle, by hardening their external cuticle that changes from a whitish or yellowish to a brown color. These hardened dead females containing eggs are called cysts which are dispersed in the soil after harvest and represent the tool for spreading the infestation. Therefore, *Globodera* and *Heterodera* spp. are commonly called cyst-nematodes (CNs). *Meloidogyne* spp. induce the formation of few discrete giant or nurse cells by blocking their cell division; nematode secretions induce hypertrophy and hyperplasia of the surrounding tissues, thus leading to the formation of tumor-like galls or knots on roots. Female bodies are completely immersed in these galls; however, they lay eggs in an external gelatinous matrix (egg mass), which is clearly visible outside the roots. For these characteristics, these nematodes are called root-knot nematodes (RKNs).

The first task the researcher has to face is to collect and isolate these parasites from their environment, may it be roots or soil, for identification, characterization, and quantification of infestation levels. Different life stages may be required for this task: eggs, egg-containing formations, motile juveniles, motile or sedentary adults, etc. Chapter 1 first describes how to obtain samples from soil and from different plant tissues. Specific sections focus on nematode extraction procedures of vermiform stages or cysts from soil, extraction of nematodes from roots and seeds. As the most preferred stage for artificial inoculations is the invasive motile vermiform stage, one section is dedicated to the description of the Materials and Methods for obtaining J2s from cysts and egg masses.

Collecting nematodes from fields is not easy; it requires expertise, labor, and a lot of time, and is impossible in certain seasons. Therefore, most research centers in nematology must be equipped with furniture useful to artificially rear nematodes after their collection from host plants grown under field conditions. Chapter 2 describes the procedures for rearing nematode in greenhouse, procedures that can be very different according to the required family of nematodes. Moreover, this chapter describes the techniques used to axenically rear nematodes in vitro. Once populations of nematodes are produced, it is necessary to keep them alive for future inoculations or other uses. One section of the chapter is dedicated to the different ways of conservation, which depend on the nematode life stages chosen for conservation.

The responses of plants to nematode attack are extremely variable, although they can be broadly divided into responses to sedentary endoparasitic nematodes (Chap. 3) and responses to migratory and semi-endoparasitic nematodes (Chap. 4), having the last ones an intermediate parasitism between sedentary and migratory nematodes. Within a determined plant crop species, some cultivars may show resistance to nematode infection reducing population density for next crops, others may be tolerant in that they can bear the infection without severe damages and yield loss, others may be very good hosts, being more or less damaged by the attack, and multiplying the field population for the next crops. Usually, resistance traits are found in wild varieties of a determined crop species that cannot be crossed with edible or otherwise usable varieties. The most studied and used resistance gene, the *Mi-1* gene, was introgressed from the wild relative of tomato, *Solanum peruvianum,* into the cultivated tomato, *Solanum lycopersicon*, in the 1940s, using embryo rescue to obtain a hybrid of these normally incompatible species. *Mi-1*-carrying tomato plants are resistant to the attack of the so-called tropical species of *Meloidogyne* (*M. incognita, M. javanica, M. arenaria*). All present fresh-market and processing tomato originate from the progeny of this single F1 plant and following extensive breeding. Most times, resistant traits are searched in germplasm collections of hundreds of varieties of the chosen crop species. Plants are defined as resistant when they provoke a reduced reproduction level of the challenging nematodes. Nematode reproduction rates are preferred in assessments of resistance because of the complexity to quantitatively measure specific symptom development and damage levels compared with the relative time sparing and accuracy of reproduction measurements. Chapter 3 describes the techniques to carry out screenings for resistance to RKNs and CNs, as well as how to assess that a plant is tolerant to nematodes. Moreover, resistant cultivars may be attacked and infested by resistance-breaking nematode populations, called virulent. In this chapter, methods are described to assess the virulence of determined nematode populations and recognition of different nematode virulence groups. In Chap. 4, methods of screening for resistance are listed concerning many families of migratory and semi-endoparasitic nematodes, such as *Pratylenchus, Radopholus, Nacobbus, Ditylenchus, Tylenchulus* spp. and others.

Although identification of different families is rather straightforward, discrimination of species within one genus has always been a difficult time-wasting but necessary task, because plant resistance can be strictly specific to determined species or even pathotypes within one species. Classical methods using microscopy and morphometric measurements have been used until recently; however, results obtained by such methods were subjected to long discussions among the different laboratories, especially when groups supposedly containing tens of species were examined. Molecular techniques that exploit phenotypes/genotypes consisting in electrophoresis profiles of total proteins or isozymes, or of DNA fragments obtained by DNA recombinant techniques, have increasingly and successfully been used to identify nematode species. However, these innovative techniques have generally been applied to recognition of the most economically important nematode families, such as RKNs and CNs. Chapter 5 describes all the most up-to-date molecular methodologies for the identification of RKN species. If the readers are interested in molecular identification of *Heterodera* cysts, *Xiphinema* spp., and *Globodera* spp. and pathotypes, they can refer to lots of papers of mine in google scholar on the topic.

Nematodes in the soil can also be used to determine soil biodiversity along with soil micro-biome. In Chap. 6, the Denaturing Gradient Gel Electrophoresis (DGGE) separation technique is described in which DNA from the whole soil community is extracted and 18 S rDNA amplified by PCR. The amplification products are electrophoretically separated and the profiles analyzed by bioinformatics. Higher the diversity higher the health of a determined ecological niche. Nematodes are associated with soil micro-biome also in terms of microbial specific attachments to the surface of phytonematodes. Chapter 7 describes the association of classical and microbiological with advanced molecular techniques for the study of microbial attachments to the cuticle of nematodes in soil. Such a study is important to determine the effects of nematode-attached microbes on plant defense responses. Root exudates can actively alter the composition of micro-biome in the rhizosphere, affect egg hatching of nematodes, and attract or repel nematode invasive forms moving in the soil. The effects of root exudates can be detected in vitro in laboratory to evaluate the comparative attractiveness of different plants or cultivars of the same plant species. Chapter 8 describes some examples of different types of these in vitro assays and other assays to reveal the effect of root exudates on the hatching rate of nematodes.

Nematode infections have always been very difficult to control despite the fact that plants have innate immunity to react to biotic challenges. The existence of a plant immune system has convincingly been acknowledged by the scientific community not much time ago, thanks to a paper published on Nature in 2006.[1] Plant protection strategies against nematodes thus far have mainly been based on disruption and eradication of the parasite populations in the soils by volatile and non-volatile organic nematicides. These strategies ended up destroying the soils and not the nematodes. However, most stakeholders involved in nematode management still think that the only way to control nematode invasions is to destroy parasite population in soil with more-or-less toxic compounds, ignoring that a plant

[1] Jones, J.D.G. & Dangl, J.L. (2006). The plant immune system. *Nature* 444, 323–329. DOI: https://doi.org/10.1038/nature05286

immune system exists and can be activated and strengthened. The fashion, strongly supported by pesticide companies and only recently limited by EU agriculture policies, to spread in agricultural soils poisons to kill nematodes has led to the degradation of vast cropping areas. Regeneration of soils goes through a re-thinking of the strategies used to limit nematode infection to plants. The market of bio-stimulants is consistently enlarging in agricultural good practices; however, the use of activators of plant immune system is revealing to be very promising as an environmentally friendly method to control nematode attacks. The most effective activators which can be commercially available to farmers are antioxidant compounds and formulations containing biological control agents (BCAs), i.e., arbuscular mycorrhizal fungi (AMF), opportunistic fungi (e.g., *Trichoderma* spp.), and plant growth promoting rhizobacteria (PGPR). Chapter 9 indicates the methods to provide plants with salicylates and four commercial BCA-containing formulations as an effective strategy of control for sedentary endoparasitic nematodes. The success of these methods is subjected to many variables that have to be screened, such as the age/weight of the treated plants in relationship with the dosage of the provided activator, the application method, the specific plant-nematode interaction, the texture of soil, etc. In the chapter, data of the suitable dosages to be applied to vegetable crops for a consistent reduction of RKN/CN infections are also given expressed per gram of plants at treatment and are the results of many years of experimentation. In these control methods, dosages with respect to plant age are extremely important as low dosages may increase infection factors and dosages in excess may be toxic to plants. Activation of plant immune system triggers at the first attempts of nematodes to build their feeding sites a series of metabolic events that promote the generation of reactive oxygen species (ROS), such as superoxides and hydrogen peroxide. When this oxidative burst is not limited and become persistent, a so-called hypersensitive reaction (HR) occurs in roots that can cause oxidative degeneration of cells, cell death, and necrosis of the tissue surrounding the head of the invasive juveniles. This type of response is characteristic of plants carrying dominant resistant genes when attacked by avirulent populations of sedentary endoparasitic nematodes; moreover, a similar response is observed in plants immunized (primed) by pre-treatments with defense activators. Chapter 10 describes methods for ROS detection and hydrogen peroxide quantification. Chapter 11 describes methods for detection of Pathogenesis Related (*PR*) genes that are the executioners of immunity in plant-RKN interactions. Provision of salicylic acid and structural homologues at suitable concentrations to tomato plants, by soil-drench or foliar spray, induces the so-called Systemic Acquired Resistance (SAR) by which plants limit the spread of infection by RKNs. Some *PR*-genes (*PR-1, PR-5*) are markers for SAR, and detection of an over-expression of these genes indicates that SAR is activated and plants are immunized against RKNs.

Epigenetic changes are the focus of recent investigations that have established their importance in determining the outcome of nematode infection to plants. Repression of plant defenses in compatible plant-nematode interactions is most likely caused by gene silencing due to DNA methylation. On the contrary, DNA de-methylation seems to be associated with the enhanced expression of defense genes in immunity against nematodes. Chapter 12 deals with epigenetic analyses of plant response against sedentary endoparasitic nematodes. Analyses to detect miRNA regulation, DNA methylation, and histone modifications are thoroughly described.

Cultivation of resistant plants, other practices of crop cultivation, or the use of toxic compounds induce modifications of nematode communities in the soil which undergo to a selection for surviving in a hostile environment. All these challenges subject nematodes to stressful states which can be monitored by some techniques described in Chap. 13; in this chapter, detection of nematode stress state is carried out by identification and quantification of Heat Shock Proteins.

Lastly, two genera of nematodes, such as *Heterorhabditis* and *Steinernema*, are used as biocontrol agents against insects and called entomopathogenic nematodes (EPNs). Chapter 14 describes the laboratory procedures and tests on these nematodes, such as their isolation, count, culture, identification, pathogenicity, virulence, and environmental tolerance.

This book has the peculiar characteristic to make available the knowledge of most of the techniques and methodologies for working with plant parasitic nematodes, from the classical to the most recent and innovative ones. The expertise in both types of techniques is required by a research center or diagnostic laboratory that wants to work with nematodes. In the past, the knowledge, expertise, specialized personnel, and equipments for extracting nematodes from soil or plant samples to identify and characterize them were present in some specific research centers, such as the Institute of Italian CNR where I have been spending all my scientific activity, an Institute that had the name of Institute of Agricultural Nematology until about 2000. Unfortunately, such an expertise is rapidly disappearing in modern laboratories; that is why, in the first part of the book, I collected the experiences of some colleagues that are now retired to ensure the description of these methods and the required materials for the use of present and future young researchers from all over the world who will be keen on nematological studies. My hope is that this book can spur more and more students and young researchers toward this topic that allows them to get insight into one of the most intimate relationships between living organisms, that deeply and markedly transforms plant root in its histology, biochemistry, metabolism, and genetics by nematode action.

Bari, Italy ***Sergio Molinari***

Contents

Contributors

PAMELA ABBRUSCATO • *Nuova Genetica Italiana, Villa Guardia, CO, Italy*
MAHFOUZ M. M. ABD-ELGAWAD • *Department of Plant Pathology, Nematology Laboratory, National Research Centre, Giza, Egypt*
SANDRA BREDENBRUCH • *Epidemiology and Pathogen Diagnostics, Julius Kuehn Institute, Braunschweig, Germany*
REGINA M. D. G. CARNEIRO • *Embrapa Recursos Genéticos e Biotecnologia, PqEB Parque de Estação Biológica, Brasília, DF, Brazil*
MONICA CELI • *Dept STEBICEF, University of Palermo, Palermo, Italy*
VALDIR R. CORREIA • *Instituto Federal de Educação, Ciência e Tecnologia do Tocantins, Dianópolis, TO, Brazil*
RENATO CROZZOLI • *Institute of Agriculture Zoology, Central University of Venezuela (UCV), Maracay, Venezuela; Azzano Decimo, Italy*
KAREN DE KOCK • *Department of Biotechnology, Faculty of Bioscience Engineering, Ghent University, Ghent, Belgium*
AHMED ELHADY • *Laboratory of Insect and Nematode Management, Corteva Agriscience Research, Center Eschbach, Eschbach, Germany*
MARGHERITA FESTA • *CNR - Istituto di Biofisica, Genova, Italy*
NICOLA GRECO • *Institute for Sustainable Plant Protection, National Research Council of Italy (CNR), Bari, Italy*
TRACY HAWK • *Department of Plant Sciences, University of Tennessee, Knoxville, TN, USA*
HOLGER HEUER • *Laboratory of Insect and Nematode Management, Corteva Agriscience Research, Center Eschbach, Eschbach, Germany*
TAREK HEWEZI • *Department of Plant Sciences, University of Tennessee, Knoxville, TN, USA*
RENATO N. INSERRA • *Florida Department of Agriculture and Consumer Services, Division of Plant Industry, Gainesville, FL, USA*
TINA KYNDT • *Department of Biotechnology, Faculty of Bioscience Engineering, Ghent University, Ghent, Belgium*
PAOLA LEONETTI • *Institute of Sustainable Plant Protection, Research Unit of Bari, IPSP-CNR, Bari, Italy*
PEITONG LI • *Department of Plant Sciences, University of Tennessee, Knoxville, TN, USA*
BARBARA MANACHINI • *Department of Agricultural, Food and Forest Sciences, University of Palermo, Palermo, Italy; Dept SAAF, University of Palermo, Palermo, Italy*
JASPER MATTHYS • *Department of Biotechnology, Faculty of Bioscience Engineering, Ghent University, Ghent, Belgium*
VANESSA S. MATTOS • *Embrapa Recursos Genéticos e Biotecnologia, PqEB Parque de Estação Biológica, Brasília, DF, Brazil*
SERGIO MOLINARI • *Institute for Sustainable Plant Protection (IPSP), National Research Council of Italy (CNR), Bari, Italy*
SELIN OZDEMIR • *Department of Plant Sciences, University of Tennessee, Knoxville, TN, USA*
VINCE PANTALONE • *Department of Plant Sciences, University of Tennessee, Knoxville, TN, USA*
ROLAND N. PERRY • *Department of Biological and Environmental Sciences, School of Life and Medical Sciences, University of Hertfordshire, Hatfield, Hertfordshire, UK*

ZORAN S. RADAKOVIĆ • *Laboratory of Nematology, Wageningen University and Research, Wageningen, The Netherlands*
CAIO F. B. SOUZA • *Embrapa Recursos Genéticos e Biotecnologia, PqEB Parque de Estação Biológica, Brasília, DF, Brazil*
MEG STATON • *Department of Entomology and Plant Pathology, University of Tennessee, Knoxville, TN, USA*
OLIVERA TOPALOVIĆ • *Department of Biology, Section of Terrestrial Ecology, University of Copenhagen, Copenhagen, Denmark; Rheinische Friedrich-Wilhelms-University of Bonn, INRES—Molecular Phytomedicine, Bonn, Germany*
ALBERTO TROCCOLI • *Institute for Sustainable Plant Protection, National Research Council of Italy (CNR), Bari, Italy*
MIRELLA VAZZANA • *Dept STEBICEF, University of Palermo, Palermo, Italy*
SOBHAN BAHRAMI ZADEGAN • *Department of Plant Sciences, University of Tennessee, Knoxville, TN, USA*

Chapter 1

Sampling and Extraction of Plant Parasitic Nematodes

Nicola Greco and Renato Crozzoli

Abstract

The study of nematodes requires availability of nematode specimens and their population densities in plants and soil. This can be achieved using adequate sampling schemes and extraction methods. In this chapter, the most common and suitable sampling and extraction procedures and equipment are described. These include the use of Baermann's funnels, Cobb's decanting and sieving, floating methods such as the Oostenbrink method and Fenwick can, elutriators such as Seinhorst methods, centrifugation methods including that of Coolen, and mechanical and enzymatic maceration. The combination of different methods for cleaning the nematode suspensions is described, such as Cobb's sieving with Baermann's funnels or centrifugation, and for cysts combining Seinhorst's elutriator or Fenwick can with the alcohol methods. Methods for extraction of eggs and/or juveniles of cyst and egg mass forming nematodes, to be used as inoculum or to ascertain egg viability, are also described. Only little information is also noted on the use of molecular tools to identify and quantify nematode populations in soil and roots.

Key words Elutriators, Extraction methods, Floating apparatus, Mechanical and enzymatic maceration, Nematode extraction, Nematode sampling

1 Introduction

Studies on plant parasitic nematodes, as a first step, require the availability of a certain number of specimens of the target nematode. This need can be achieved first by identifying the environment in which the target nematode lives (soil, plant parts), followed by extraction and possibly its reproduction and conservation for further investigation and availability during seasons or years when nematodes under natural conditions are in insufficient numbers for many studies. As most plant parasitic nematodes live or survive in the soil, the first plant nematologists to design methods to extract them from this environment, early in the last century, were Baermann [1] and Cobb [2]. The method described by Cobb, known as decanting and sieving, basically consisted of mixing a soil sample in

Nicola Greco and Renato Crozzoli authors are now retired.

Sergio Molinari (ed.), *Plant-Nematode Interactions*, Methods in Molecular Biology, vol. 2756,
https://doi.org/10.1007/978-1-0716-3638-1_1,

water, leaving the mixture to decant, with heavier soil particles decanting earlier, and then sieving the uppermost water–soil suspension through a series of sieves of different apertures to retain nematodes of different sizes. Generally, the water nematode suspension obtained is not clear because it contains small soil particles. The method of Baermann [1] to extract nematodes directly from the soil consists of a sieve placed on a funnel in which water is added to reach the bottom of the sieve; the nematodes move from the soil to the water and are collected periodically. With this method, usually the water nematode suspension is rather clear and easy to observe under a stereoscope at a magnification of 25–40×. Further progress was made by Fenwick [3] who developed a floating apparatus, known as the Fenwick can, to extract cysts of *Heterodera* and *Globodera* spp. from dried soil. Later, to extract nematodes from soil, elutriators were designed by Oostenbrink [4] and Seinhorst [5]. In 1954, Young [6] found that incubating wetted roots or other plant parts in a small (0.5 L) loosely closed container was useful for extracting endoparasitic migratory nematodes. Similar results can be achieved with the mistifier chamber of Seinhorst [7]. Another important method is the centrifugation method of Jenkins [8], useful for extracting most kinds of nematodes (motile and nonmotile) from soil and plant parts. In many laboratories, centrifugation is also used to clean nematode suspensions obtained by Cobb's sieving and decanting methods and different elutriation methods. The centrifugation method has the advantage of obtaining a clean nematode suspension after just a few minutes and is not affected by environmental conditions during the extraction procedure.

Hereafter, we report on the most common and useful nematode sampling and extraction methods; for more extensive information, the reader is referred to specialized publications [9, 10]. Some equipment, such as that necessary for centrifugation, can be rather expensive. Therefore, especially when handling small numbers of samples, it may be convenient to adapt your work to the equipment already available to you. If, instead, thousands of samples are handled, then implementing a well-equipped nematology laboratory is recommended.

2 Materials

2.1 Soil Sampling

Required equipment are soil corers such as those in Fig. 1. Two of them are made of a stainless steel tube (half to 1 inch or 3/8 diameter, about 1 m long) adapted by an artisan, or a similar one bought in the market. A steel rod to remove soil from the soil corer is necessary especially in wet and clayey soil; moreover, a 4–5 L pail, plastic bags large enough to hold 2–3 kg soil, labels, a notebook,

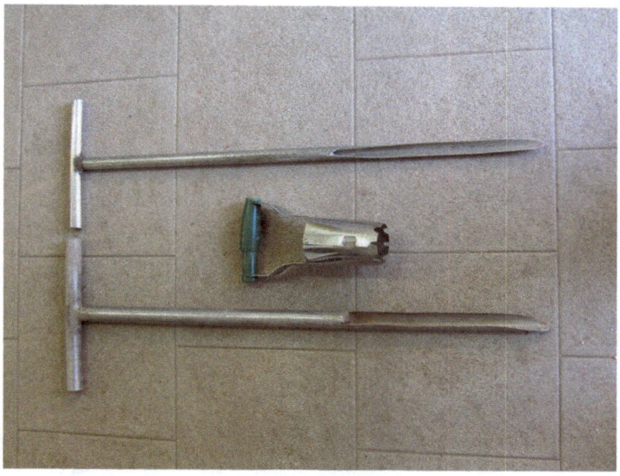

Fig. 1 Example of different soil cores for collecting soil samples, with kind permission of SIN

and, if possible, an insulated box (ice chest) in warm weather to transport samples to the laboratory.

2.2 Plant Sampling of Aerial Parts (Annual Plants)

For it scissors, knives, plastic bags of about 3–4 L capacity, cool box necessary for transporting samples to laboratory during warm period, and bowls are necessary.

2.3 Plant Sampling of Belowground Parts

The minimum equipment should consist of a hoe or a rather large soil auger, shovel, scissors, knives, plastic sacks, labels, cool box necessary especially during warm weather, bowls, and scale.

2.4 Plant Sampling from Trunks of Host Plants (Bursaphelenchus xylophilus and B. cocophilus)

What you need for sampling is a portable driller, better if powered, pitsaw or belt saw, a devise to eliminate bark, and other materials used for sampling in general.

2.5 Nematode Extraction of Vermiform Stages from Soil

Have available the Oostenbrink elutriator (Fig. 2); four sieves of 30 cm diam. and 45 μm pore size; plastic bowls with a capacity of 4 L; decanting tray with cross piece; extraction sieve (16 cm diam) with supporting wires; clamping ring for securing the nematode filters. Watch glass (6 cm diam); two nematode filters (or equivalent); extraction dishes; if necessary, covering material to cover the extraction dishes; a bench free of vibrations on which to keep the tray (at constant, non-extreme temperature; about 20 °C); 100 mL capacity beakers; clean water.

2.5.1 Nematode Extraction According to Baermann's Funnels

The most important materials you need are several common plastic (or glass) funnels (Fig. 3) of 10 cm diam. having the stem inserted in a rubber tube of about 10 cm long; a clip to close the rubber tube

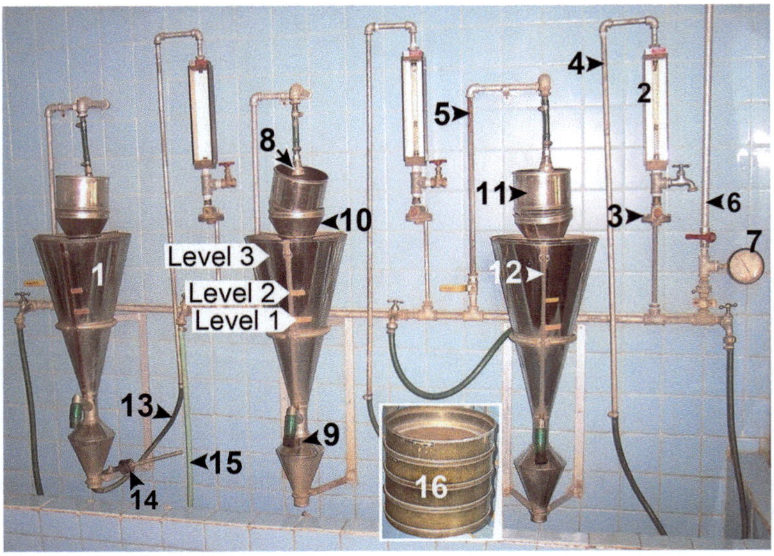

Fig. 2 Description of Oostenbrink's apparatus installed at Zoology Institute, Central University of Venezuela, Maracay. 1, Elutriator; 2. Flowmeter; 3, Valve; 4, Copper pipe 12 mm diam.; 5, Copper pipe 15 mm diam.; 6, Main water supply; 7, Main flowmeter; 8, Nozzle on rubber tube; 9, Bottom plug said outlet; 10, Funnel with baffle plate; 11, Top sieve with conical plate sieve; 12, Gauge plastic or glass; 13, Rubber tube; 14, Inlet pipe bottom plug; 15, Inlet pipe; 16, Set of four sieves (45 μm aperture). (Grateful collaboration of Guillermo Perichi (Universidad Central de Venezuela, Maracay, Venezuela))

and avoid leaching; small artisan made plastic sieves of 6–8 cm diam. and having a rather large net (2–3 mm aperture) to hold one to two layers of cheesecloth or facial tissues with soil or other plant parts on them; cheesecloth; facial tissues; a small weighing scale and/or graduated beaker or cylinder to measure the sub-sample to submit to nematode extraction; a laboratory plastic bottle; single (metallic) or multiple (woody) funnel stands; a room or even a cabinet with air-conditioning and with nearly no vibration.

2.5.2 Nematode Extraction According to Cobb

For Cobb's decanting and sieving technique, have available at least a set of 20 cm diameter stainless steel sieves (Fig. 4) having different net apertures, such as 710 μm (25 meshes), 250 μm (60 meshes), 150 μm (100 meshes), 100 μm (140 meshes), 74 μm (200 meshes), 45 μm (325 meshes), and 20 mm (500 meshes). A number of 4–6 L pails according to amount of work; plastic or, preferably, glass beakers of 100–200 mL capacity; a scale and a larger beaker to measure the amount of soil sub-sample, usually 400–500 g or cm^3; a wood or other material rod or spoon (to mix the soil in water).

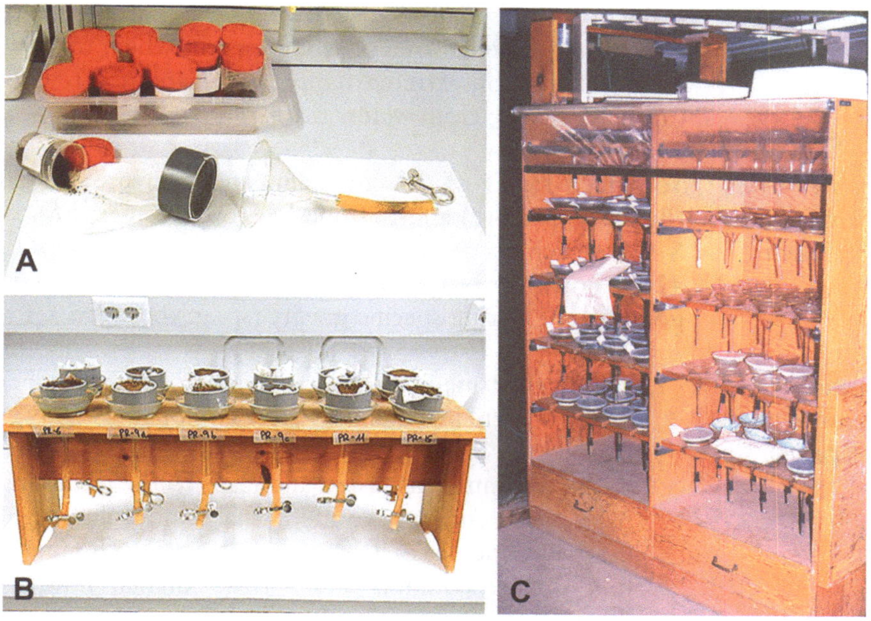

Fig. 3 Baermann's funnels. (**a**) Different parts: glasses containing soil sub-samples, sieve with rather large aperture (2 mm) on which the soil is directly put or enveloped in a paper tissue, funnel with the stem inserted in a rubber tubing, clip to close the tubing. (**b**) Arrangement of funnels on a woody stand. Note the sieve containing soil on the funnel and the clip closing the rubber tubing. (**c**) Arrangement of funnels in a cabinet. (**a** and **b** originally from **c**. Ornat, Barcelona, Spain, with kind permission of SIN)

Fig. 4 Stainless steel sieves of 20 cm diam. and 7–8 cm high, having aperture of different sizes, used in Cobb's sieving and decanting method. (From ref. 18, with kind permission of SIN)

2.5.3 Nematode Extraction According to Oostenbrink

The Oostenbrink elutriator is shown in Fig. 2. Four sieves of 30 cm diam. and 45 μm pore size; plastic bowls with a capacity of 4 L; decanting tray with cross piece; extraction sieve (16 cm diam) with supporting wires; clamping ring for securing the nematode filters. Watch glass (6 cm diam); two nematode filters (or equivalent);

extraction dishes; if necessary, covering material to cover the extraction dishes; a bench free of vibrations on which to keep the tray (at constant, non-extreme temperature; about 20 °C); 100 mL capacity beakers; clean water.

2.5.4 Nematode Extraction According to Jenkins' Centrifugation

You need sieves of 20 cm diam. and apertures of 20 meshes (840 μm aperture) and 270 meshes (50 μm aperture); buckets of 8–9 L capacity; a centrifuge that can hold tubes of about 50 mL at 420 *g*, bottles or beakers of about 100 mL; sugar and a densimeter good for measuring specific gravity (sp. g.) between 1.1 and 1.2.

2.5.5 Nematode Extraction According to Cooken's Centrifugation

1. The needed materials are composed of two benches of hard plastic material or stainless steel equipped as in Fig. 5. Both benches are installed on a channel in which the soil remaining in the column is discarded and leads to the decanting ditches. Also, both benches have an internal removable frame on which is set a plastic net of 1 cm diam. aperture. The first bench has four mixing columns and a large kitchen funnel of about 20–25 cm diam. with a stem of 3–4 cm diam, adapted to contain firmly a sieve 11 cm diam. and 8 cm high having a net of about 4–5 mm aperture. Each column consists of a Plexiglas cylinder of about 11 cm diam. and 36 cm height, with a funnel-shaped base ending in a tube 7–8 cm height and about 1.5 cm diam. The second bench (avoid that the first bench while operating causes vibration to the second bench) is to lie on it four plexiglass sieves of 11 cm diam. and 8 cm height (enough to contain about 500 mL of the magnesium solution–nematode suspension from the second centrifugation) and a net of 5 or 10 μm aperture and four under sieves having a hollow slightly larger than the diameter of the sieve in which must be laid the sieve without causing leaks when pouring in it the nematode suspension from the second centrifugation of the sample. Each base of the sieve has laterally an outlet hole (2–3 mm diam) closed with a rubber stopper.
2. Table centrifuges (Fig. 6a) with a rotor having four swinging arms that can hold four centrifuge tubes each of 650 mL capacity (Fig. 6b) and having the bottom as a spherical bowl. If the tube holder does not have the internal bottom shape similar to the tube, it is necessary to insert between the tube holder and the tube a piece of plexiglass or similar materials, of the same diam. of the tube, having the bottom modeled as the inner bottom of the tube holder and the upper part as the spherical bowl of the tube. If the tube does not adhere exactly into the tube holder, the tube will be deformed just during the first centrifugation.
3. Magnesium sulfate (not for chemical laboratory use, but for industrial use).

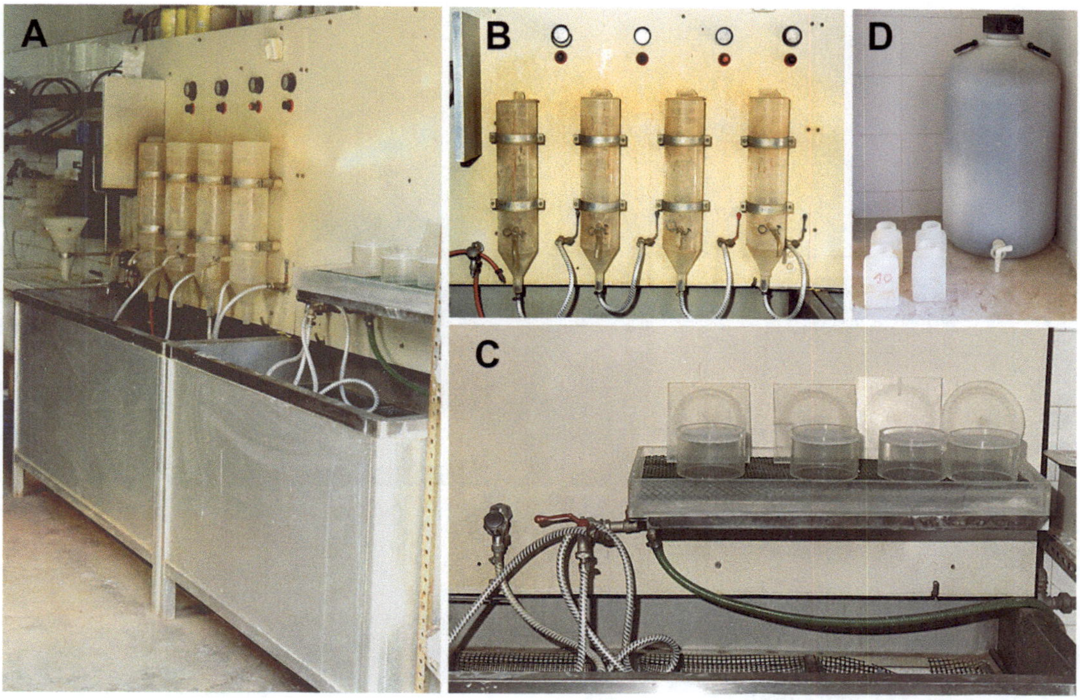

Fig. 5 Benches to extract nematodes with Coolen's method. (**a**) general view of the two benches with the mixing columns on the second bench, (**b**) close-up view of the mixing columns; (**c**) close-up view of the second bench; (**d**) tank containing the magnesium sulfate solution and four half liter plastic bottles. (From ref. 18, with kind permission of SIN)

4. A tank (15–25 L capacity) for the magnesium sulfate solution at 1.16 specific gravity (sp. g.), or otherwise as suggested in particular cases, and four 0.5 L beakers or cylinders for measuring the magnesium sulfate solution.
5. Two densimeters, one to measure sp. g. between 1.1 and 1.2 and the other between 1.2 and 1.3.
6. Four 250 mL beakers to draw soil–water mixture from the mixing columns. Four to eight laboratory bottles cut to contain 0.5 cm^3 soil.
7. A 2–3 L beaker to sieve the soil–water mixture before pouring it in the mixing column.
8. About 12–16 beakers of 100 mL, plastic bottles of 30–50 mL, or 100 mL beakers for storing temporarily the nematode samples.
9. About eight small plastic funnels with short stem to help in pouring the nematode gathered on the small sieves into the 100 mL beakers or in the small bottles.
10. About 12–20 buckets (multiple of the four tubes of the centrifuge), in which 500 cm^3 or 500 g soil is first mixed with

Fig. 6 Bench centrifuges. (**a**) two centrifuges with a scale to balance centrifuge tubes in between; (**b**) particular of the centrifuge tube holder; (**c**) aluminum tubes of 650 mL and tube stand. (From ref. 18, with kind permission of SIN)

1–1.5 L water and two of about 10 L capacity to prepare the magnesium solution.

11. Two vibromixers (Fig. 7) (laboratory model EL with a vertical vibration, vibrating frequency 50 t/min, amplitude 0–3 mm), one just on the left side of the first bench to resuspend in the magnesium sulfate the sediment of the first centrifugation. Between the vibromixer and the attachment wall is inserted a disk of rubber (to act as anti-vibration) and the other attached to the stem of the large sieve for an easier sieving of the water–soil mix before pouring it in the mixing columns.

12. Air compressor, better if a rather large one, installed in a separate room to reduce the noise, connected with pipes next to each column whose airflow is regulated by an airflow meter and a valve.

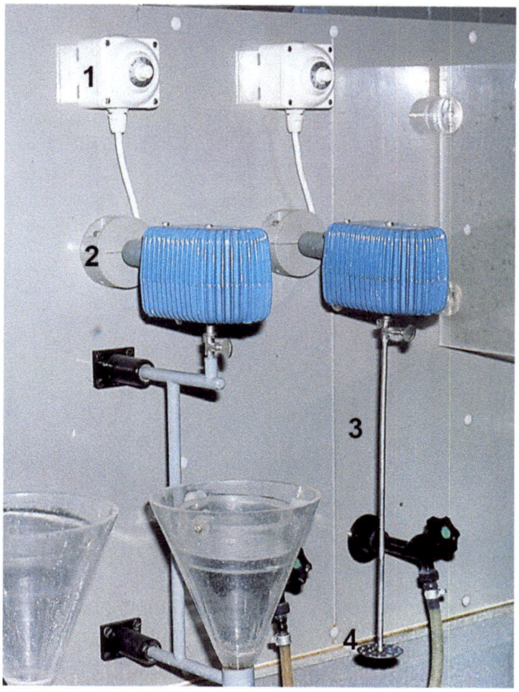

Fig. 7 Vibromixers. On the right, the vibromixer used to suspend kaolin in the water before the first centrifugation and suspend the sediment in the magnesium sulfate solution before the second centrifugation; on the left, the vibromixer connected to the stand of the large funnel, to favor the sieving of the soil–water suspension. Note above the electrical potency (1) to regulate the vibration, anti-vibrating dowel (2) to reduce transmission of vibrations to the benches, the steel rod of the vibromixer (3), and the vertical vibrating disk (4). (From ref. 18, with kind permission of SIN)

2.5.6 Nematode Extraction According to FLOTAC Basic Technique

The required materials are as follows:

1. A FLOTAC apparatus, composed of the base, the reading disk, the key, and the microscope adaptor (Fig. 8). It contains two 1 mL flotation chambers for microscopic examination and counting of the nematode eggs and juveniles in the suspension, each divided into 12 sections.
2. Infested soil, pail or bowl, centrifuge, centrifuge 11 mL tubes, centrifuge adapters for flotation chambers, sodium hypochlorite solution (common bleach), 1 mm aperture sieve, flotation solution (magnesium sulfate is good), mixer, stereoscope for observation at 20–40×.

2.6 Extraction of Cysts from Soil

2.6.1 Extraction of Cysts with Fenwick Can

You need a Fenwick can (Fig. 9). This is a simple apparatus that consists of a frustum of cone on top of which is cast a funnel stand to hold a funnel in which can be nested a sieve having an aperture of about 1 mm size. On the bottom, the apparatus has an outlet closed with a stopper during operation and removed for emptying and cleaning of the apparatus at the end of each extraction.

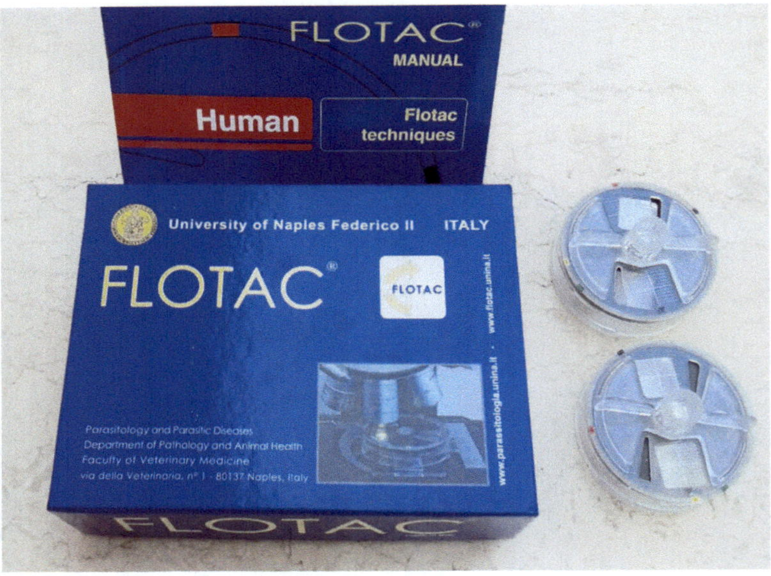

Fig. 8 A FLOTAX box showing its content: the manual and two chambers

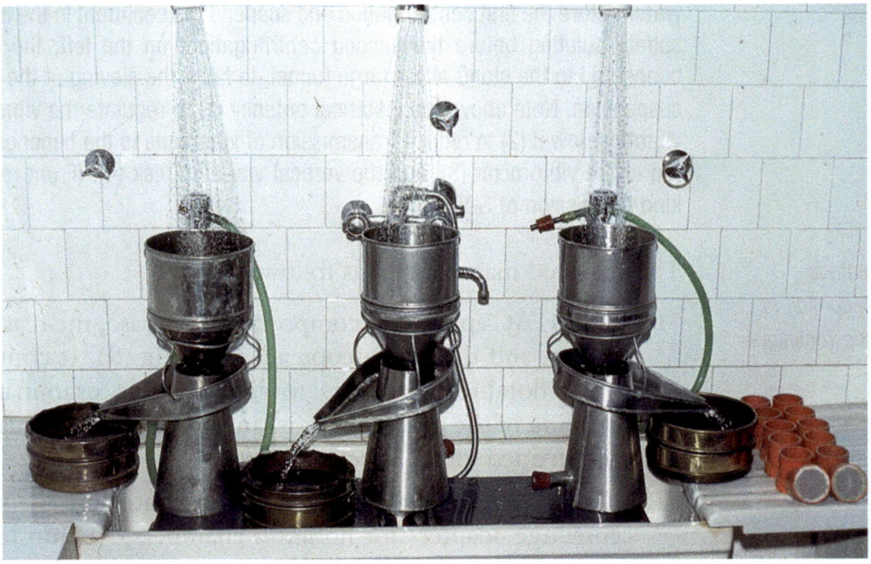

Fig. 9 Fenwick cans to extract cysts from dried soil. Note lower part having a frustum of cone with a collar inclined and the upper part formed of a funnel in which is nested a cylinder with the base closed by a net or a steel having apertures of about 1 mm. Laterally and near the base of the can is visible an outlet to empty the can of the remaining soil at the end of the extraction process, but is maintained closed by a rubber stopper while working. (From ref. 18, with kind permission of SIN)

Other required materials:

1. Washbasin having two to three showers, one per operating apparatus; a 25 mesh (710 μm) aperture sieve of 20 cm diam. and a sieve of 60 meshes (250 μm aperture) per working apparatus.
2. 40–50, or more, aluminum pans or similar containers to dry soil.
3. A warm room maintained at about 40 °C, in which to dry the soil samples.
4. 40–50, or more sieve about 8 cm diam, 8 cm high and having a net of about 200 μm apertures; small plastic bottles (35–50 mL capacity) to store cysts.

2.6.2 Extraction of Cysts with Seinhorst Elutriator

You need a Seinhorst elutriator (Fig. 10). It is made of a metallic tube about 6–7 cm diam. casted at its upper part to an upside-down frustum of cone 20 cm high and having the smaller base of the same diameter of the tube and the larger base of about 25 cm diam. In turn, the larger base of the frustum of cone is casted to a cylinder of the same larger diam. and 20 cm high. Around the upper part of the apparatus, there is a collar inclined downward of 45°. A rubber tube, about 50–60 cm long, is nested on the lower part of the collar to collect cysts and other soil particles in a 175–180 μm aperture sieve or in a pail whose bottom has been substituted with a net of the same aperture. At 10 cm from the bottom of the tube, there is an outlet with a valve to which is nested another rubber tube to collect cysts remaining in the upper part of the apparatus at the end of the process. At the bottom of the apparatus is located an inlet for a water flow regulated by a flowmeter. Moreover, on its base, the apparatus has an enlargement, for the decanted soil particles, closed with an outlet plug that is operated by a hand lever. This outlet plug serves for emptying and cleaning of the apparatus.

Other materials required are as follows:

1. 2 mm aperture kitchen sieve (20 cm diam.), 175–180 μm aperture sieve (20 cm diam.).
2. 15 cm diam. special filters (Binzer Munktell Filter GMBH, Battenberg, Germany).
3. 10 cm diam. funnels, small plastic sieves about 6 cm diam. and 7 cm high (Fig. 11a) and having an aperture of about 100 μm (small cysts) or 200 μm (large cysts), a thermostatic stove with airflow to dry cysts and soil debris at 50–60 °C.
4. A 25 mesh sieve and white paper, small colorless steel sieves (Fig. 11a) of about 5 cm diam. and 2.5 cm high, a warming electrical plate (Fig. 11b).

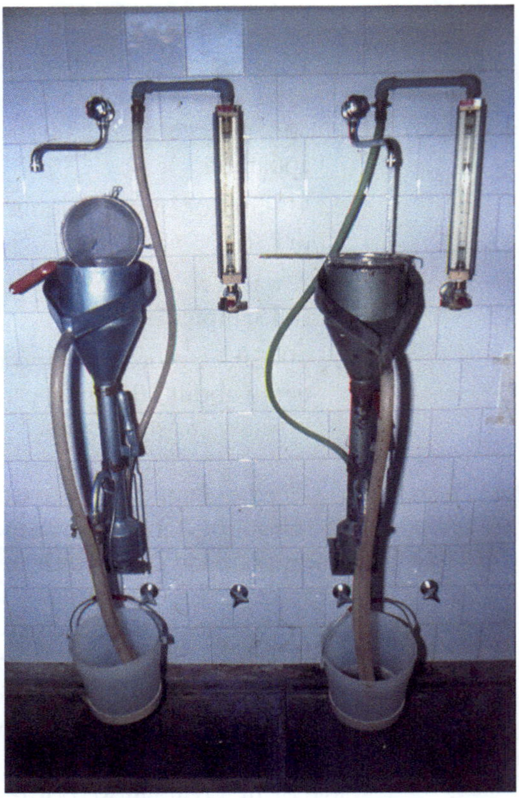

Fig. 10 The Seinhorst apparatus to extract cysts from wet soil. Functions are similar to the Oostenbrink elutriator (Fig. 4); however, a single sieve per apparatus is enough to collect cysts. Note on the inferior part that the two plastic tubes end here in a pail whose base has been substituted with a net of 215 μm aperture. (From ref. 18, with kind permission of SIN)

5. At least two flasks of 100 or 200 mL, acetone, alcohol, or 1.25 sp. g. magnesium sulfate solution, watch glass of 15 cm diam. and light concavity; some round painting brushes (No. 1 or 2).
6. Laboratory knife or a spatula to rake cysts; small vials made of synthetic (polyallomer) and transparent material (such as those used by virologists), about 1 cm diameter and 5–6 cm high (Fig. 12) devices to crush cysts, such as that according to Seinhorst and Ouden [11] methods (Fig. 12a).
7. Beakers of about 100 mL, magnetic stirrer, Pasteur pipette, or calibrating pipette (Fig. 13) calibrated to the desired amount of liquid suspension.
8. Counting slides (Fig. 14), such as Peter's slides, which contain just 1 mL liquid suspension, or other devices.

Fig. 11 (**a**) On the left, plastic sieves in which cysts and soil debris are transferred from the 60 mesh 20 cm diam. sieves of the different apparatus and used to be dried in a warm cabinet. On the right, smaller steel sieves for further drying of the cysts before being separated in alcohol. (**b**) Small stainless steel sieves, containing cysts and soil debris, on a heated plate for further drying of the cysts before being separated in alcohol in a flask. (From ref. 18, with kind permission of SIN)

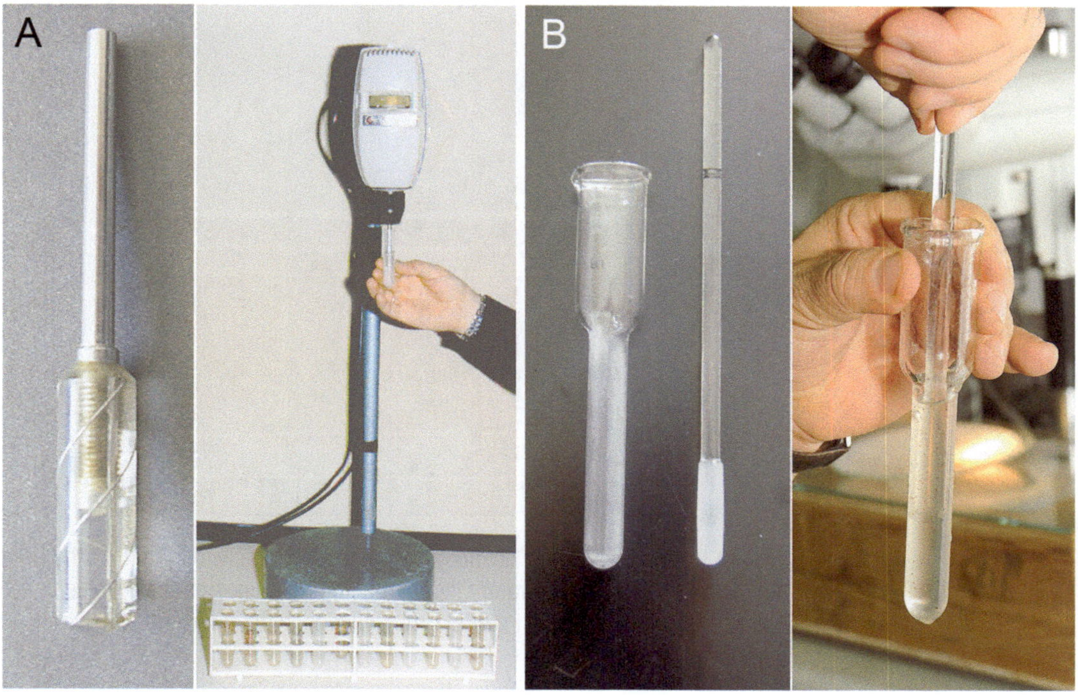

Fig. 12 (**a**) The Seinhorst and Ouden cyst crusher. On the left alone and on the right while operating. Below on the right, a tube stand with tubes containing cysts to crush. (**b**) crushing cyst with an old tissue crusher. (From ref. 18, with kind permission of SIN)

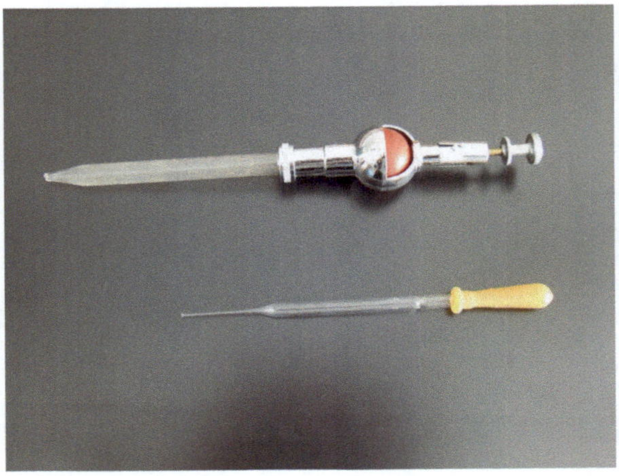

Fig. 13 Pasteur's and calibrating pipettes to draw the required nematode suspension to count nematodes. (From ref. 18, with kind permission of SIN)

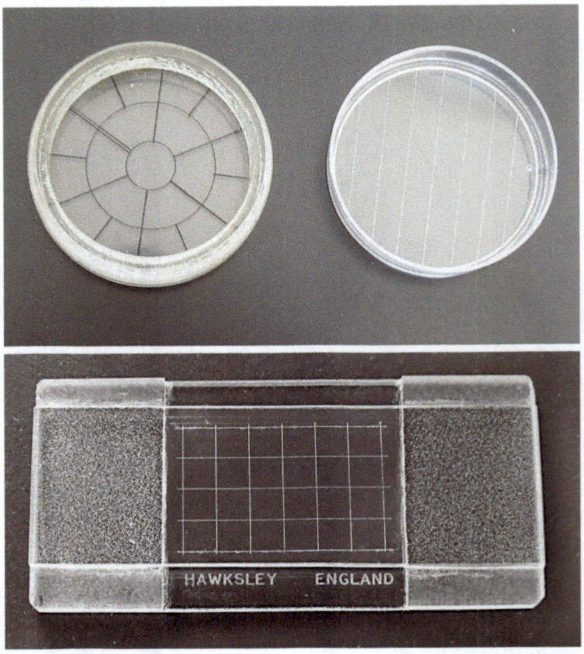

Fig. 14 Up, two dishes for counting nematode in suspension. That on the right has been designed in laboratory just by tracing right lines on the reverse of the plastic Petri dish. Below a Peter's 1 mL counting slide. (From ref. 18, with kind permission of SIN)

9. A stereoscope equipped (Fig. 15) with an incident light that can be orientated on the observing field of the filter on the glass watch, where cysts are to be picked up and counted, when on the stage.
10. A counter (Fig. 16).

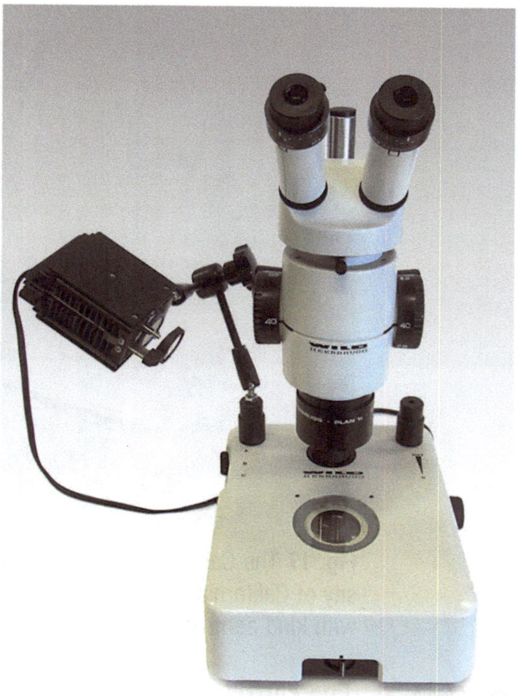

Fig. 15 A stereoscope equipped with a devise for reflected light, incorporated in the base, useful for counting nematodes in water suspension. The illumination must allow to observe nematodes in transparence. Laterally is a devise for incident light necessary to observe cysts and females on the roots or cysts on filters. (From ref. 18, with kind permission of SIN)

Fig. 16 A scaler to take record of the nematodes counted. (From ref. 18, with kind permission of SIN)

Fig. 17 The California elutriator used at the Department of Nematology, University of California, Riverside, USA, to extract cysts from dried soil. (From ref. 18, with kind permission of SIN)

2.6.3 Extraction of Cysts with California Elutriator

The elutriator (Fig. 17) consists of a cylinder 18 cm in diam. and 33 cm high. On the upper side, at 6 cm from the ridge, the cylinder has a hopper 9 cm wide, for the water overflow on the sieves, while on the lower side, at about 10 cm from the bottom, there is an inlet connected inside with a device creating six horizontal water stream receiving water from a tap connected to the domestic water pipeline with a plastic or rubber tube. The rate of water entering the apparatus can be regulated via a flowmeter and/or a valve. The California elutriator should be installed on one side of a sink, and complemented with sieves and materials as for the Fenwick can apparatus.

2.7 Extraction of Nematodes from Plant Roots

1. Sieves of 22, 45, 100, 250, and 710 μm aperture; apparatus and other materials according to different procedures.
2. Incubation according to Young [6]: infected roots; half liter glass jars with cap, number depending on workload; a room at about 20–22 °C; 50 mL plastic or glass bottles to store samples until observation.
3. Mist chamber according to Crow et al. [12] (Figs. 18, 19, 20, 21, 22, and 23). The apparatus consists of a box made of a polycarbonate sheet of 0.63 cm thick. In the box, there is a shelf at 29 cm from the base with 40 holes (10 frontally by 4) to insert 15 cm diam. glass funnels whose stems of 15 cm long end beneath the 250 mL flasks. The floor of the box is made of galvanized metal on a wood frame and has a drain in the bottom; American coffee filters (outside USA similar materials can be used); around the holes on the shelf, a plastic ring

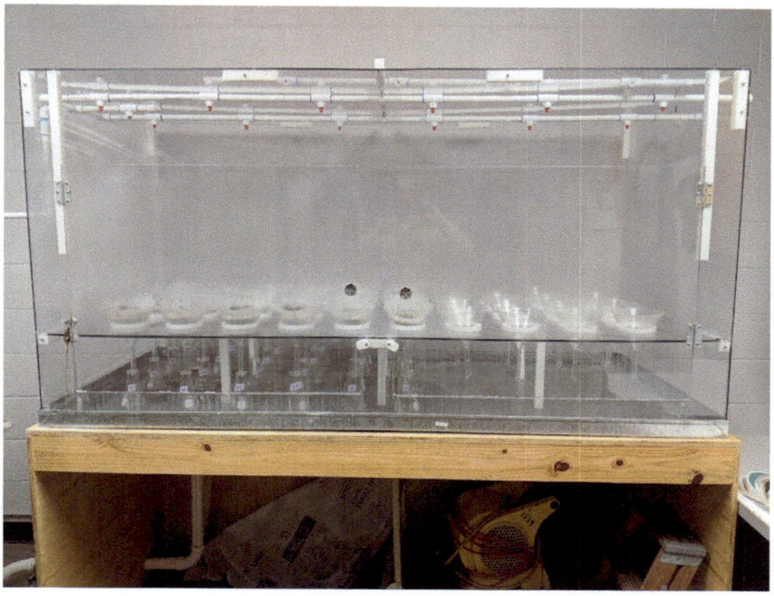

Fig. 18 The University of Florida Nematode Assay Lab mist chamber. (From ref. 12, with kind permission of Journal of Nematology)

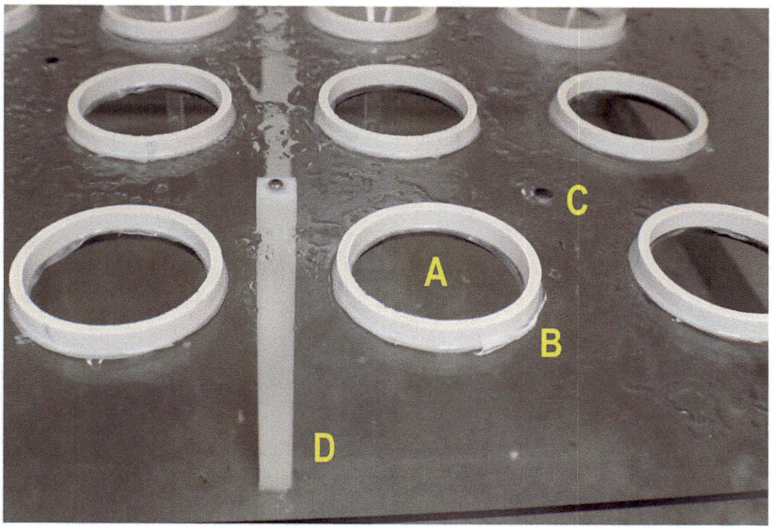

Fig. 19 The University of Florida Nematode Assay Lab mist chamber. (**a**) funnel hole cut in the center shelf; (**b**) water flow prevention ring; (**c**) drain hole; (**d**) shelf support. (From ref. 12, with kind permission of Journal of Nematology)

(0.63 cm high and 10 cm diam.) is glued to avoid water leaking along the funnel stems in the flasks and at the same time to hold the funnels firmly in place; on the funnel, there is a similar ring on the base of which is glued a rather large net (window net) to hold the plant materials. On the shelf, there are holes to drain

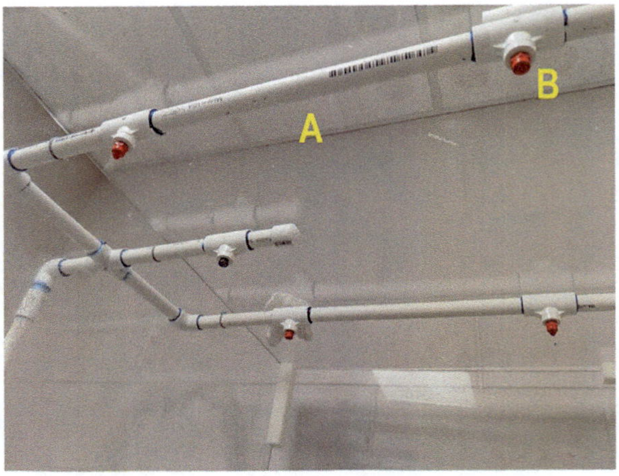

Fig. 20 The University of Florida Nematode Assay Lab mist chamber. (**a**) misting rail; (**b**) superfine mist emitters. (From ref. 12, with kind permission of Journal of Nematology)

Fig. 21 The University of Florida Nematode Assay Lab mist chamber. Funnels with sample supporting ring. Samples are placed onto coffee filters on the sample support rings. Funnel necks inserted slightly into the 250 mL flasks. (From ref. 12, with kind permission of Journal of Nematology)

water and below the ceiling of the boxes, there are superfine mist emitters.

4. Maceration and filtration according to Stemerding [13]: electric blender. That used by the authors is shown in Fig. 24, which can operate also with a 250 mL jars having four internal and vertical indentations to interrupt the whirl created by the blades when operating, for a better maceration. A kitchen

Fig. 22 Samples placed onto coffee filters on the sample support rings. Funnel necks inserted slightly into the 250 mL flasks. (From ref. 12, with kind permission of Journal of Nematology)

Fig. 23 The University of Florida Nematode Assay Lab mist chamber. Washed roots on the funnel (in this particular case, four grass plugs per funnel) in the mist chamber and incubated for 72 h. (From ref. 12, with kind permission of Journal of Nematology)

blender will work but should be as close as possible to that described. Large jars are not very good and would take longer time for maceration. Equipment to clean the suspension: Baermann's funnel (Fig. 3), Oostenbrink dishes, centrifuge, beakers (100–250 mL), and small bottles (50–100 mL).

5. Application to *Rotylenchulus reniformis*. Required materials: household blender; set of sieves (250, 149, 74, 44 μm pore aperture); beakers or glass containers of 100 mL capacity; scissors; scale; 45% sucrose solution (not fundamental);

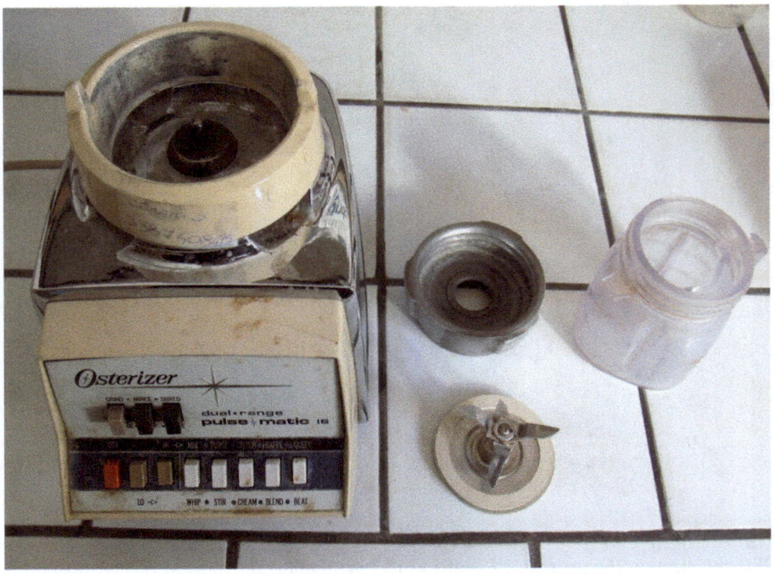

Fig. 24 A mini blender, for grounding roots, and its accessories (250 mL jar, closing disk with blades, and gasket to avoid leaking from the jars and the cap to close the jar). (From ref. 18, with kind permission of SIN)

centrifuge holding 50 mL tube (not fundamental); 1% sodium hypochlorite solution; sieves of 25 μm pore.

6. Centrifugation method of Coolen [14] (Fig. 25): a centrifuge and a bench as the second bench of Coolen's method and a vibromixer (Figs. 5c, 6, and 7), magnesium sulfate solution at 1.16 sp. g., a densimeter, and kaolin.
7. Nematode extraction according to FLOTAC basic technique: material as in Subheading 2.5.6, plus a blender to grind roots.
8. Extraction of *Tylenchulus semipenetrans*: colloidal silica at 1.16 sp. g. The original colloidal product is denser and must be diluted with water and the density checked with a densimeter.
9. Enzymatic maceration: heavily galled roots, water bath, 1 L plastic bottle, macerating solution made of 150 mL Pectinex® Ultra SP-L and 350 mL of Celluclast® (both from Novo Nordisk, Denmark), 500 mL of water, temperature-controlled water bath, blender, 250 and 710 μm sieves.

2.7.1 Staining of Roots

To stain roots, you need infected roots, phenol, lactic acid, glycerine, sodium hypochlorite, chlorine acid, stains (acid fuchsin, phloxine B, food stain, or erioglaucine dye as suggested by Omwega et al. [15]), glass beakers to boil the stain solution, and small plastic bottles (50–100 mL).

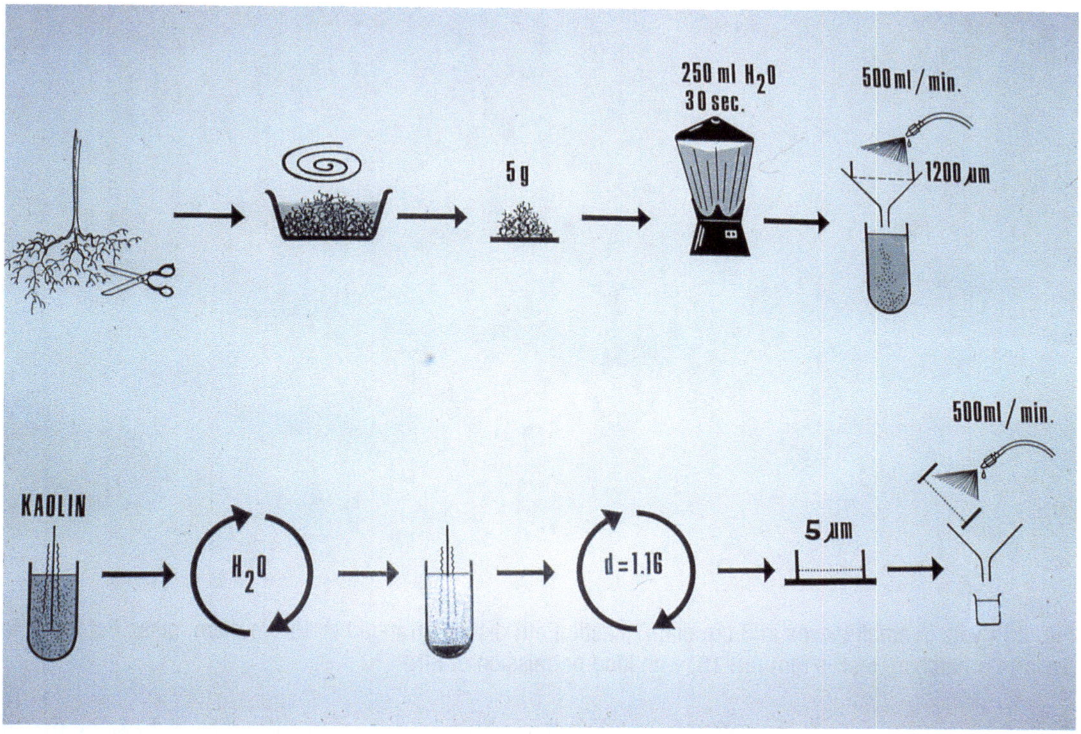

Fig. 25 Scheme of Coolen's centrifugation method to extract nematodes from soil

2.8 Extraction of Nematodes from Seeds

Baermann's funnels on which to hold the sample on a sieve with large aperture, 200 µm or even 1–3 mm; cheesecloth, small aperture sieve (44 µm or better 10 µm if dealing with thin nematodes) to concentrate the nematode suspension if necessary.

2.9 Extraction of Eggs from Egg Masses

To proceed have infected roots, scissors, sodium hypochlorite (common bleach is fine), square section bottles of half liter, automatic shaker to allocate several bottles at the same time, sieves of 20 cm diam. and 710 µm, 74 µm, and 10 µm aperture.

2.10 Collection of Second-Stage Juveniles from Cysts and Egg Masses (Hatching Tests)

Necessary equipment is composed of cysts and egg masses of the nematodes; several 2 cm diam. and 2 cm high sieves having a net of about 100 µm aperture for egg masses or small cysts or 200–250 µm aperture for large cysts. The sieves are made of a plexiglass tube 2 mm thick cut in pieces of 2 cm long. In the base of the tube, internally a circular hollow about 1.5 mm height is made with a lathe. The net is then cut in circles and kept firmly in the bottom of the sieve with a ring that fits exactly in the hollow when pushed in. Such a sieve can be disassembled for cleaning. Many 3 cm diam. plastic Petri dishes (depending on workload); plastic or glass Petri dishes that can contain several small sieves. Petri dishes of 15 cm diam. can contain nine small Petri dishes, each containing a sieve (Fig. 26).

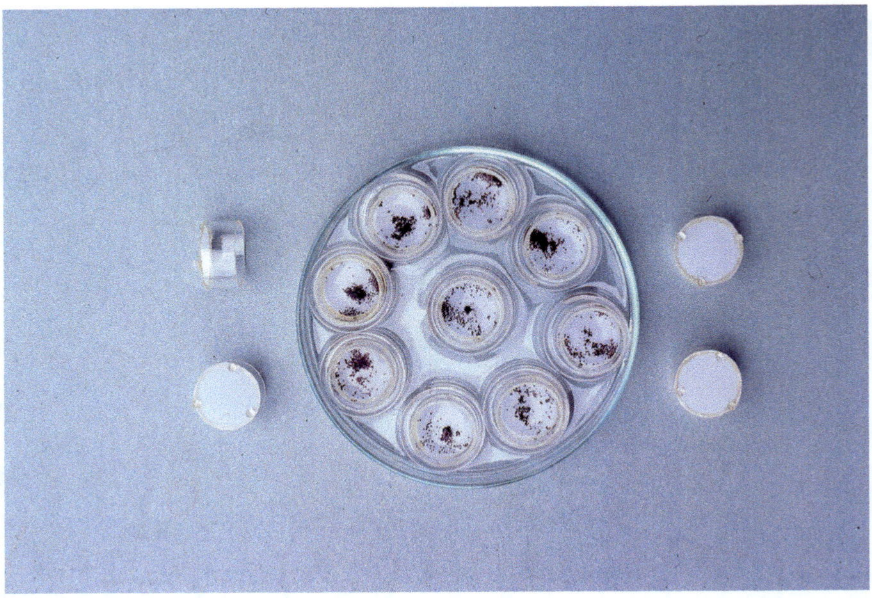

Fig. 26 Cysts in small sieves in 3 cm diam. plastic Petri dishes arranged in 15 cm diam. glass Petri dish to perform a hatching test. (From ref. 18, with kind permission of SIN)

Sheets of laboratory filter paper; root diffusate of host plants of the cyst nematode species. Egg masses will hatch easily also in distilled water but using diffusate of a host plant would result better. An apparatus to extract living cysts from soil; a stereo microscope to collect cysts from soil debris or to detach egg masses from roots infected by *Meloidogyne* species; thermostats to set temperatures in the range 15–28 °C.

3 Methods

3.1 Soil Sampling

The sampling procedure adopted depends very much on the project aim. For qualitative sampling aimed at identifying the causal agent of the plant symptoms observed, the procedure consists of taking a soil sample from the rhizosphere of the diseased plant, discarding the top 5 cm soil, or a sample of roots or aerial plant parts to be sure that there is a clear and constant association of the nematode presence with the observed symptoms. However, it is suggested to collect samples also from apparently healthy adjacent area; if the same nematode species and similar densities are found, the extracted nematodes may not be the causal agent of the crop decline. When, instead, it is necessary to estimate the nematode population density affecting the entire field, to estimate expected yield losses and to determine the need for implementing control programs, the collected samples must be representative of the entire field or, more generally, of the field area investigated, such as an experimental pot, microplot, or plot.

Most plant parasitic nematodes are soilborne. Therefore, to ascertain the nematode species occurring in a field, and their effect on the crop based on a measure of the population size for economical considerations and crop management options, it is important to collect soil samples representative of the area under study from which nematodes will be extracted and counted. In the soil, nematodes are not evenly distributed across the field, having an aggregated distribution. This occurs because by themselves nematodes move only a few meters per year and their distribution across a field relies mostly on passive movements due to farm machinery, irrigation and rainwater, and other soil movements. Moreover, several nematodes are sedentary and all eggs laid during their life cycle remain together. Finally, most plant parasitic nematodes feed on roots, and, therefore, they can be found mostly in the soil profile where roots are more abundant: 0–30 cm deep in the case of annual plants and 0–50 cm deep in perennial crops, although some can be found 1–2 m deep if roots are able to explore such a soil profile. Thus, representative soil samples must be a composite of several (50–100) soil cores taken uniformly across the sampling area, having the same recent cropping history and soil texture.

3.1.1 Soil Sampling in Fallow Field

The aim of such sampling is to ascertain the presence and density of the nematode population, to estimate the possible impact on the following crop [9, 16, 17], and, if necessary, to adopt the most appropriate management measures. As nematode soil population densities usually decline with time, it is suggested that sampling is done just a few weeks before cropping, especially during summertime when the nematode decline in the soil is expected to be greater. However, if the aim is to estimate the number of eggs in cysts of *Globodera* and *Heterodera* spp., then the soil can be sampled up to 1–2 months before cropping in spring. A special case is that of sampling experimental plots. It is preferable that a soil sample is collected soon after working the soil for sowing, but wait for 7–15 days after a heavy rain.

Then proceed as follows:

1. Before sampling, have the field worked in two crossed directions and up to 30–40 cm deep.
2. Divide or stratify the entire field in different sampling sections, each of about 1 ha or less. Be sure that each section has the same soil texture and was cultivated with the same crops at least during the previous 2 years.
3. Select the most appropriate sampling scheme. This must consider collecting generally about 50 cores/ha and 100 cores/ha

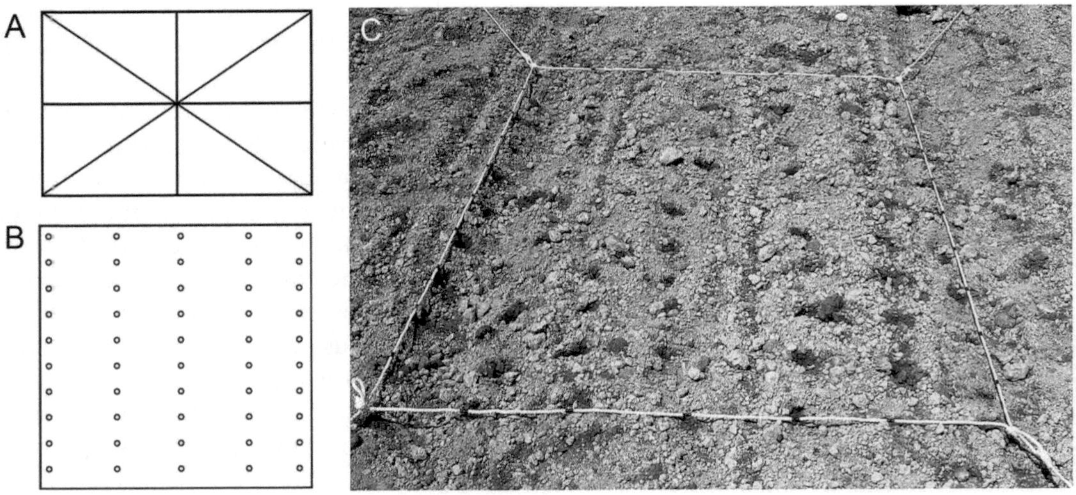

Fig. 27 Schemes for sampling noncropped fields. From ref. 18, with kind permission of SIN

in case of sampling for quarantine purposes. Collect soil along 10 rows 10 m apart and collect cores every 10–20 m along the row [18] (Fig. 27).

4. Each core is emptied in the pail with the help of the core emptier. Each core must be taken up to 30 cm deep if the next crop will be an annual and 40 cm deep if a perennial.
5. Pour the soil (about 2–4 kg) in the plastic bag, close the bag tightly with a wire or string attached to the label, and number the samples.
6. On the block notes, take note of the sample number, owner name, locality, crop present at sampling, soil texture, and other useful information. Also, we suggest to have a GPS to take note of the exact geographical coordinates.
7. Clean the pail and soil corer before collecting another sample.
8. Leave the samples in the car trunk during cool months and in a cool box during warm periods of the year until returning to laboratory.

3.1.2 Soil Sampling in Cropped Field

Follow the same procedures as in bare soil, but take the soil cores from the soil rhizosphere along the plant rows in order to obtain the same number of cores per ha. In case of perennials, take the soil cores from under the plant canopy and up to 30 cm deep. If roots are present, do not separate them from the soil.

3.1.3 Soil Sampling in Experimental Pots, Microplots, and Plots

1. For pots, at experiment termination, turn the pot upside down, separate all roots, mix thoroughly the soil, and take all of it (small pots of 10–14 cm diam) or about 1 kg, i.e., at least double the size of the sub-sample to be processed for nematode extraction.

Fig. 28 Example of microplots used at: (**a**) first author institute (CNR, IPSP, Bari, Italy), 1982; (**b**) At Servicio Agrìcola y Ganadero, La Serena, Chile, 1988; (**c**) USDA-ARS, Salinas, CA, USA, 1978; (**d**) Dept of Plant Pathology and Genetic, Raleigh, NC, USA, 1982

2. From microplots (0.3–1 m diam. size) (Fig. 28), collect at least 20 cores 30 cm deep each using a soil narrower corer, 3/8 of an inch diam.

3. While collecting soil from plots, soil border effects must be avoided; therefore, it is suggested to collect the samples, before planting and at end of the experiment, from the central part of each plot. At the end of the trial, samples must be collected along the planting rows if it is more important to have information on the reproduction of the nematode and both along the plant row and between rows if it is important to have information on the possible impact of the nematode population on the following host crop. As a guide to where to collect soil cores, we suggest preparing a square or rectangular scheme made of a rope having on each side attached rings of black tape, as in Fig. 29. This scheme is firmly set in the center of each plot, and then the soil cores are collected following the position indicated by the black tape, for a total of about 40–60 cores per plot.

Fig. 29 Sampling experimental plots in Italy

3.2 Plant Sampling

Many nematode species live in plant tissues, during the entire or part of their life cycle, including sedentary (root-knot *Meloidogyne* spp., false root-knot *Nacobbus* spp., cyst *Globodera* spp., and *Heterodera* spp.) and migratory (*Ditylenchus* spp., *Pratylenchus* spp., *Radopholus similis*, *Aphelenchoides* spp., *Anguina* spp.) endoparasites or semi-endoparasites (*Tylenchulus* spp., *Rotylenchulus* spp.). Therefore, samples of different plant parts must be taken to extract specific nematodes.

3.2.1 Plant Sampling of Aerial Parts (Annual Plants)

This is the case of sampling for the presence of *D. dipsaci*, *D. angustus*, *Anguina tritici*, and several species of *Aphelenchoides* in their multiple host plants (see specialized articles or books for biology and range of host plants of the nematode species under study). Depending on the nematode species, time of the year, plant species, and nematode species, it may be necessary to collect samples of the entire plant or just of some plant parts, such as leaves along the plant stem or just from the basal plant, spikes, flowers, and seeds.

Proceed as follows:

1. Collect at least 0.5–1 kg leaves and stems from around 40 plants per ha, evenly distributed across the sampling field of 1 ha.
2. Put plant materials in the plastic bag (3–4 L capacity), close the bag tightly, identify the sample with a label or a waterproof pen, and store the bag in the cool box.
3. When in the laboratory, cut with a knife or scissors the collected materials in about 0.5–1 cm long pieces and mix them well on a bench or preferably in a bowl containing water (mixing in water

is easier but must be done just before extracting nematodes; otherwise, nematodes may come out of the plant material before the extraction process and be lost).

4. Take one or two sub-samples each of 5–10 g and extract nematodes from them without delay (*see* **Note 1**).

3.2.2 Plant Sampling of Belowground Parts

Most of the plant parasitic nematodes feed on roots and are attached to them either just during feeding or throughout their development cycle. Therefore, to detect endoparasitic nematodes, it is necessary to collect root samples and extract nematodes from them to ascertain species and population densities, to evaluate if the observed symptoms are caused by the extracted nematodes or by different factors, or to use them for taxonomic, anatomical, biochemical, or molecular studies. Follow the suggestions listed below:

1. If an entire field must be sampled, collect about 20 plants per ha in case of large annual plants (potato, sugar beet, tomato, eggplant, etc.) and at least double that for rather small-sized plants (such as winter cereals, garlic, onion, chickpea, oil radish, grasses, and clovers), evenly distributed across the field.
2. Then, with a knife or scissors, remove the top plant and slowly wash the roots under running tap water, to eliminate soil particles. Be aware that in case of roots infected by species of cyst-forming nematodes, mature females and cysts may detach easily and get lost. Then with the scissors, cut roots in 0.5–1 cm long pieces, mix them on a table or in a bowl containing water (*see* suggestion for above plant parts), and collect one to two sub-samples each of 5–10 g according to the method you are going to use to extract nematodes.
3. Extract nematodes from sub-samples without delay (*see* **Note 2**).

3.2.3 Plant Sampling from Trunks of Host Plants (Bursaphelenchus xylophilus *and* B. cocophilus)

To detect the pine wood nematode (PWN) *B. xylophilus* in trunks of pines and derived materials, which typically involves quarantine issues, the reader is encouraged to consult protocols prepared by the Regional Plant Protection Organization (RPPO) and national phytopathological services. Moreover, for a complete understanding of the issues related to the PWN and its sampling, we suggest consulting Schröder et al. [19], while here we describe the sampling procedures when trees and wood materials to sample have already been identified.

When inspecting forests for the presence of PWN, in areas still known as noninfected, select trees showing weakness. Also sample pines in areas at risk of infection, such in parks, near urban centers, industrial areas, or important national and international highways.

For quarantine purposes, it may be necessary to sample pine trunks as consignments on ships, on lorries, or at wood factories.

What you need for sampling is a portable driller, better if powered, a pitsaw or belt saw, a devise to eliminate bark, and other materials used for sampling in general.

1. If the plant to sample does not show symptoms of PWN infection, select plants that appear weak or showing activity of its insect vector (*Monochamus* spp.), and among these, select five plants per sampling area. If possible, the sample should be taken from the trunk at different heights and also from canopy branches.
2. If the plants are symptomatic, select five plants but in this case, it is sufficient to collect samples from the trunk only at chest height.
3. Then with the driller, take samples of sawdust in different positions of the plants or just from three different directions at the same height, in order to collect at least 60 g of sawdust per plant and 300 g per sampled areas to process for nematode extraction.
4. In case of dead plants, it is suggested to take three disks, each 3–5 cm thick, at different heights. The disks are then first reduced in small pieces with a hatchet or a scalpel and then hashed in a machine in the laboratory before nematode extraction.
5. Mix the five sub-samples, put it in a plastic bag to avoid loss of humidity, label it, and keep it in the car or better in a cool box.

A similar case is that of sampling for *B. cocophilus* from coconut and oil palm. *B. cocophilus* is found in tropical environments, closely associated with palms. In coconuts palms, it causes the "red ring disease" and in oil palm the "short leaf syndrome." The palm weevil, *Rhynchophorus palmarum*, is the principal vector of the nematode, and life stages of *B. cocophilus* can be found in the gut, body cavity, and the region of the ovipositor of the weevil.

In coconut palm, *B. cocophilus* infestation occurs more commonly in trees 2.5–10 years old, with greatest incidence in those 4–7 years old. The nematode is not found in the first 3–5 cm from the trunk periphery, but it does occupy the entire central portion of trunk tissues. In a longitudinal cut, a pair of reddish-colored, parallel bands can be observed. These bands can extend to the entire length of the trunk or occupy only a portion of it, either from the crown downward or from the base upward to the top [20].

In oil palms with the short leaf syndrome, nematode specimens are found in the floral primordia and at the base of the spear leaf. In palms with initial symptoms, nematodes are not found in the stem or in the meristem; in palms with advanced or intermediate symptoms, it is found at the base of the spear leaf and leaves, and stem, but not in the petioles, flower peduncles, roots, and soil [21].

This information is useful to select the plant parts to sample for this nematode as suggested hereafter:

For coconut palm, look for 4–10-year-old plants with red ring disease symptoms, then cut into the lower stem, or bore into the stem with an increment borer. To survey for infected weevils, cut down a 4–10-year-old palm with symptoms, leaving the cut faces exposed. Cover the exposed faces with fronds from the fallen palms as open sun inhibits weevil visiting. After 24 h, return to the site and collected weevils on or near the cut faces [20]. Martinez [22] observed that, in coconut plantations showing red ring disease, approximately 42% of the *R. palmarum* specimens were infested with *B. cocophilus.*

For oil palm, the use of a drill is not recommended because necrosis in the trunk may be very limited or absent and, even in the presence of necrosis, the number of nematodes in this area can be extremely low or absent. Therefore, it is recommended to take entire pieces of stems [23].

With the drill method, it was not possible to confirm the presence of the red ring nematode in plants with external symptoms, and the nematode only was recovered from samples taken from inside the stem in a low density. The drilling method is faster, but the ring is not always observed leading to underestimation of the population level [24]. Therefore, it is necessary to take samples from different plant parts (*see* **Note 3**).

3.2.4 Transportation and Storing of Samples

While transporting, storing, and handling samples before processing to extract nematodes, be aware that you may be carrying live organisms that are very sensitive to environmental stresses and handling. Therefore, pay attention to reduce any disturbance to them that may cause death or drastic population reduction.

When transporting nematodes by car, use cool boxes and always park in a shaded place; during sunny hours, the temperature increases in the car, due to greenhouse effect, which may greatly suppress nematode populations in the samples and reduce the chance to detect them. Also, do not stack samples on each other; this causes pressure on nematodes and reduction of oxygen necessary for the nematode to remain active.

In the laboratory, store the samples in a cool room at 4–6 °C having several shelves, handle samples with care, and put on each shelve only a layer of samples. Take samples out of the storage room, mix it thoroughly, take a sub-sample, and return the left sample to the storage room for further needs. You may need to process again the samples for confirmation of the results, in case of problems during the first processing, or because you decide to reproduce some nematodes for future studies. Throw away the leftover samples only after having processed and checked results carefully. While handling samples, never use warm or chilled water. Finally, before getting rid of the samples, sterilize all the leftover soil and plant parts; steaming them for a few hours is suggested.

3.3 Nematode Extraction from Soil

As anticipated, several methods and apparatus have been designed and modified during time to adapt to different needs. Some can serve to extract different nematodes from soil or plant parts and other only a few nematodes or nematode stages, such as cysts and eggs. At the end of each extraction process, clean the apparatus and other materials used very well and if possible with hot water, to kill all nematodes that may have remained attached to them. This suggestion should be mandatory when dealing with quarantine samples.

In a nematology laboratory equipped for extracting nematodes, it is necessary that soil residues after the extraction process be conveyed in a series of two to three contiguous decanting ditch (each of 80–100 cm square section and 1.5 cm) deep next to each other so that water move from the first to the other, in which soil particles of different sizes will decant, before reaching the city sewage system. These decanting pitches should be easy to inspect and those containing soil deposit emptied periodically according to the amount of laboratory work. Moreover, all tubes draining water must be rather large (60–70 mm diam.) and sloped toward a channel or a larger collecting tube connected with the pitches. If all drains go into a channel, this must be covered by a stainless steel grid for easy inspection and cleaning.

3.3.1 Extraction of Nematode Vermiform Stages by Baermann's Funnel, Sieving and Decanting, and Elutriation Methods

First of all, crumble soil clods and sieve the soil sample with a rather large net sieve (2–4 mm aperture) to eliminate small stones and other inert materials. Then mix the sample thoroughly and take a sub-sample of the size required by the selected method of extraction, usually from 50 to 500 g or cm^3 and put directly on/in the extraction equipment or in beakers or other container ready for processing later. For more reliability of the extraction results, extracting two to three sub-samples per sample, especially when using small sub-samples (50 g or cm^3), is suggested.

Extraction by Baermann's Funnels

To be active, nematodes need to be covered by a film of water. Therefore, this method takes advantage of the hydrotropism of active nematode stages. Moreover, quiescent and anhydrobiotic nematodes return active when they resume their water body content, making this method also useful to extract surviving stages of the nematodes in dried soils and seeds. This is a rather simple method [1] and results in a nematode–water suspension clear and easy to observe under a stereoscope at 25–40× magnification and is still widely used, alone or in combination with other methods. The method has received several modifications; for more insights, see references [25, 26].

The procedure is as follows:

1. Assemble the funnel on stands as in Fig. 3 [18] in a conditioned room and set temperature at 20–25 °C, according to nematode species.
2. Close the rubber tube with a clip.
3. Fill the funnel with tap water up to the base of the sieve.
4. Put a double layer of the cheesecloth or facial tissue on the sieve and pour on it the amount of plant material (5–10 g) or soil (50 g or cm^3).
5. As soon as the edges of the cheesecloth or paper tissues are wet, turn each on the outside of the sieve adhering to it or on the soil or planting material.
6. Check soon and again at least when entering and leaving the laboratory each funnel to be sure that the water level is about 2–3 mm above the soil or plant material level and that below the sieve there are no air bubbles. If necessary, refill the funnel by adding water with a laboratory bottle containing tap water, between the funnel and the sieve.
7. Leave the funnels for 24 or 48 h. In case of extraction of *Meloidogyne* spp. and other nematodes laying eggs in egg masses, we suggest to leave the soil on the funnel for a longer period (7–10 days) to allow eggs of the nematodes to hatch and migrate into the funnel stem; otherwise, the number of the nematodes extracted will be underestimated.
8. At the end of the process, most active nematodes will have migrated into the water and settled at the base of the funnel stem. Open the clip rapidly and collect 10–15 mL of the nematode–water suspension in a small beaker or bottle (50–100 mL capacity).
9. Finally, adjust nematode suspension to the same volume, collect a small amount (1–5 mL), pour in a counting slide or plate, and count nematodes under a stereoscope at 25–40× magnification or process them for more studies.
10. If the size of the nematode suspension is more than 50 mL, leave the bottles or beakers decanting for a few hours or overnight on a bench without vibration, especially when small nematodes are expected to be present, and slowly and carefully remove the top excess water of the suspension. This operation can be made in two steps separated by at least 1 h to avoid removing too much water at once which can mix the basal part of the suspension and cause loss of nematodes (Fig. 30) (*see* **Note 4**).

Fig. 30 A modification of Baermann's funnel. Sieves of about 8–10 cm diam. and 1.5 cm high, having three to four 2 mm high legs to allow nematodes to freely move in the water. The aperture of the sieve net is of 90 μm for long nematodes and 50–60 μm for short nematodes. (From ref. 18, with kind permission of SIN)

Extraction by Cobb's Decanting and Sieving Technique

This method is useful just to extract nematodes from soil, including cysts, and is based on the different nematode and soil particle sizes and densities and their decanting time. Therefore, experience can indicate the most appropriate decanting time and sieve aperture to use.

Then follow the procedures reported hereafter:

1. Pour the soil sub-samples in the pail, add about 1–2 L tap water, and leave the soil undisturbed for a few minutes for sandy soil or longer for soil containing a higher percentage of clay.
2. Add more water to at least five times the soil volume and mix with a plastic or wood rod or spoon until all the soil has been homogenously dispersed in water.
3. Leave the water–soil suspension to decant for 1–2 min; use shorter time for very sandy soil and larger nematodes and longer time for heavy soil and small nematodes.
4. Then insert the 25 mesh (710 μm aperture) sieve in the 100 (149 μm) (long nematodes)–200 (74 μm) (most adult tylenchids) mesh sieves. Hold them in your left hand at 45° angle and slowly pour on them the water suspension in the pail on the two sieves and collect the filtering suspension in another pail (pail 2) and leave in the first pail (pail 1) all soil deposited in

the bottom of the pail, and spray water on the 25 mesh sieve with a gently stream of tap water or a laboratory bottle to wash nematode on it in the second (narrower aperture sieve); now add more water in pail 1, mix, decant, and sieve again through the same two sieves. This operation can be repeated two to three times.

5. At this stage, take away the larger sieve and clean it to be used for processing the next soil sample.
6. Thereafter, try to gather all nematodes and soil particles on the second sieve in a corner by using again a gentle stream of tap water or a laboratory bottle. If the nematode–water suspension contains still much soil particles, collect it in another clean pail by turning upside down the sieve and pouring on its reverse water; leave it to decant again for about 30 s, filter again through the second filter, and collect again all material in an angle of the sieve by pouring water (easier with a laboratory bottle and if clean enough, collect this suspension in a 100–200 mL beaker).
7. While repeating this operation, the sieving water must also be collected in pail 2.
8. Now repeat the same procedure with the soil suspension in pail 2, but using a small aperture sieve, such as a 325 mesh (44 μm aperture) (*see* Fig. 31 for details).

It is suggested that the nematode–soil fractions collected from the two sieves are kept separately if they will be observed directly under a stereoscope or if they will be subjected to further cleaning with Baermann's funnel, while they can be mixed if the further cleaning will be by centrifugation. It is not convenient to superimpose all the three sieves especially with heavy soil (*see* **Note 5**).

Extraction by the Oostenbrink Elutriator [4]

The method is used for the extraction of active nematodes from different substrates that can host nematodes and can be suspended into a water suspension, but not eggs. Depending on the type of soil, the sample size should be of 100-max 1000 cm^3. The method is based on the difference in size, shape, and sedimentation rate between nematodes and soil and finer substrate particles. In the cone of the apparatus, the finer and lighter particles float, while the heavier ones settle to the bottom of the apparatus. The suspension retained in the extraction cone is drawn through a lateral tube and sieved by passing it through a set of sieves. What is retained on the sieves can be observed directly; however, it may be difficult to observe the nematodes properly because of the presence of large amount of residues of soil or other substrates. In this case, the sample is cleaned, taking advantage of the hydrotropism of the nematodes; the sample is further processed using either the cotton filter or Baermann's funnel method. In this way, the substrate

Fig. 31 Procedure to extract nematodes with Cobb's sieve. (**a**) a pail containing soil; (**b**) a pail containing soil and water; (**c**) sieving the soil–water suspension through the 25 mesh (710 μm pore) sieve in pail 1; (**d**) soil debris remained on the 25 mesh sieve to be discarded; (**e**) sieving the water suspension in pail 1 or 2 through a 200 mesh (74 μm pore) or 325 mesh (44 μm pore) sieve that will be sprayed with water and then nematodes collected in a beaker. (Courtesy of N. Sasanelli (e.g., IPSP, Bari, Italy))

particles are retained on the filter, while the nematodes pass through it into the water below. Proceed according to Fig. 2, as follows:

1. Take a sub-sample (optimum 100–200 cm^3) soil.
2. Close the side outlet with a rubber stopper and the bottom with the plug. Fill the funnel with water up to level 1 (point in which the water level touches the funnel) and flow water from below at the rate of 1000 mL/min.
3. Using the top nozzle, wash the sample through the top sieve into the funnel and continue until the water level has reached

level 2 (2/3 of the funnel). The sample must be washed completely and slowly. Close the nozzle and reduced undercurrent to 600 mL/min.

4. Moisten the four 45 μm sieves to avoid clogging of the mesh and place them under the side outlet.
5. As soon as the water has reached level 3, the side outlet is switched off and the suspension flows over the sieves. Tap the side of the sieves for an easier sieving.
6. Immediately wash residues from sieves into a container (100–200 mL capacity beaker is appropriate). Tilt the sieves and wash both sides with small amounts of water (can be used a sprayer). Carefully wash all parts of the funnel and sieves.

For the next procedural steps (cleaning of the sample with nematode filters) (Fig. 32), you need milk filters or disposable filters, clamping rings, extraction sieves with two parallel bars at the bottom, settling tray, cross support, and watch glasses and then:

1. Place two nematode filters with a clamping ring on an extraction sieve. Milk filters or disposable tissues can also be used (it is necessary before standardizing the use of alternative filters to compare efficiency with filters for nematodes or milk filters). Moisten with the spray to eliminate possible air bubbles between the filters. Place the sieve in a settling tray filled with water and with a cross support and place a watch glass over the filters to avoid breaking them when pouring the suspension.
2. Carefully pour the suspension from the container onto the watch glass of the filter. The last amount of suspension remaining in the container should be shaken well and quickly poured onto the watch glass. Remove the watch glass and rinse it. Lift the sieve and remove the retaining ring. Identify the sample.
3. Place a dish containing ±100 mL of clean water in a place free from vibrations. Carefully place the sieve with the wet filter on an angle.
4. After an extraction period of 16–48 h (24 h is the most indicated according to our experience), the sieve can be removed and the nematode suspension poured into a 100 mL container for identification and counting. Population size is indicated as per 100 mL of soil.

The Oostenbrink elutriator is excellent for the extraction of small and medium sized nematodes; for the extraction of large nematodes (e.g., *Xiphinema* and *Longidorus*), it is not efficient following the above procedure. Isolation of these nematodes requires an increase in the rate of upward water flow. Nematode filters hinder the passage of large nematodes in suspension;

Fig. 32 Cleaning nematode suspension samples obtained with the Oostenbrink elutriator. (**a**) materials necessary; (**b,c**) two nematode filters with a clamping ring in an extraction sieve (moisten the filters with the sprayer to remove air bubbles from between the filters); (**d**, **e**) sieve in a water-filled decanting tray with cross piece; (**f**) a watch glass on the filters; (**g**) carefully pouring the suspension from the bowl onto the watch glass on the filter; (**h**) shows how to remove the watch glass, rinse it, and as soon as all the water is poured on the filter, lift the sieve and remove the clamping ring; (**i**) an extraction dish in an area free of vibrations. (Grateful collaboration of G. Perichi (Universidad Central de Venezuela, Maracay, Venezuela))

however, personal experience of the second author indicates that the use of a single filter can be effective, since large nematodes can also pass through it.

For extraction of longidorids, take a sub-sample (optimum 100–200 mL) soil. Close the side outlet with a rubber stopper and the bottom with the plug. Fill the funnel with water up to level 1 (point in which the water level touches the funnel) and flow water from below at the rate of 1300 mL/min.

After washing the sample into the funnel (and as soon as the water reaches level 2), the undercurrent must be reduced to 800 mL/min. As soon as the water level reaches level 3, the side outlet is unplugged and the suspension passes over the two 180 μm sieves. It is possible to capture large, medium, and small nematodes in a single procedure, using the flows of 1300 and 800 mL/min and the set of four 45 μm sieves instead of the 180 μm ones. In this way, all nematodes are captured regardless of size. What is retained in the 45 μm sieves is washed and passed through two 180 μm sieves to a second plastic container. Small- and medium-sized nematodes end up in the second bowl and large nematodes are retained on 180 μm sieves. (*see* **Note 6**).

3.3.2 Centrifugation Methods in Nematode Extraction

Jenkins [8] modified the method of Caveness and Jensen [27], based on centrifugation, to separate nematodes from other soil particles after Cobb's sieving and decanting process. The results were so good that this method is still used in many laboratories, including that of the first author. The different changes to the original method refer mainly to the procedure followed before centrifuging. The centrifugation method takes advantage of the different specific gravities of the various particles in a suspension and their separation when centrifuged in a solution having a given specific gravity (generally 1.16–1.18). The method extracts all nematode groups and stages, including longidorids if the right solution is used, it is rather rapid and the final nematode suspension is very clear, and the nematodes are easy to observe and count under a stereoscope. However, if nematodes are extracted for morphological studies or inoculation purposes, this may not be the best method. In some laboratories, the soil sample is first processed with an elutriation method and then all or part of the elutriate is sieved through a small aperture sieve (325 meshes = 45 μm aperture) and then centrifuged. The method also reduces most of the operator subjectivity of the procedure. However, in the market, it may be difficult to find a centrifuge and accessories exactly suited for nematode extraction. If not, contact a good artisan for the necessary adaptations. Hereafter are first reported methods as suggested by Jenkins [8], later of one of the most important modifications, according to Coolen [14], and then the FLOTAC basic technique just published [28].

Nematode Extraction According to Jenkins

The suggested procedure is as follows:

1. Wash soil samples of 100 to 500 mL through a 20 mesh sieve into an 8 L bucket and made up with water to 6 L.
2. Stir the mixture well and then allow standing for about 30 s before sieving through a 270 or 325 mesh sieve. Do this twice and collect the residue on the sieve in a beaker and discard the remaining soil.
3. Transfer the residue in the beaker into two 50 mL centrifuge tubes and spun for 4 to 5 min at about 1750 rpm. Then pour off the supernatant liquid and add a sugar solution (1 lb. cane

sugar/L water, approximately equivalent to a 1.18 sp. g. solution), thoroughly mix with the sediment, and spin the tubes for 0.5 to 1 min at the same speed.

4. Now pour the supernatant containing the nematodes onto a 270 or 325 mesh sieve (we suggest a 20 μm aperture sieve or better 10 μm to avoid loss of eggs and small nematodes) and wash the sugar solution away with a gentle stream of water on the sieve.
5. Gather nematodes first in an angle of the sieve maintained inclined to 45° with your left hand and then collected in beakers or bottles, ready for observation and counting.
6. The process recovers all free-living nematodes (*see* **Note** 7).

Nematode Extraction According to Coolen

The procedure [14], hereafter reported with slight modifications, was especially designed to extract nematode eggs from both soil and plant parts but is effective with all kinds of nematode species and stages, although it may require some adjustments. If the procedure has been conducted well, the final nematode suspension is clear and the nematodes are easy to observe and count. The method is very suitable to extract slowly moving nematodes such as criconematids. However, some nematode species may appear distorted and therefore, are not useful for morphometric studies. After all material is ready, an operator can process up to 12 samples in half an hour using at the same time two centrifuges.

For the procedure, see also scheme in Fig. 33.

1. Prepare the magnesium sulfate at 1.16 sp. g. Therefore, dissolve 450 g of it per L in hot water (70–80 °C) for easier dissolving. Leave to cool and then check the density by immersing the densimeter in the solution. Realize that after cooling, the solution will become denser. If the density of the solution is not that required, adjust it by adding more salt or water. Magnesium sulfate is used instead of sugar because the solution has less viscosity and in it algae will be more difficult (but not impossible) to select and grow.
2. After thoroughly mixing the sample, take a sub-sample of 500 cm^3 or 500 g and pour it in the buckets and add 1–1.5 L tap water, leave it for a few minutes (sandy soil) or 0.5–1 h for silty or rather clayey soil, and mix with a plastic or wood rod or large spoon. We suggest to determine the water content of each sample so that the nematode content can be referred to soil dry weight.
3. Sieve all the content in the bucket through the large funnel on bench 1 in the 2–3 L beaker underneath, with the help of a flat water nozzle.

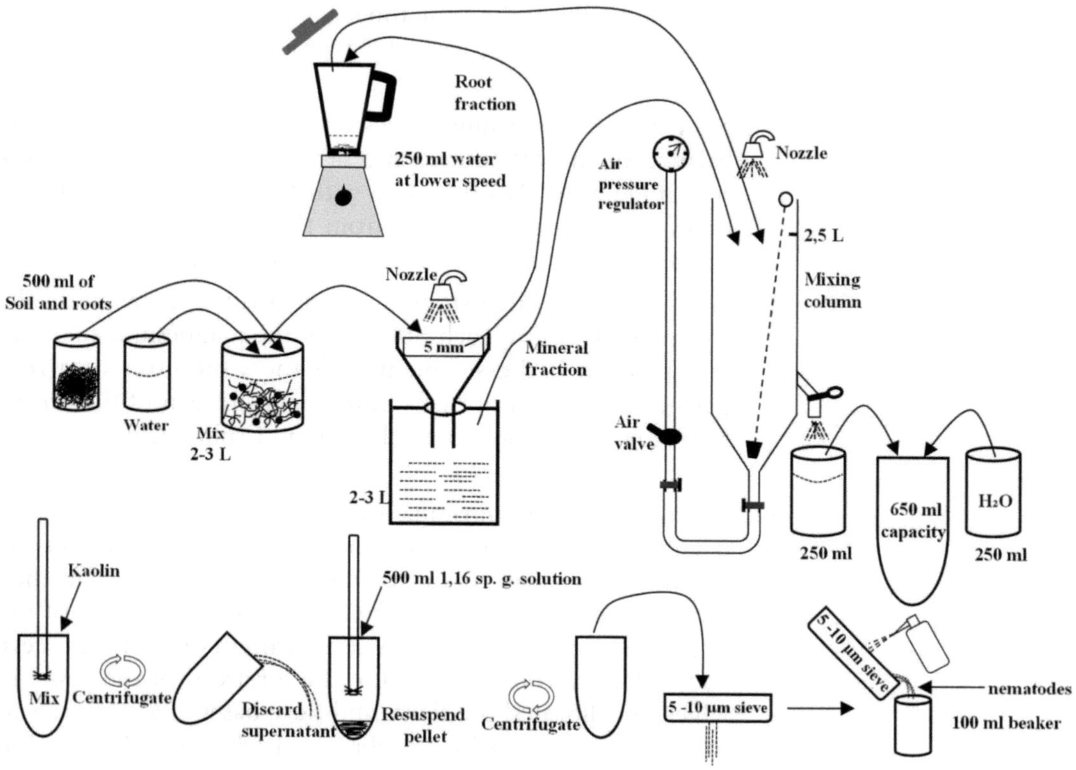

Fig. 33 Scheme to extract nematodes from roots with the centrifugation method of Coolen

4. If on the large sieve roots remain, collect and homogenize them in a blender (Fig. 24).
5. Close from inside the hole in the bottom of the column with a rubber stopper fixed at the end of a 40 cm long steel wire 1 mm thick and externally from another rubber stopper in which has been introduced a rigid plastic tube (4–5 cm diam.) inserted into a flexible plastic or rubber tube connected to the pressurized air duct.
6. Now pour in the column both the soil and root mixer of the same sample and adjust water volume to 2.5 L.
7. When all four columns are ready, at the same time, pull out of the column the internal bottom stopper while starting bubbling air in it via the bottom pipe opening gradually the valve of the pressurized air duct. Then continue to open the valve until all the soil mixture in the column is thoroughly bubbling (look carefully to be sure that the soil in the bottom of the column is also mixing).
8. Leave the air bubbling in the columns for 5 min and then, while operating, draw in a beaker from the outlet in the lower part of the column 250 mL of the suspension and pour in a centrifuge tube; with the same beaker, take another 250 mL clean tap

water and pour in the same tube, add kaolin up to 5 g per tube (the amount varying with soil type), and mix well with the vibromixer.

9. Now put two centrifuge tubes on the two pans of the scale, balance them with water, and arrange in the two opposite centrifuge arms. Do the same with the other two centrifuge tubes containing soil mixers from the other mixing columns and leave the centrifuge to spin at 1500 rpm for 5 min.
10. When the centrifuge stops, from each tube, throw away the supernatant, add to each 500 mL of the magnesium solution, resuspend in it the sediment with the vibromixer or a rod, balance the weight of tubes in the opposite arm of the centrifuge now using the same solution, and spin again for 5 min more at the same speed.
11. When the second centrifugation is terminated, pour the supernatant containing nematodes on the 5–10 μm sieve on the second bench, leave them to stand for about 1 m, and then remove the rubber stopper from the hole in the base of the sieve and leave the solution to sieve through the 5–10 μm sieve and collected from a channel in the forefront of the vessel in which are allocated the sieves and sieve bases in a container to which is connected with a rubber or plastic pipe. This solution could be reused after adjusting the density.
12. Now with a gentle water stream, wash away the magnesium solution, and with a flat water nozzle, wash all nematodes on the sieve in the small bottles or beaker. The suspension can now be used to observe and count nematodes.
13. Before starting again, wash thoroughly all materials and equipment (*see* **Note 8**).

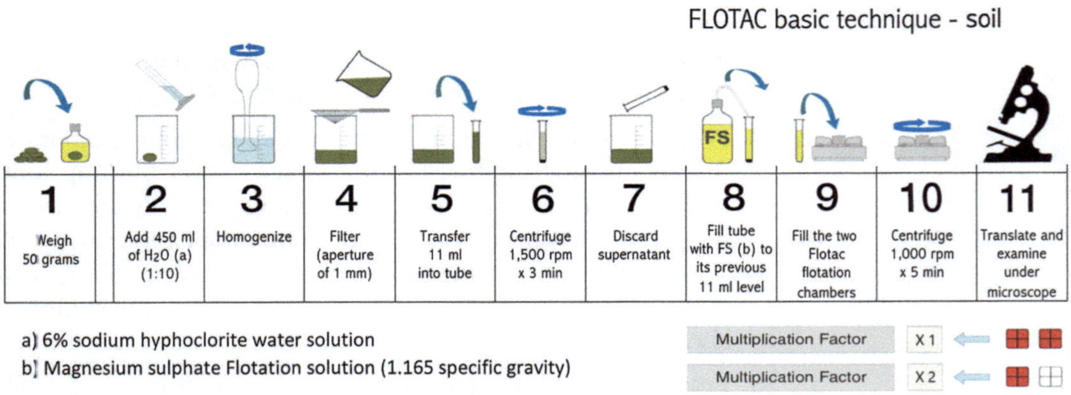

Fig. 34 Scheme showing the steps of the FLOTAC basic technique for extraction of nematodes from soil. (From ref. [28], with kind permission by the authors)

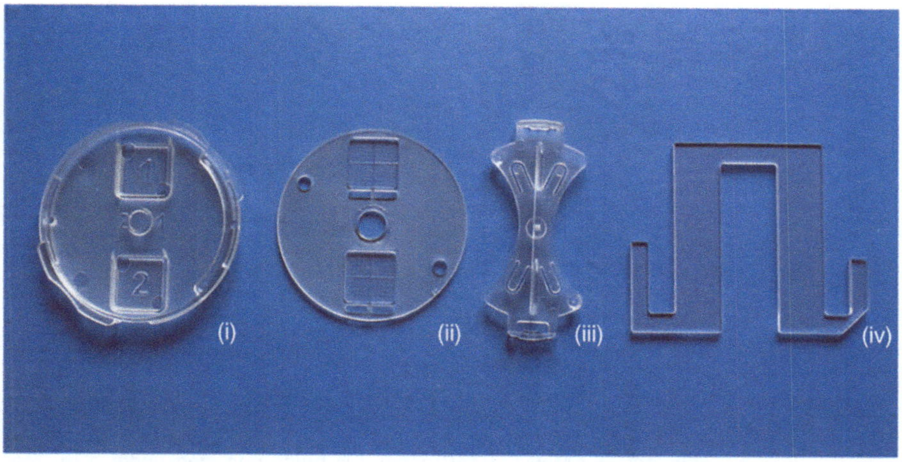

Fig. 35 Mini-FLOTAC components. (i) Base; (ii) reading disk with two ruled grids; (iii) key; and (iv) microscope adaptor. (From ref. [50], copyright Springer Nature)

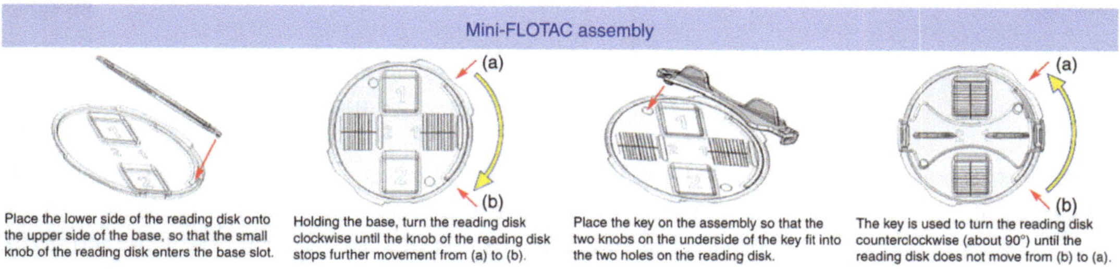

Fig. 36 Step-by-step assembly of the Mini-FLOTAC. (From ref. [50], copyright Springer Nature)

Nematode Extraction According to FLOTAC Technique

Procedure according to Troccoli et al. [28] and scheme in Fig. 34:

1. Sieve the soil sample through a 2 mm aperture wire sieve.
2. Place 50 g soil in a 2 L beaker or bowl, add sodium hypochlorite solution (common bleach) to 450 cm^3 volume, and stir very well with a hand electric stirrer.
3. Filter the soil suspension through a 1 mm aperture wire sieve and collect in another beaker.
4. Transfer 11 mL of the filtered suspension in a conic centrifuge tube and spin for 3 min at 170 *g* (1500 rpm).
5. Discard the supernatant, add in the centrifuge tube the separation magnesium solution at 1.165 sp. g. up to 11 mL, resuspend in it the pellet settled in the bottom, and then pour into the two flotation chambers of the FLOTAC apparatus (Figs. 35 and 36).
6. Now centrifuge the FLOTAC apparatus (use the adapter to your centrifuge) for 5 min at 120 g (1000 rpm).

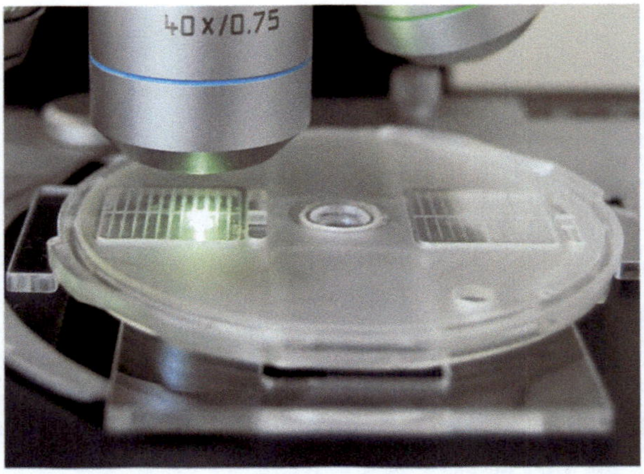

Fig. 37 The Mini-FLOTAC apparatus being analyzed under a microscope. The microscope is focused on one of the two flotation chambers, each of which is divided into 12 sections by the ruled grid on the reading disk to facilitate counting of nematodes in the sample. (From ref. 50, copyright Springer Nature)

7. Finally, move the chambers on the stage of an optical microscope (Fig. 37) and count eggs and J2 (*see* **Note 9**).

3.3.3 Extraction of Cysts from Soil

Cysts are the survival and final life stage of nematodes of the family Heteroderidae, of which the genera *Heterodera* and *Globodera* are the most economically important. In general, they detach from host plant roots and survive in the soil for years (up to 20 years), and species of *Globodera* attacking potatoes are also found on tubers with which they can be dispersed at long distance. Each cyst may contain several hundred eggs, but their content reduces year after year naturally or because of parasitism of other soil microorganisms, especially fungi. Therefore, to ascertain numbers of eggs of cyst nematode species, it is necessary to extract cysts from soil, crush them, and determine the egg content that is instrumental to predict their impact on the yield of following host crop plants.

So far, several methods have been designed to extract cysts from soil but probably the most used is the Fenwick can. Therefore, details on the use of this method are provided, together with some information on the other methods.

Fenwick Can

The method [3] is based on the fact that dried cysts float easily in water and other liquids especially when, after harvest of the host crop plants, the internal structures of the adult females, from which the cyst derives, are completely decomposed or with the aging their egg content is reduced. This method is rapid, and the apparatus does not require much space and while not operating can be left in a storage room. The method is suitable to extract cysts from dried soil and therefore the eggs in cysts can be killed. To obtain living cysts, the described procedures must be slightly modified.

Then proceed as follows:

1. Take the soil sample from the storage room, break clods if any, pour into the small aluminum trays, and leave in a warm room (greenhouse or outside in summer) until the sample is well dried.
2. Mix the dried soil thoroughly, measure 100 or 200 g and pour into a beaker or directly into the Fenwick can, and store the remaining soil until cysts have been extracted, crushed, and their egg content determined.
3. Fill the apparatus with tap water until water gently overflows, nest a 25 mesh sieve on a 60 mesh sieve, and arrange the sieve so that the overflowing materials sieve through them. For easier sieving, insert a space between the nested sieves and the sink. Then pour the sample into the sieves on top of the apparatus, open the shower on top of the apparatus, and regulate the water operating on the tap valve. The water–soil mixture must overflow from the apparatus onto the nested sieves slowly and without creating turbulence and obstruction of the sieves; should this occur, reduce the water flow.
4. When the water overflowing from the apparatus is rather clean, the extraction process can be considered complete; generally, it takes 3–5 min per sample. Then turn off the water, clean the collar of the apparatus with water to collect any attached cysts on the sieves, remove the stopper from the outlet on the base of the apparatus, draw out all soil deposited in the bottom, and wash it clean.
5. Spray water with a nozzle or water stream all over the 25 mesh sieve, and discard all material remaining on this sieve; now spray water on the 60 mesh sieve to eliminate finer soil particles and then try to collect all cysts and soil debris first in a corner of the sieves and then in a beaker or directly on special sieves, with the same nozzle or a laboratory bottle, for further cleaning (*see* **Note 10**).

Seinhorst Cyst Extraction Apparatus

It is used for the extraction of cysts from moist soil samples (usually 600 g) [29]. It is recommended for the extraction of *Globodera*, *Heterodera*, and *Punctodera* cysts if it is necessary to evaluate the "living" content of the cysts, assuming that no further cleaning of the cysts from soil debris is made using rather high temperature and chemicals (alcohol or acetone). The method is based on the different sedimentation rates between the cysts and the soil particles. In the Seinhorst apparatus, an ascending water stream causes the cysts to float, while the heavier soil particles settle. Empty cysts float and are carried by the water to the upper edge of the overflowing apparatus and are collected in a sieve, while heavy, full cysts remain afloat in the upper half of the apparatus and are recovered through an outlet on one side, kept closed during the operation, and are opened at the end to let the cysts in the upper half of the apparatus to be collected in the same sieve.

Follow the steps described hereafter:

1. Pour the sample in the kitchen sieve that must be completely submerged in the water.
2. Shake the sieve horizontally until the sample passes completely through the sieve into the elutriator. When the entire sample has passed into the elutriator, wait for 5 min.
3. Carefully rinse the collar with water to ensure all cysts are washed onto the 175–180 μm aperture sieve. Open the side outlet, so that suspension passes through the 175 μm sieve. Run for 2 min with gentle water flow. Open the outlet at the bottom and clean the elutriator with water. Later, Seinhorst substituted the sieve with a pail of which the bottom was cut and substituted with a net of 100–200 μm, depending on cyst nematode species and cyst size.
4. Therefore, collect both soil–water mixture overflowing and from the outlet when open, into the pail via two different plastic or rubber tubes tied to each other (Fig. 9).

Cysts retained on the sieve can be recovered in different ways (*see* **Note 11**).

California Elutriator

Another simple method to extract cysts from soil is the use of the California elutriator [30]. In general, the apparatus works with dried soil but can also extract cysts from moist soil. Therefore:

1. Measure a sub-sample of 600 g.
2. Open the tap and fill the apparatus until about 10 cm from the hopper, put the sieves under the hopper to collect cysts and soil debris.
3. Pour the sub-sample into the elutriator from the top and mix it in the water with a plastic or wood rod or spoon and then open the tap to create inside the elutriator an upward water stream of about 6.4 L/min that will carry up cysts and finer and light soil debris that will overflow in the sieves.
4. After 2 min, close the tap and pour 2/3 of the soil–water mixture onto the same sieve. Repeat this procedure for 2 more min.
5. At the end, pour all the sediment in the apparatus into a large pail and leave to decant to get rid of the soil.
6. Now take the sieves from under the hopper and proceed as with the Fenwick can and as suggested in Note 12.

Other Methods

The centrifugation method of Coolen is also useful to extract cysts. The procedure is the same for extracting other nematodes from soil, but because cysts, especially newly formed cysts, are heavier than other nematode stages, in the second centrifugation, the density of magnesium sulfate solution must be 1.25 instead of 1.16. In general, the exposure of cysts to magnesium sulfate for the time required for centrifugation (6–7 min) should not harm eggs in the cysts. Moreover, when sieving the supernatant of the second centrifugation, 100–200 μm instead of 5–10 μm aperture sieves can be used, and there is no need to wait a few minutes before collecting cysts from the sieves. If the cysts must be crushed to estimate their egg content, the cyst suspension can be filtered through a large pore size paper filter and then collected with a spatula and put in a container for crushing.

In many nematology laboratories, apparatus and procedures are used that have been modified to adapt the described apparatus to local needs. However, a new apparatus has also been designed, for example, in England the fluidizing column and in the Netherlands the Schuiling centrifuge. In some agricultural services, which handle several thousand samples per year, carousels have been built on which are installed several small elutriators that are run automatically and can process around 200 samples per hour.

Finally, it is noted that cysts can also be extracted with Cobb's sieving and decanting method. In this case, after dissolving the soil in the water of the buckets, before sieving, waiting just a few seconds or as long as necessary considering the soil texture and the size of the sample is enough. Also, for sieving the soil–water mixture, a 25 mesh sieve nested on a 60 mesh sieves is sufficient. At the end of the extraction, while still nested, spray water on the 25 mesh sieve, discard what remains on this sieve, and collect the remaining cysts and soil debris on the 60 mesh sieve into a 100 mL beaker.

Cleaning of Cysts

After extraction, with most of the extraction methods, the cyst suspension is rather dirty and the cysts are difficult to separate and collect. Therefore, cysts must be separated from soil debris and collected for further use.

If the cysts with live eggs will be used, we suggest that they are extracted from partially dried soil and cysts and soil debris collected with a stream of water into 100 or 200 mL beakers. Promptly, put a special filter (Binzer Munktell Filter GMBH, Battenberg, Germany) of 15 cm diam. on a 10 cm diam. funnel and pour in the content in the beaker. While filtering, keep the funnel in your left hand, and with a finger, close the base of the stem and when the cyst–water reaches 2 cm from the edge of the filter, slightly close or open the stem of the funnel so that the water suspension in the funnel remains at the same level. Add some water in the beaker and shake and pour in the same funnel to collect any remaining cyst and

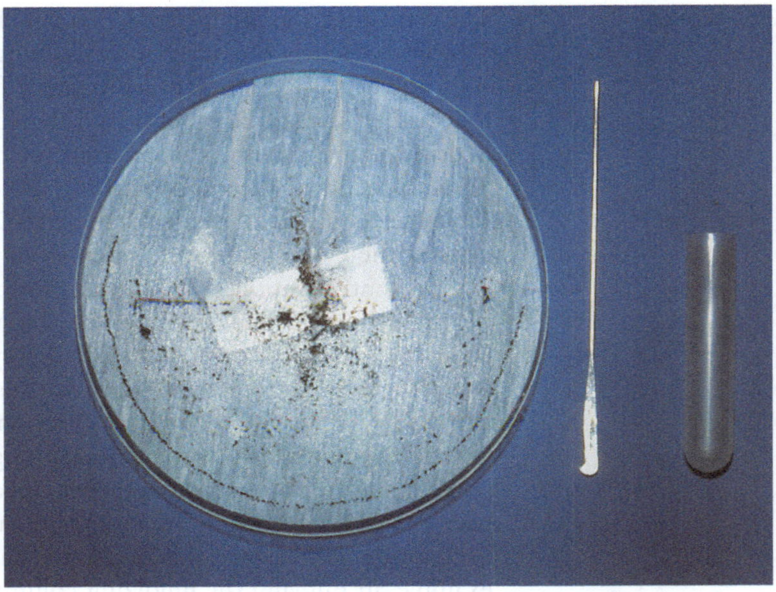

Fig. 38 A watch glass of 15 cm diam. in which is subsided a special filter of the same diam. containing cysts extracted with different apparatus. Note the cysts separated from soil debris and aligned at 1–2 cm from the edge. On the right are a lancet and tube in which cysts are transferred before being crushed. (From ref. 18, with kind permission of SIN)

debris. Then quickly open the base of the funnel hole, so the suspension will drain rapidly, leaving most of the cysts and some debris on a semicircle at 2 cm from the edge (Fig. 38) of the funnel, but some will also sink into the center of the funnel along with soil debris. Now lay the filter on a 15 cm diam. watch glass (just slightly concave), avoiding as much as possible the formation of folds; put the watch glass on the stage of a stereoscope; rotate the glass watch slightly, keeping it slightly sloping toward you; set the magnification at 6–10×; and with a wetted brush in your right hand, pick up cysts first from the semicircle and then also from the center of the filter. The cysts can be temporarily arranged in a semicircle 1 cm from the edge of the filter and collected with a laboratory knife or spatula using them as a rake. These cysts can now be used for many tests such as hatching, nematicide, and inoculation tests. However, if entire cysts will be used, the cyst egg content of five to six replicate samples should be estimated to have a measure of eggs per cyst.

When conducting pot and field trial experiments, generally an estimation of the nematode cysts and eggs, does not matter if alive or dead, is required. As the estimation of dead eggs and cysts is easier because the final nematode–water suspension is clearer, this procedure is preferred.

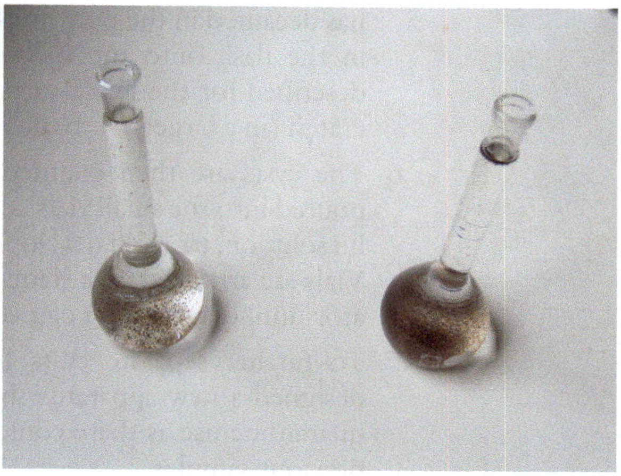

Fig. 39 Flasks for further separation of cysts from soil debris in alcohol. Note that most of the cysts are floating up to the neck of the flasks. (From ref. 18, with kind permission of SIN)

Therefore, when processing soil samples with the Fenwick can, elutriators, or Cobb's sieving and decanting method, it is preferable to start from very dry soil, but not with the Seinhorst elutriator.

The following procedure is mostly according to Seinhorst's [31] and the authors' own experience:

1. At the end of the extraction process, collect cysts and soil debris in the small plastic sieve (Fig. 11a), and put sieves in a tray and the tray in a stove at about 50–60 °C for about 5–6 h or until the mixture of soil debris and cysts is well dried.
2. Take the sample out of the stove; put a clean and dried 25 mesh sieve on a sheet of white paper (A4 size), using a flat painter brush; pour the content of each plastic sieve on the 25 mesh sieve; and with your fingers, gently press on the dried mixture to separate cysts and soil debris from each other and let them fall on the white paper beneath the sieve.
3. Transfer cysts and soil debris from the white paper in the small stainless sieves (Fig. 11a), put them on the electrical warming plate, switch on the power and regulate temperature until you hardly can keep your hand on it, and leave the sample in the stainless sieve to warm up for 5–8 min. This is to further dry cysts for easier floating in the separation liquid.
4. Then, with the help of a small funnel, pour all cysts and soil debris into a flask (Fig. 11b), fill up to 1–1.5 cm of the edge with a separating liquid (alcohol or colloidal silica, Ludox, at 1.25 sp. g.), and shake and leave on a bench until no more soil particles move upward to the collar of the flask.
5. After a few minutes (2–4 min), when all cysts and some soil debris have float on the top of the flask and most of the debris

has decanted in the bottom (Fig. 39), pour nearly all the liquid in the flask onto the special filter placed on the funnel, as described for the living cysts and the separating liquid recuperated (in a larger 1 L beaker) and reused.

6. The cysts are then counted and collected as explained and poured into the small vials containing 1.5 mL of a 0.5% formalin solution, to avoid development of fungi until crushing cysts. Vials are arranged on a stand (Fig. 12) and stored in a refrigerator until determining egg content.
7. To further separate cysts from soil debris, Seinhorst [31] designed a new apparatus that is faster but not suggested for quarantine use as there could be some carryover of cysts from previous samples.
8. During all steps of the procedure, do not forget to label different containers. The use of a sticking tape, on which to write a few identification details of the samples, is helpful as it can be removed at each step.

Because cysts of the same cyst nematode species usually contain different numbers of eggs, the egg content, but not the number of cysts, gives an exact measure of the viable nematode population size. Therefore, it is necessary that cysts are crushed and eggs counted.

In the case of a few cysts, they can be crushed putting them on a glass counting dish and crushing them with laboratory knife or scalpel, or putting cysts between mounting slides and pressing on them. However, when dealing with hundreds or thousands of eggs, these procedures are tedious and time consuming.

To speed up the work, Bijloo [32] designed a special cyst crusher made of plexiglass rod in which is inserted a steel rod to for inserting into a mandrel jaw. This crusher was later modified by Seinhorst and Ouden [11], inserting in the plexiglass rod three helicoidal nickel wires, 1 mm thick, protruding 0.5 mm (Fig. 12).

1. To crush cysts, insert this cyst crusher into the mandrel and the plexiglass rod in the vial containing cysts and run it for about 10 s at 700 rpm, moving the vial up and down slowly and pressing lightly against the crusher with your right hand (Fig. 12).
2. At the University of California, Riverside, a similar devise is used to crush cysts along with soil debris.
3. Cysts can also be crushed using a small leaf triturator (Fig. 12b). From the vial, the cysts are poured into the triturator containing about 1.5 mL water and then with the piston do several up and down strokes.

4. Crushed cysts are then poured into a beaker, and the crusher washed with a laboratory water bottle and the final volume made up to 50 mL or as desired.
5. Put the beaker on a magnetic stirrer and while stirring take an aliquot of egg suspension for counting eggs under a stereoscope using a scaler. We suggest using a 1 mL Peter's counting slide and count eggs in at least three different 1 mL aliquots of the suspension.
6. However, other counting devices can also be used.
7. Avoid using very concentrated egg suspensions as this may impact counting accuracy. Should this occur, make larger dilutions. It is suggested that each observed sub-sample contains from 150 to 250 eggs.
8. At the end of the counting and calculation, report the egg number as per 1 g of dried soil, to the amount of soil you wish, or as eggs per cyst.
9. The triturators reported here can also be used to estimate the content of viable cysts (*see* **Note 13**).

3.4 Extraction of Nematodes from Plant Roots

Many nematodes live or survive in plant parts. These include migratory and sedentary species, whose estimation can be required for a number of purposes, such as diagnosis and quarantine issues, to estimate the effects on plant growth and yield, to assess the degree of infestation, or to assess the effects of any control means or biochemical and molecular changes following given treatments.

3.4.1 Direct Observation

Some larger nematode types can be observed on the roots just by the naked eye in the field, after gently shaking to eliminate adhering soil, or in the laboratory after gently washing the roots, such as for females and cysts of *Globodera* and *Heterodera* species. To detach females and cysts of these species, collected roots can be gently washed free of adhering soil, and then shaken vigorously in a bucket containing water, the water suspension sieved through a 25 mesh sieve (710 μm aperture) nested on a 100–250 μm aperture sieve, and then debris on the 25 mesh sieve discarded and cysts and debris gathered on the 100–200 μm sieve collected.

For other nematodes, the observation of the main symptoms, such as galls in case of infection by *Meloidogyne* spp. or *Nacobbus* spp., can be a clear indication of the presence of these nematodes. However, observation under a stereoscope is easier after dipping roots in some dye solutions that stain all or part of the nematodes.

To stain egg masses of *Meloidogyne* spp. on the roots, phloxine B can be used. The procedure is as follows:

1. Immerge roots in a water solution containing 0.15 g/L phloxine B for 15–30 min and then rinse in water to eliminate excess

stain. The egg masses will retain the stain and therefore will be easier to observe under a stereoscope and also by the naked eye.

2. Acid fuchsin, instead, stains all nematode parts outside and inside plant tissue. Therefore, plant tissues, especially roots, are cut in 1–2 cm pieces and then boiled for 1–3 min in a solution containing 400 mL glycerine, 200 mL water, 200 mL lactic acid, 200 mL phenol, and 0.1 g fuchsin [33]. Boil for 1 min when using tender plant parts and 3 min when dealing with hard woody roots. To eliminate excess stain, the roots after cooling are destained in bottles containing the same solution but without fuchsin, for 1–2 days. Then roots can be observed under a stereoscope or further processed to extract nematodes (by maceration or centrifugation methods) for easier observation and counting.
3. As these stains and phenol are harmful, they should be used wearing personal safety equipment. Moreover, disposal of the remaining solutions as chemical wastes is expensive. Therefore, Thies et al. [34] developed a method in which a food coloring stain is used. The procedure described by these authors is as follows:
4. Infected roots are washed, blotted dry, cut in 1–2 cm long if large, placed in a plastic tissue staining capsule, and covered with about 100 mL of 1.5% sodium hypochlorite water solution in a 200 mL beaker. Small roots (<10 g) are left entirely and not placed in the capsules.
5. Roots are agitated in the solution for 4 min, placed in a kitchen strainer and rinsed with running water for 30 s, and then submerged in about 150 mL water for 15 min.
6. The capsule with the roots is placed in a 100 mL beaker containing a 12.5%, 33.3%, or 50% (v/v) solution of McCormick Schilling red food color available in a 250 mL bottle (McCormick & Co., Inc., Hunt Valley, MD, USA) in distilled water.
7. The solution is brought to a boil for 30 s and then cooled to room temperature and the roots rinsed with tap water, swirled for 15 s in acidified glycerine (40 mL glycerine +5 drops of 5 N HCl) brought at 40 °C, and removed from glycerine.
8. The roots are allowed to cool to room temperature, put between two glass slides (5 × 12 cm), and observed under a stereomicroscope for the colored nematode inside.

The Omwega Method

Omwega et al. [15] developed a method based on the use of erioglaucine dye that stains only the gelatinous matrix of the egg masses but not eggs, females, and roots. Besides, erioglaucine is not toxic to plant and stained plants can be transplanted in larger pots without problem. The method, developed to stain egg masses of *M. incognita* on haricot bean, is useful also to stain other egg masses of other nematodes on different host plants, especially when tested in a small container. It is as follows:

1. Twenty-one days after inoculation, bean plants are irrigated daily and for 7 consecutive days with a solution containing 50 mg erioglaucine/mL.
2. Twenty-eight days after inoculation, bean plants are taken out of the container, placed on glass sheet with diffuse bottom light, and the egg masses counted directly and their root index rated (*see* **Note 14**).

3.4.2 Incubation

Young [6] observed that endoparasitic migratory nematodes were able to come out of different plants parts if they remained wet. Therefore, Young, based on this nematode hydrotropism, designed a method to extract nematodes. The method is known as the Young incubation method and is effective with all migratory endoparasitic nematodes, such as *Pratylenchus* spp., *Radopholus similis*, *Ditylenchus* spp., *Anguina* spp., and *Aphelenchoides* spp., but some researchers use it also to obtain the second-stage juveniles emerging from eggs in egg masses outside roots.

The procedure is as follows:

1. Clean under running water the infected roots, blot them dry, weigh 5–10 g, cut them in 5–10 cm long and if rather thick or hard also lengthwise, and then put in a jar (Fig. 40).

Fig. 40 The Young incubation method to extract migratory nematode stages. Two half liter glass containers with roots, caps for the containers and a small tray (centre) containing clean roots. Labels jars to indicate sample details

2. Wet roots with a spray bottle or by immerging in water and eliminating excess water and close (not firmly) with the cap so that air easily enters into the jars (Fig. 40).
3. Every 24 h, spray water on the roots and collect the water containing nematodes. Be sure that roots always remain well wetted.
4. Continue for about 5–7 days or more if many nematodes are still collected daily (check for nematode in the water suspension under a stereoscope).
5. If too much water–nematode suspension has been collected, concentrate it by decanting in a beaker and, every hour of more, throw away the top of the suspension and repeat if still necessary to reduce the suspension to the desired volume. Also, you may sieve the suspension by nesting a 250 μm aperture sieve, to retain large particles, on a small aperture sieve (10–44 μm), depending on the size of the nematodes you expect to extract, and collect in a beaker or bottle (50–100 mL).
6. Then add water to each bottle up to the same and desired volume, take 1–3 small volume sub-samples (1–5 mL), and count nematode in them under a stereoscope.

This method does not require expensive equipment, is safe, and if run for the necessary time gives good results. In some laboratories, the method is also used to extract second-stage juveniles and males of *T. semipenetrans* from roots of citrus as soon as they hatch from eggs in egg masses. Moreover, during time the method has received modification mainly due to the root container.

3.4.3 Mistifier Chamber

The method of mistifier chambers is based on the same principle of the incubation method, but changes the root container, that in this case is a Baermann's funnel, while wetting of the roots and collection of the water–nematode suspension is automatic. The method was first designed by Seinhorst [7] but has received modifications. We believe that a good modification is that proposed by Crow et al. [12]. However, it is important that the chamber is made of material that does not oxidize while in contact with water, such as plexiglass or colorless steel.

To operate, follow these steps:

1. Set temperature of the room in the range 20–25 °C, depending on expected nematode species to extract.
2. Put coffee filters on the holding ring (probably similar materials would be fine).
3. Put plant material samples directly on the coffee filters and set the mist timer to apply mist for 10 s per hour, to maintain wetness of the plant material.

4. Leave roots and other plant materials on the funnels four 4–7 days.
5. Nematodes exiting the plant material are flushed down to the flask at each misting and decant into the bottom of the flask.
6. To reduce water volume, leave the water–nematode suspension to decant and pour away the excess water several times. Also, you may sieve the flask content though a narrow aperture sieve (10–44 μm) and collect nematodes in small (35–50 mL capacity) bottles and store in a refrigerator.
7. Thereafter, take small aliquots in counting dishes and observe and count nematodes under a stereoscope at 25–40×.
8. Store the remaining nematode suspension for further studies, and then fix nematodes if necessary or discard (*see* **Note 15**).

3.4.4 Maceration and Filtration

The method commonly used today is that proposed by Stemerding [13]. Although the nematodes in the water mixture from the blender can be observed directly under a stereoscope, this is rather difficult because of the plant debris. Therefore, to obtain clean nematode suspensions, the method is combined with another such as Baermann's funnel or centrifugation.

The procedure is as follows:

1. After sampling, the plant material is washed under running water, cut in 0.5–2 cm long pieces, mixed in a bowl containing water, blotted dry, and a 5–10 g sub-sample put in the 250 mL jar and the jar filled 2/3 with tap water.
2. Put the jars on the blender and macerate. In general, macerating 10 s at 8000 rpm (the lowest speed), 5 s at intermediate speed followed by 5 s at more rpm, and then at the two previous speeds would be fine. However, before processing experimental samples, one should check the results obtained with your equipment, especially the best amount of planting material and the speed × time combination. The best combination, depending also on the plant material, is that resulting in the largest number of nematodes extracted and the fewest damaged nematodes.
3. Then clean nematodes using the mentioned methods. However, combining with Baermann's funnels or Oostenbrink dishes will extract only active nematode stages.
4. Then concentrate and store the final nematode suspension in the small bottles.

When extracting nematodes from roots containing swollen stages, such as with root-knot and cyst nematodes, reduce the maceration time and use the lowest two speeds.

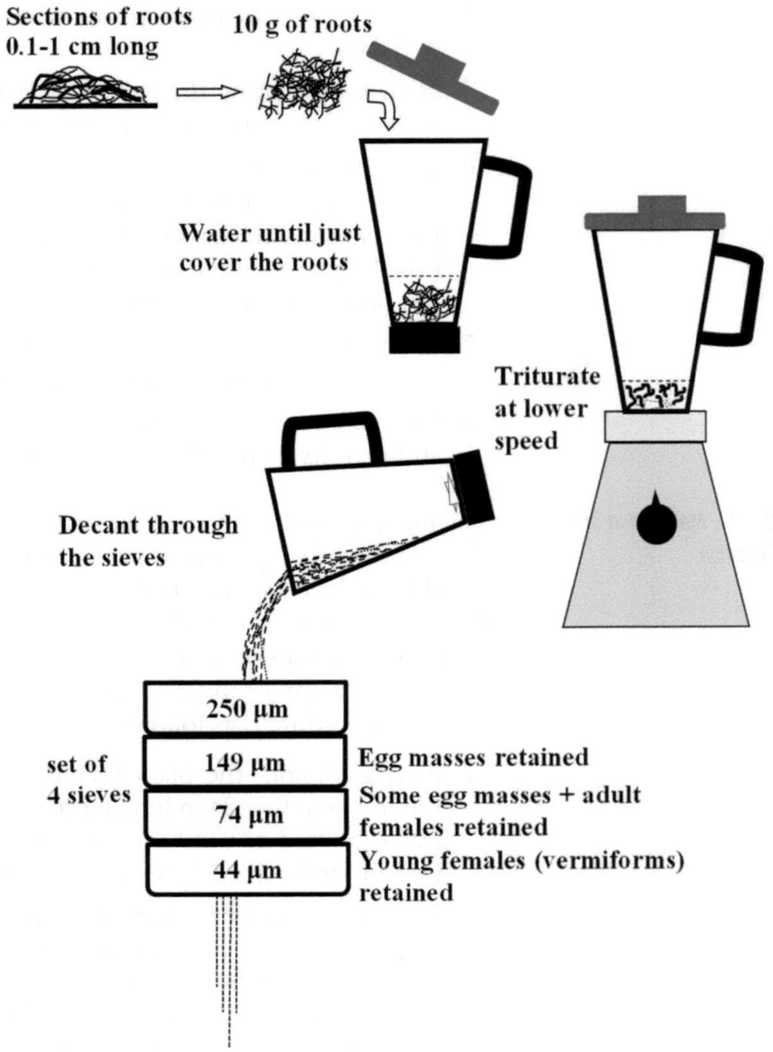

Fig. 41 Scheme to extract different life stages of the reniform nematode *Rotylenchulus reniformis*

3.4.5 Application to Rotylenchulus reniformis

The J2 and following J3 and J4 of *R. reniformis* are not parasitic and retain the cuticle of the previous stages after molting. Young females penetrate the root cortex and establish a permanent feeding site. The anterior part of the body remains embedded in the root, whereas the posterior portion protrudes from the root surface and swells during maturation of the reproductive system, assuming a kidney-like shape, and loses the ability to move. While males and young females (vermiform) can be recovered from soil with other extraction methods, to extract females inserted into root tissues and their egg masses, a more reliable procedure is necessary. Follow the method described hereafter.

1. Gently wash infected roots (Fig. 41), cut with scissors in 0.5–1 cm long pieces, weigh and place 10 g in the blender, add water just to cover the roots, and ground for 8–10 s at the minimum speed of the blender.
2. Sieve the mixture from the blender through a sieve set of 250 μm, 149 μm, 74 μm, and 44 μm pore size, and spread carefully with water. Sedentary females and some egg masses are collected on the 74 μm, while the remaining egg masses are collected on the 149 μm pore sieve. Young vermiform females can also be collected on the 44 μm pore sieve. These suspensions can be observed directly under the microscope or can be further cleaned by separating the nematodes from the plant debris. Therefore, Ganji et al. [35] described a rapid technique to clean samples, for which the sedentary female–root debris mixture is suspended in a 45% sucrose solution in a 50 mL tube, mixed and centrifuged at 450 g for 5 min. In this sucrose concentration, the majority of sedentary females will remain suspended in the solution, while heavier root debris will form a pellet.
3. At the end of centrifugation, sieve the supernatant through a 75 μm pore sieve and wash thoroughly with a water nozzle. Finally, collect nematodes on the sieves in a beaker.
4. If you need individual eggs, place the egg masses collected on the 149 μm pore sieve in a closed bottle containing 1% sodium hypochlorite solution, shake for 3 min, and sieve through the 74 and 25 μm pore sieves. Gently wash the eggs retained on the sieve 25 μm pore and recover them in a beaker for counting.

Depending on the type of roots (woody, fibrous, succulent), it is suggested to first conduct some tests to find the suitable blending time. If the blending time is insufficient, many females and egg masses will not be extracted, and if it is too long, they will break.

3.4.6 Centrifugation

Method of Coolen

This method is also excellent to extract nematodes from plant materials, especially from roots [14]. The preparation step is exactly as in the maceration described previously and requires the same materials, but the final nematode suspension collected, if the procedures were performed correctly, is very clean and can be observed easily under a stereoscope.

Proceed as follows:

1. Macerate the plant materials as explained before.
2. Pour the suspension in the 650 mL tube of the centrifuge, add kaolin (5 g would be enough), and then add water up to 500 mL and balance tubes on the opposite centrifuge arm with water.

3. Run the centrifuge at 1500 rpm for 5 min and discard the supernatant.
4. Add to each tube 500 mL of 1.16 sulfate solution, resuspend in it the sediment using the vibromixer, balance the opposite tube, and now with the same magnesium sulfate at the same density, run the centrifuge again as before.
5. At the end of the centrifugation, pour the supernatant containing nematodes in the 5–10 μ sieve as described in Subheading 3.3.4 in this chapter, collect the nematode suspension in the 50 mL bottle, and store it in a refrigerator until observation and counting under a stereoscope at 25–40× magnification.

To be sure that the amount of kaolin used is enough, after discarding the supernatant of the first centrifugation, check to see whether the sediment is firm and well covered with kaolin; if not increase the amount of kaolin. If you realize that too much kaolin was added, then reduce it.

FLOTAC Technique

This method has been tested also to extract nematode stages from plant roots with good results. The procedures differ in the first steps from that used to extract nematodes from soil and are reported hereafter.

1. Place a sample of 5 g roots in the 250 mL jar of a kitchen blender and add 6% sodium hypochlorite solution to 200 mL and a drop of wetting agent.
2. Ground it thoroughly and then filter through a 1 mm aperture wire sieve in another container. Then add more (800 mL) hypochlorite rinsing the blender jar and the mesh wire to obtain a final dilution rate of 1:200.
3. Mix the root suspension well, pour an aliquot of 11 mL in the conic centrifuge tube, and then follow the same procedures as for extracting nematodes from soil.

3.4.7 Extraction of *Tylenchulus semipenetrans*

As mentioned earlier, the incubation method and mistifier chamber are often used to extract second-stage juveniles and males of root-knot nematodes and of the citrus slow decline nematode *T. semipenetrans* from roots of different *Citrus* species. However, these methods extract only active stages but not swollen sedentary stages. In this case, maceration of roots combined with centrifugation is used. However, maceration may be detrimental to female tonicity as they become wrinkled, reduced in size, and plasmolized by the hypertonic solution of sucrose and magnesium sulfate, with female body content becoming indistinguishable from other debris, especially when swollen stages have been damaged by the blades of the blender. This problem can be overcome by macerating at given times and centrifuging in silica gel (Ludox) at 1.16 sp. g.

The following procedure is according to Greco and D'Addabbo [36] when using *Citrus* roots:

1. Wash gently the root sample in running water, cut in about 0.5 cm long pieces, put them in a bowl, submerge with tap water, mix gently with your hand, blot dry, and take a sub-sample of max 10 g.
2. Put the roots in the 250 mL jar of an electrical mini blender (Fig. 23) and add 150 mL water.
3. Macerate for 15 s at 8000 rpm and sieve through a 250 μm sieve nested on a 10 μm aperture and spray water on the larger pore sieve and gather all nematodes in a corner on the small pore sieve with a tap water spray. If the 10 μm sieve is not available, use a 20 μm sieve (easily available in specialized stores).
4. Then with a water nozzle or a laboratory plastic bottle, pour all roots on the small pore sieve in the 650 mL centrifuge tube, add 500 mL of water and 8 g kaolin, and mix with a vibromixer.
5. Balance the opposite tubes with water and centrifuge at 1600 g (about 2200 rpm) for 5 min.
6. Discard the supernatant water in the tube, substitute it with 500 mL of the colloidal silica (1.16 sp. g.), with the vibromixer, resuspend pellets in the colloidal silica, balance the tubes with the colloidal silica, and centrifuge again as before.
7. Pour the supernatant in the centrifuge tube containing nematodes on the 5–10 μm aperture sieves, rinse with water, and collect in 100 mL beakers or 50 mL bottles using a fine nozzle.

This procedure is very good for extracting at the same time eggs, second-stage juveniles and males, and all juvenile and adult female stages. However, when compared to other methods, the best results are obtained with nematode female stages. As numbers of eggs and second-stage juveniles in the water suspension are much greater than females, when counting nematode specimens, use small suspension aliquots for eggs and second-stage juveniles (three 1 mL samples) and larger (three 5 mL samples) for different female stages. It is also possible to process root samples using a single centrifugation in colloidal silica with very good results, but the nematode suspension is not as clear as with double centrifugation.

3.4.8 Enzymatic Maceration for Extraction of Sedentary Nematode Stages in Biochemical and Molecular Studies

In several biochemical and molecular studies, it may be necessary to have undamaged nematodes that may occur dissecting roots with a needle or scalpel or macerating in a blender. In this case, a gentle method must be used to separate swollen nematodes from plant tissue.

This can be obtained by using pectolytic and cellulolytic enzymes that digest cell walls and free nematodes. The method described hereafter is according to Molinari [37] and is especially designed to extract entire females of *Meloidogyne* spp. from heavily galled roots of annual plants.

The suggested protocol is as follows:

1. Wash roots free of adhering soil and cut them in pieces 0.5–1 cm long, put a sample of 5–20 g in 1 L plastic bottle, and add enough macerating solution to abundantly cover roots so that they float in the solution.
2. Transfer the bottle in the water bath set at 30–35 °C, shake gently, and leave for a minimum of 2–3 h to a maximum of 10–12 h.
3. Then sieve the contents of the bottle on a 710 μm aperture sieve nested on a 250 μm sieve and rinse with water.
4. Nematode females and debris on the 250 μm sieve are then gathered in a beaker (100 mL).
5. Nonmacerated roots on the 710 μm sieve are transferred in the small jars of the blender, add physiological solution at 1%, homogenize for 15 s at the lowest speed, and collect mixture in the same beaker.
6. If only a few females are needed, these can be picked up directly from the nematode suspension in the beaker, while if hundreds or thousands of females are needed, all the water suspension can be processed with Coolen's method to obtain a clean nematode suspension [14].

However, when extracting nematodes for studies at biochemical or molecular level, be aware that a nematode enters stress conditions as soon as it is taken out of its usual environment: soil, plant, and in vitro culture. Also, nematodes undergo stress when stored at temperature far from the optimum for development and reproduction.

3.5 Extraction of Nematodes from Seeds

Many nematodes survive in seeds and their extraction and enumeration is important for quarantine purposes and also to decide if seeds must undergo some treatment before sowing, to avoid nematode dispersion. The most appropriate method to extract nematodes from seeds is Baermann's funnel method or one of its modifications. However, more details are given by Singh et al. [38].

To extract nematodes:

1. Take a sample of seeds and then a sub-sample of 5–10 g, clean it under running tap water if dirty, and then distribute on the small sieve and place on the funnel. Put seeds directly on the sieve net of the small aperture, or, otherwise, put first on the sieve a piece of two to three layers of cheesecloth and then seed on it.
2. Then add water into the funnel to nearly submerged seeds.

3. Leave seeds on the funnel for 4–5 days or until germination, but remove water suspension from funnels every day, check for nematodes, and substitute with fresh water.
4. Then sieve all nematode–water suspensions through a small aperture sieve, collect nematodes in beakers or directly in small bottles, and store in a refrigerator until the entire or part of the nematode suspension is observed and nematodes counted.

We suggest to put on the funnel only a small layer (1–3 mm) of seeds or use more funnels per sub-sample. Moreover, if dealing with rather large seeds, try to crush them a little to promote exit of dormant nematode stages.

3.6 Extraction of Eggs from Egg Masses

In many instances, it is necessary to extract eggs from egg masses especially of root-knot nematodes. This may occur whenever inoculum is needed, and assess the reproduction rate of a nematode or the infestation degree of plants. Although several methods have been used, we suggest using the Hussey and Barker [39] method as it is mostly used, is very simple, and does not require expensive equipment. Also, the eggs are disinfected.

Follow the next protocol:

1. Prepare the sodium hypochlorite solution at 0.5% or 1%.
2. Wash roots free of adhering soil, cut in 1–2 cm long pieces in a bowl containing water enough to submerge the roots, mix, have them blotted dry, and take a sub-sample of 10–20 g if roots come from experiments and 30–50 g if roots are to obtain eggs for inoculation. In this last case, there is no need to have the roots blotted dry.
3. Put root sub-sample in the half liter bottle, add in each bottle about 40 mL of the hypochlorite solution (just to submerge roots), and shake vigorously by hand, or setting the bottles on an automatic horizontal shaker, for 3 min.
4. Then pour the content of each bottle on the 710 μm sieve nested on both the 74 μm sieve and 10 μm sieve, and spray water on the 710 μm sieve to wash away the sodium hypochlorite and flush eggs on the below sieves, discard roots on the large aperture sieve, spray more water on the 74 μm sieve, and finally collect eggs retained on the 10 μm sieve.
5. If collecting eggs for inoculation, the content of several bottles can be sieved at once and then collected in rather large glass cylinders, flasks, or beakers (2–3 L capacity). At the end, estimate the eggs in the entire suspension. Therefore, make rather large dilutions and count numbers of eggs and second-stage juveniles in three 1 mL aliquot and calculate the amount of the eggs available.

6. If the roots are from experimental tests, sieve them separately and store them in 50 mL bottles before counting.
7. Wash all materials and start again (*see* **Note 16**).

3.7 Collection of Second-Stage Juveniles from Cysts and Egg Masses (Hatching Tests)

In several studies, such as investigations on viability of eggs in cysts exposed to different treatments, for inoculation or diagnostic purposes, it is necessary to obtain second-stage juveniles of cyst nematodes (*Globodera* spp., *Heterodera* spp.) or other nematodes forming egg masses, especially the root-knot nematodes (*Meloidogyne* spp.) but also of some cyst nematode species that form large egg masses. This goal can be reached by performing a hatching test using either mature cysts or egg masses. The following procedures are adopted at the former institute of the first author [40]:

1. Collect root diffusate of the host plant or prepare the chemical solution specific for each cyst nematode species (*see* also Chap. 3 in this book).
2. Collect cysts from soil or egg masses from roots under a stereoscope and put about 50–100 cysts or 10–20 egg masses per sieve.
3. Put the sieve in the 3 cm diam. Petri dishes, add about 3 mL of the hatching agent, transfer the small dishes in the larger dishes (Fig. 26), and set temperature at optimal temperature for hatching. In general, set temperature at 25–28 °C for egg masses and cysts of warm season root-knot nematode and cyst nematode species and at 20–22 °C for cool season species of both root-knot and cyst nematode species.
4. Every 5–7 days, renew the hatching solution in the small Petri dishes and count nematodes that hatched in the previous hatching period in the old solution. In doing so, you may use counting slides (especially useful in the presence of many second-stage juveniles) or transfer the 3 cm diam. dishes into larger plastic Petri dishes (5 cm diam) on the bottom, of which several parallel lines with a needle to serve as a guide for counting under a stereoscope at 25× magnification have been drawn externally.
5. Most of the juveniles will hatch during the second to fourth week of incubation, while hatching will be negligible after 7–8 weeks.
6. If it is necessary also to determine numbers of non-hatched eggs, then remove cysts from the small sieve, put them in the small polyallomer vials, and crush as reported for extracted cysts in this chapter.
7. To dissolve the egg mass matrix, put egg masses in a vial containing sodium hypochlorite at 1% and shake or in the 250 mL jars of the blender and homogenize for 3–4 min.

Then count the non-hatched eggs and add them to the number of emerging juveniles to have the total number of eggs in cysts or egg masses at the beginning of the hatching test. Then express data as % of hatch or % non-hatched eggs compared to the total number of eggs, whichever fits to your study.

When investigating the effect on eggs of different substances or temperatures, it is assumed that non-hatched eggs are dead. However, not all eggs hatch at the same time, and some can be in a dormant stage and will hatch much later. The use of Meldola's blue stain to differentiate live eggs has shown to be as good as a hatching test in case of nondormant eggs and better than when using cysts in diapause [41].

3.8 Final Considerations

Sampling and extraction methods described above are the most common and can be useful in most situations. However, some particular needs may require adaptations and/or integration of more methods. We have explained the principles on which the methods presented are based on. Therefore, once a researcher has become familiar with them, we trust that he/she will easily find the best solution.

In plant nematology, and particularly in classical plant nematology, several small tools and even more complex equipment may not be found in the market. Therefore, the researcher must be prepared to build them or have them built by an artisan under his guidance. This is the case of the small pore (5–10 μm) sieves used in Coolen's centrifugation method [14]; sieves used for drying cysts and soil debris before further separation in alcohol, for performing hatching tests; cyst crusher according to Seinhorst and Ouden [11]; or adaptation of commercial centrifuges to nematode extraction.

Finally, the extraction methods described above require time, expertise in nematode diagnosis, and sometimes complex apparatus. Today, molecular taxonomy and systematics are making impressive progress and have become instrumental in the classification and identification of nematodes. A number of studies have shown the possibility of identifying nematodes directly in the soil and some also of estimating nematode population densities and certainly will become more and more important in the future. For more details, we suggest to read the publication by Powers et al. [42] who demonstrated that the internal transcribed (ITS) region is a marker for nematode; Molinari et al. [43] who used superoxide dismutase of isoelectric focusing patterns to differentiate pathotypes of *Globodera* spp.; many articles on identification of nematodes, and some for identification and quantification of nematodes in soil, such as Ciancio and Loffredo [44] with a fluorogenic molecular probe for *M. incognita*; Niu et al. [45] for the four main species of *Meloidogyne* using long isothermal amplification (LAMP) assay; Min et al. [46] for *M. incognita*, *G. rostochiensis*, and *H. glycines* by real-time PCR; Toumi et al. [47] for *H. avenae* and *H. latipons* using qPCR; and Baidu et al. [48] for *Pratylenchus penetrans* using qPCR.

It is likely that much more research on this topic will be published soon. Certainly, their findings will be very helpful especially for identification and population density assessment in soil and roots for nematode dynamics and management purposes.

4 Notes

1. Several nematodes survive as quiescent second-stage juveniles (*Anguina tritici* in winter cereal kernel), fourth-stage juveniles (*D. dipsaci* in seeds of broad been, alfalfa seeds and hay, onion, clover, and *D. gigas* in seeds of broad bean), or adults (*Aphelenchoides besseyi* in rice seeds). Therefore, when searching for these nematode–crop combinations, instead of collecting the entire plants, sampling can be limited to seeds, although the mentioned nematodes can be found in all the aerial plant parts and in plant residues after harvest. Then thresh spikes, collect seeds, and store them until processing to extract nematodes. Nematodes quiescent in seeds or other plant materials may survive for years, and therefore, although nematode populations decline with time, seed samples can be stored at room temperature even for months before extracting nematodes. Seed samples can be collected also for quarantine purposes from seed stocks stored in different containers.
2. Some nematodes may attack plants at crown level, both tap roots and basal top plant. This is the case for carrot and sugar beet infected by the stem and bulb nematode *D. dipsaci*. In these plants, the nematodes attack the top 3–4 cm of the tap roots and the basal 4–5 cm of leaves in carrots and a little more in sugar beet. Therefore, of the collected plants, take separate sub-samples of the top tap roots and basal leaves before cutting them in small pieces and mixing as suggested earlier. *D. dipsaci* is also known to damage bulbous plants such as Alliaceae, tulip, jacinth, and *Narcissus*. In these plants, both aerial plant parts and bulbs are damaged. Therefore, after collecting plants, separate bulbs and leaves, cut them, and take separate samples from stems, leaves, and bulbs. Discard roots as this nematode does not attack them. We suggest that each bulb and top plant is first cut in two halves longitudinally of which one half is discarded and the remaining half is cut as suggested, mixed, and one to two sub-samples used to extract the nematode. For potato, tubers can be attacked by root-lesion nematodes (especially *Pratylenchus penetrans*), stem and bulb nematodes (*Ditylenchus dipsaci* and *D. destructor*), and several species of root-

knot nematodes, such as the warm season species *M. arenaria*, *M. javanica*, *M. incognita*, and the cool season species *M. hapla*, *M. chitwoodi* and *M. fallax*, the last two species considered quarantine organisms. However, while the two species of *Ditylenchus* attack all parts of the tuber, the species of *Meloidogyne* and *Pratylenchus* can be found only in the external 4–6 mm part where xylem and phloem vessels are found. Therefore, the parts of the tuber to sample are also different. To obtain a sample to detect the mentioned *Ditylenchus* species, we suggest to randomly select from half to 1/3 of the potato tubers per plant, clean them free of soil, and cut each tuber longitudinally, one half is then cut horizontally in slice 1 mm thick, and the slices cut in smaller pieces; alternatively, cut the half tuber in several segments longitudinally and then slice transversally taking the segments in one hand. Then mix together the sliced tubers and take one to two sub-samples of 5–10 g or more each. Try to extract nematodes from sub-samples soon, possibly the same day. In case of infection by *Meloidogyne* spp. or *Pratylenchus* spp., the half tubers are used to cut the external 5–6 mm part that is then cut transversally to obtain small pieces, and these are mixed and one to two sub-samples collected.

3. Sometimes, it may be necessary to check woody sawdust, residues, or hand-manufactured articles, such as packaging materials. In this case, collect small pieces of them, grind, take a sample, and tightly close in a labeled plastic sac for transportation to the laboratory.

4. The method is also good for extracting *B. xylophilus* and *B. cocophilus*. One of the most common modifications consists of using plastic or glass Petri dishes of 10–12 cm diam. in which is placed a sieve of 2 cm less in diameter (Fig. 30); this sieve has three to four legs of 2 mm height to favor nematode crossing the sieves and move within the water in the Petri dishes. In general, these sieves are made of a plastic ring 2 cm high closed at one end with a net having the required aperture, such as 90 μm for long nematodes and 60 μm for short nematodes. This modification is useful for seeds and other plant materials and to further clear the nematode–water suspension obtained with Cobb's decanting and sieving method. Store nematode suspensions in a refrigerator and count nematodes without delay. For long-term conservation, hot formalin is added to the suspension to have a final concentration of 2.5–4%, but counting nematodes in formalin is unhealthy and, therefore, counting should be limited to the minimum necessary and just in a very aerated room with rapid air recharge. Instead of the sieve, a sheet of a common metallic net of 1–3 cm aperture can be used to forge as a small cone trunk bowl with the larger base

up, to be adaptable to the internal conical part of the funnel. The modeling can be made by hand or better using a die. The use of a metallic net may cause problem to longidorid nematodes. Be sure that the water level in the funnel always remains in touch with the soil or plant material and check that air bubbles do not form; if this occurs, nematodes will be impeded to move into the water of the funnels. Instead of facial tissues, square pieces of toilet paper can also be used that are large enough and are unscented.

5. Especially when extracting small nematode stages, the final nematode suspension contains too much soil and organic matter particles, making the suspension too dirty and difficult to observe and count nematodes. Therefore, the suspension needs further cleaning, and this is usually done by putting the final nematode suspensions on Baermann's funnel for 24 h or centrifuging. Whenever mixing soil, wear gloves. When sieving the water soil suspension, if this does not sieve well because of the many small soil particles, leave the soil suspension to decant longer and/or reduce the size of the soil sub-samples. An experienced operator will soon learn how to make the best selection.

6. This is a very good extraction method used in many nematology laboratories, including that of the second author for nearly 50 years. However, with this method, it is not recommended to use more than 200 cm^3 of soil (100 cm^3 or less is the adequate amount). When washing large samples (>200 mL), in the funnel, it is necessary to increase the volume of ascending water to prevent nematodes from being dragged to the bottom by the large number of soil particles or other materials. This also means that a greater amount of dirt remains in the suspension. For nematodes that are extremely sensitive to mechanical damage or friction (e.g., Trichodoridae), the top sieve is not used. Instead, the sample is pre-suspended with a shaker or hands before and brought into the funnel by pouring it over your hand at the top of the funnel. You can also place the soil in the funnel and wash gently without causing friction with your hands or agitator. Four 45 μm sieves is the adequate number for retention of most nematodes; however, for small and slowly moving nematodes, such as *Paratylenchus* or juvenile stages of some Criconematidae, it is better to use another extraction method such as the centrifuge or to increase the number of sieves. Once the sieve is placed on the plate, it must not move; otherwise, the suspension becomes dirty. Observe if water evaporation occurs; if so, add small amounts through the edge so that the filter does not lose contact with the water in the tray. You can also cover the dishes to prevent the evaporation of water.

7. At the end of the first centrifugation, be sure that the sediment remains firmly attached to the bottom of the centrifuge tube and does not mix with the supernatant when pouring it off. If not, add about 0.5 g kaolin (less with soil containing high clay content and more with sandy soil) and check that at the end of the centrifugation, the sediment is well separated from the supernatant. Also, be sure that no kaolin remains attached on the internal wall of the centrifuge tubes; it may remain suspended in the sugar solution and results in a dirty final nematode suspension. At the beginning of the second centrifugation, be sure that all sediment in the tube is well dispersed in the sugar solution; the use of glass or transparent plastic tubes and vibromixers or rods for stirring is helpful.
8. When extracting from soil nematodes producing egg masses, the recovery of the eggs is improved adding in each column 150 mL of a 6% sodium hypochlorite solution (common bleach) to have a concentration in the column of about 0.22%. This solution dissolves egg masses that are then dispersed in the soil–water mixture in the column and recovered. This improvement is particularly useful when extracting nematodes soon after uprooting plants, when only a small proportion of juveniles have already hatched from the eggs. Also, adding bleach will cause foaming of the soil mixture in the mixing column; therefore, before bubbling air, add a few drops, or as many as necessary, of an antifoam. However, when adding bleach in the tube, be sure that you are working in a very well-aerated room. If the centrifugation method is used to extract cysts, the sp. g. of the magnesium solution must be 1.25; otherwise, new and full cysts will not float and get lost in the sediment during the second centrifugation. Also, in case of soil containing longidorids, it is highly recommended to use colloidal silica (Ludox^{R} AS, DuPont de Nemours. It is a material used to make surfaces anti-slippery) at 1.16 sp. g. Silica does not exert osmotic pressure that would damage nematode specimens and get lost. Originally, Ludox is denser than 1.16 sp. g. and, therefore, must be diluted with tap water.
9. FLOTAC basic technique. This is an innovative technique designed for uro-copromicroscopic diagnosis of protozoal and helminthic infections of humans and animals [49, 50] and has been patented by the Department of Pathology and Animal Health, University of Naples Federico II, Italy. The method is very simple and also accurate to extract nematode eggs and other stages of the root-knot nematode, *M. incognita*, from soil and plant roots [28]. The technique is used at the Centre for Monitoring of Parasites (CREMOPAR) (Eboli, Salerno, Italy) and at the WHO Collaboration Centre of the University of Naples. Detailed information can be found at

www.flotac.unina.it and at www.parassitologia.unina.it. CREMOPAR is pleased to welcome, free of charge, colleagues wishing to be familiarized with the technique.

10. The amount of material should be proportioned to the extent of the workload. In many cases, it is necessary to extract living eggs and therefore the soil must not be dried in the warm room, but the sample can be left at room temperature in a shaded area until the excess water has evaporated and the soil does not stick when pressed between fingers. However, these cysts are heavier than those in well-dried soil. Moreover, and especially when collecting soil samples soon after harvest of the host crop, new cysts are heavier than old cysts, and they may sink into the sediment in the apparatus, if operated as described, and be lost. To avoid this problem, in the laboratory of the first author, the procedure has been slightly modified. A rubber or plastic pipe is connected to one side to the tap and the other to a rigid pipe (4–5 mm internal diameter) inserted into the rubber stopper that closes the outlet of the apparatus. With this modification, it is possible to create in the apparatus also an upward stream of water, operating similar to an elutriator.
11. Large samples can be handled in a standardized way. Compared to extraction methods where dry soil is used, this method captures fuller cysts. Another advantage is that it is not necessary to dry the sample. However, large amounts of water are needed, and the equipment is expensive and requires space to install two elutriators, suggested to speed up the work. For some species of nematodes with small cysts such as *H. carotae*, *H. cruciferae*, and *H. urticae*, it is recommended to use the 100 μm sieve. On the other hand, when the cysts are very large (e.g., *H. schachtii*), the 250 μm sieve can be used, allowing the sample obtained to be cleaner.
12. The flow rate of the water inside the elutriator must be regulated according to soil type (more when dealing with rather sandy soil and less when working with rather clayey soil). Also, you may need to increase the upward water stream if working with soil that may contain newer cysts that are heavier. See also suggestions given for the Fenwick can.
13. The first separating liquid to be used and still used in some laboratories is acetone. As exposure to acetone is a health hazard, Seinhorst [31] substituted it with denatured alcohol. However, denatured alcohol is also hazardous as it is methylated. Therefore, as done in the laboratory of the first author, we suggest to use pure alcohol (95–96%). When using acetone or alcohol, we suggest to work under a fume hood and leave samples on the filter until acetone and alcohol have completely

evaporated and then add in the center of the filters a few drops of tap water to keep the filters wet; if samples are not furtherly processed, soon they can be covered with another watch glass and kept on a laboratory table. When counting cysts on the filter, be sure that the filter is wet enough; if not, it will be difficult to pick up cysts with the brush. The same problem may occur if in the center of the filter there is too much water, because cysts may float. The authors found that when dealing with newly formed lemon-shaped cysts, about 20% of them can sink in the flask and be lost. Therefore, as a separating liquid, a 1.25 sp. g magnesium sulfate solution is used. This solution is a little sticky but will result in a greater recovery of the fullest cysts. When separating cysts, do not leave the cyst suspension in the flask for a long time; otherwise, cysts may sink and be lost. Paper filters used in the final step are characterized by allowing a fast flow of the separating liquid. Finally, it is necessary that the person in charge of crushing cysts becomes familiar with the method before undertaking routine work. Optimal crushing must result in a limited number of damaged eggs and second-stage juveniles and of eggs sticking to each other.

14. If there are females and cysts of cyst forming nematodes on the roots, these detach easily from the roots. This use of the red food color is also effective to stain egg masses. For it, the blotted dry roots are placed in a 500 mL beaker containing 10% or 20% (v/v) of the same red food color for 15 min, rinsed in tap water, and blotted dry. Then egg masses can be observed directly or under a stereoscope at 20× magnification. The staining with the food red coloring stain, as shown by photos provided by the authors, appears as good as that of fuchsin and phloxine B and do not pose health, environmental, and economic issues. For information on modification of some methods for adaptation to specific needs, you may wish to see also Byrd et al. [51]. Realize that phloxine B does not stain egg masses of *M. artiellia*. As phenol is dangerous, it has been eliminated from the acid fuchsin method without loss of efficiency.

15. The obtained water–nematode suspension is clear. This mistifier box, designed to extract *Meloidogyne* second-stage juveniles from golf course grasses, also can be useful to extract many endoparasitic nematodes. This method is also effective to extract *B. xylophilus* and *B. cocophilus*. For the extraction of nematodes, 15 g of minced tissue is placed in a blender with 50 mL of water and processed for 2 min. The contents of the blender are then poured into a dish and left for 20 min for the nematodes to emerge. The nematodes are then recovered by

sieving. Generally, advantage is taken of the level of activity of the nematode during extraction methods [52].

16. Although this is a very good method, the eggs extracted are much less infective [53, 54], probably because of the deleterious effect of sodium hypochlorite and that eggs isolated from egg masses can be damaged while mixing into the soil. To minimize this effect, when collecting eggs for inoculum, we suggest using a sodium hypochlorite solution at 0.5% and not at 1%. The 1% solution should be preferred whenever studying dynamics of populations or reproduction rates of the target nematode. When studying the effects of treatments on the nematodes, in case of severe root infection at the end of the trial, if large galls are developed on the roots, it is likely that many egg masses remain embedded into the roots. Then we suggest macerating the roots in jars of a mini blender in which water has been substituted with a 1% sodium hypochlorite. Then sieve through a 10–20 μm aperture sieve and rinse with water, to wash away the chemical solution, and centrifugate as explained. Maceration and hypochlorite will dissolve egg masses.

Acknowledgments

Most of the figures in this chapter are from the first author already published in Greco and Carletti (2014) [13]. They are here reused with the kind permission of the publisher, the Italian Society of Nematology (SIN), to which we are very grateful. A special thank goes to Prof. Philip Roberts, Riverside, CA, USA, for critically checking the manuscript and the precious suggestions, and to Dr. Alberto Troccoli, Bari, Italy, for helping with final formatting of most of the figures.

References

1. Baermann G (1917) Eine einfache Methode zur Auffindung von Ankylostomum (Nematoden) Larven in Erdproben. Genesk Tijdschr Ned-Indië 57:131–137
2. Cobb NA (1918) Estimating the nema population of the soil. Agriculture Technical Circular of the Bureau of Plant Industry, U.S Department of Agriculture, No. 1. 48 pp.
3. Fenwick DW (1940) Methods for the recovery and counting of cysts of *Heterodera schachtii* from soil. J Heminthol 18:155–172
4. Oostenbrink M (1960) Estimating nematode populations by some selected methods. In: Sasser JN, Jenkins WR (eds) Nematology. University of North Caroline Press, Chapel Hill, pp 85–102
5. Seinhorst JW (1956) The quantitative extraction of nematodes from soil. Nematologica 1: 249–267
6. Young TW (1954) An incubation method for collecting migratory endoparasitic nematodes. Plant Dis Rep 38:794–795
7. Seinhorst JW (1950) De betekensis van de toestand van de grond voor het opreden van aantasting door het stengelaaltje (*Ditylenchus dipsaci* (Khün) Filipjev). Tijdschr PlZiekt 56: 289–348

8. Jenkins WR (1964) A rapid centrifugal-flotation technique for separating nematodes from soil. Plant Dis Rep 48:692
9. Been TH, Schoemaker CH (2006) Distribution patterns and sampling. In: Perry RN, Moens M (eds) Plant Nematology. CABI, Wallingford, pp 302–326
10. van Bezooijen J (2006) Methods and techniques for nematology. Wageningen University, Wageningen. 112 pp.
11. Seinhorst JW, den Ouden H (1966) An improvement of the Bijllo's method for determining the egg content of *Heterodera* cysts. Nematologica 12:170–171
12. Crow WT, Habteweld A, Bean T (2020) Mist chamber extraction for improved diagnosis of *Meloidogyne* spp. from golf course bermudagrass. J Nematol 52:1. https://doi.org/10.21307/jofnem-2020-96
13. Stemerding S (1964) Een mixer-wattenfilter methode om vrijbeweeglijke endoparasitaire nematoden uit wortels te verzamelen. Versl. Meded. Plziektenk. Dienst Wageningen, No 141 (Jaarbock 1963), 170–175
14. Coolen WA (1979) Methods for the extraction of *Meloidogyne* spp. and other nematodes from roots and soil. In: Lamberti F, Taylor CE (eds) Root-Knot nematodes (*Meloidogyne* species): systematics, biology and control. Academic Press, London, pp 317–329
15. Omwega CO, Thomason IJ, Roberts PA (1988) A nondestructive technique for screening bean germ plasm for resistance to *Meloidogyne incognita*. Plant Dis 72:970–972
16. Seinhorst JW (1965) The relationship between nematode density and damage to plants. Nematologica 11:137–154
17. Shoemaker CH, Been TH (2006) Plant grow and population dynamics. In: Perry RN, Moens M (eds) Plant Nematology. CAB International, Wallingford, pp 275–301
18. Greco N, Carletti B (2014) Estrazione, allevamento e conservazione dei nematodi. In: Ambrogioni L, d'Errico FP, Greco N et al (eds) Nematologia Agraria Generale e Applicata. Società Italiana di Nematologia (SIN) Firenze, Italy, pp 115–153
19. Schröder T, McNamara DG, Gaar V (2009) Guidance on sampling to detect pine wood nematode *Bursaphelenchus xylophilus* in threes, wood and insects. EPPO Bull 39:179–188
20. Esser RP, Meredith J (1987) Red ring nematode. Nematology Circular N. 141. Florida Department of Agriculture and Consumer Services, Division of Plant Industry. 4 pp
21. Calvache H, Mora S, Guevara LA (1995) Anillo rojo – hoja corta en palma de aceite. Tecnología disponible en Colombia. Palmas 16:211–218
22. Martínez RGJ (1970) *Rhynchophorus palmarum*, vector of the red ring nematode in Venezuela. (*Rhynchophorus palmarum* L. (Coleóptera, Curculionidae) portador del nematodo del "anillo rojo" en Venezuela). Revista Fac Agron Maracay 5:81–85
23. Crozzoli R (2014) La nematologia agrícola en Venezuela. Universidad Central de Venezuela, Facultad de Agronomía, Ediciones de la Universidad Central de Venezuela, Maracay. 447 pp. CD
24. Sarria GA, Riascos-Ortiz O, Medina HC et al (2022) Molecular identification of *Bursaphelenchus cocophilus* associated to oil palm (*Elaeis guineensis)* crops in Tibu (North Santander, Colombia). J Nematol 117
25. Cesarz S, Schulz AE, Beugnon R et al (2019) Testing soil extraction efficiency using different variations of the Baermann-funnel method. Soil Org 91:61–72
26. Saikai K, Vanjala S, Oduori C et al (2021) Revisiting the modified Baermann extraction method: extraction efficiency of *Radopholus similis* using different extraction materials. Nematology 23:1215–1218
27. Caveness FE, Jensen HJ (1953) Modification of the centrifugation-flotation technique for the isolation and concentration of nematodes and their eggs from soil and plant tissues. Proc Helminthol Soc Wash 22:87–89
28. Troccoli A, D'Errico G, D'Addabbo T, Sasanelli N, Bosco A, Maurinelli MP, Rinaldi L, Cringoli G (2022) The FLOTAC basic technique as a new extraction method for root-knot nematodes (*Meloidogyne* spp.) from soil and roots. Front Parasitol. https://doi.org/10.3389/fpara.2022.1000673
29. Seinhorst JW (1964) Methods for the extraction of *Heterodera* cysts from not previously dried soil samples. Nematologica 10:87–94
30. Caswell E, Thomason IJ, McKinney HE (1985) Extraction of cysts and eggs of *Heterodera schachtii* from soil and assessment of extraction efficiency. J Nematol 17:337–340
31. Seinhorst JW (1974) Separation of *Heterodera* cysts from dry organic debris using ethanol. Nematologica 20:367–369
32. Bijloo JD (1955) Een eenvvoudige methode voor het bepalen van de inhoud van grote aantallen cysten van *Heterodera*-soorten. Meded LandbHoogesch OpzoekStns Gent 20:291–300
33. Southey JF (1970) Laboratory methods for work with plant and soil nematodes. In: Technical Bulletin. Ministry of Agriculture and

Fisheries & Food No. 2, 5th edn. Her Majesty's Stationery Office, London, 148 pp

34. Thies JA, Merril SB, Corley EL (2002) Red food coloring stain: new safer procedures for staining nematodes in roots and egg masses on root surface. J Nematol 34:179–181

35. Ganji S, Wubben MJ, Norton JJ (2013) Two simple methods for the collection of individual life stage of reniform nematode, *Rotylenchulus reniformis*. J Nematol 45:87–91

36. Greco N, D'Addabbo T (1990) An efficient procedure for extracting *Tylenchulus semipenetrans* from *Citrus* roots. J Nematol 22:590–593

37. Molinari S (2009) Bioassay on plant-nematode interaction. In: Narwal SS (series ed), Sanpietro DA, Catalàn CAN, Vattuone MA et al. (co-eds) Plant bioassay. Science Publisher, Enfield, pp 293–326

38. Singh RK, Pandey SK, Chattopadhyay (2020) Detection and diagnosis of seed-borne and seed-associated nematodes. In: Kumar R, Gupta A (eds) Seed-borne diseases of agricultural crops: detection, diagnosis & management. Springer Nature Singapore Pte Ltd, pp 201–229. https://doi.org/10.1007/978-981-32-9046-4_8

39. Hussey RS, Barker KR (1973) A comparison of methods of collecting inocula of *Meloidogyne* spp., including a new technique. Plant Dis Rep 57:1025–1028

40. Greco N, Brandonisio A, De Marinis G (1982) Investigation on the biology of *Heterodera schachtii* in Italy. Nematol Medit 10:201–214

41. Kroese D, Zasada IA, Ingham RE (2011) Comparison of Meldola's blue staining and hatching assay with potato root diffusate for assessment of *Globodera* sp. egg viability. J Nematol 43:182–186

42. Powers TO, Todd TC, Burnel AM et al (1997) The rDNA internal transcribed spacer region as a taxonomic marker for nematodes. J Nematol 29:441–450

43. Molinari S, Greco N, Miloslav Z (2010) Superoxide dismutase isoelectric focusing patterns as a tool to differentiate pathotypes of *Globodera* spp. Nematology 12:751–758

44. Ciancio A, Loffredo A, Paradies F et al (2005) Detection of *Meloidogyne incognita* and *Pochonia chlamydosporia* by fluorogenic molecular probes. EPPO/OEPP Bull 35:157–164

45. Niu J-h, Guo K-x, Jian H et al (2011) Rapid detection of *Meloidogyne* spp. by LAMP assay in soil and roots. Crop Prot 30:1063–1069

46. Min YY, Toyota K, Sato E (2012) A novel diagnostic method using the direct quantification of major plant-parasitic nematodes in soil by real-time PCR. Nematology 14:265–276

47. Toumi F, Waeyenberg L, Viaene N et al (2015) Development of qPCR assay for quantitative detection of *Heterodera avenae* and *H. latipons*. Eur J Plant Pathol 143:305. https://doi.org/10.1007/s10658-015-0681-0

48. Baidoo R, Yan G, Nagachandrabose S (2017) Developing a qPCR assay for direct identification and quantification of *Pratylenchus penetrans* in soil. Plant Dis 101:1432–1441. https://doi.org/10.1094/PDIS-01-17-0117-RE

49. Cringoli G, Rinaldi L, Maurelli MP, Utzinger J (2010) FLOTAC: new multivalent techniques for qualitative and quantitative copromicroscopic diagnosis of parasites in animals and humans. Nat Protoc 5:503–515. https://doi.org/10.1038/nprot.2009.235

50. Cringoli G, Maurelli MP, Levecke B, Bosco A (2017) The mini-FLOTAC technique for the diagnosis of helminth and protozoan infections in humans and animals. Nat Protoc 12:1723–1732. https://doi.org/10.1038/nprot.2017.067

51. Byrd DW, Kirkpatrick T, Barker KR (1983) An improved technique for clearing and staining plant tissues for detection of nematodes. J Nematol 15:142–143

52. Schuiling M, Van Dither JBM (1981) 'Red ring disease' in the Paricatuba oil palm Estate, Para, Brazil. A case study. Z Angew Math Phys 91: 154–169

53. Di Vito M, Greco N, Carella A (1986) The effect of *Meloidogyne incognita* and importance of the inoculum on yield of eggplant. J Nematol 18:487–490

54. Gomez-Gonzales G, Cruz-Lachica I, Màrquez-Zequera I et al (2021) *Meloidogyne enterolobii* egg extraction in NaOCl versus infectivity of inoculum on cucumber. J Nematol 57:1. https://doi.org/10.21307/jofnem-2021-057

Chapter 2

Rearing and Conservation of Nematode Populations Under Greenhouse or Axenic Conditions

Nicola Greco and Alberto Troccoli

Abstract

Once a nematode has been identified, to conduct studies for screening programs or pathogenicity tests, it is necessary a supply of large numbers of nematodes from field crops or reproduced and stored to be used in periods of the year when they are not available from fields. Therefore, nematodes must be reared in greenhouse or under in vitro conditions and stored for future needs. In this chapter, suggestions are given on how to obtain nematodes from fields and reproduce most of them on host plants in greenhouse (mainly *Meloidogyne* spp. and *Globodera* spp.) or in vitro. Reproductions in vitro include:

- On suitable callus of host plants (*Pratylenchus* spp., *Ditylenchus* spp.)
- On fungal cultures mainly of *Botrytis cinerea* or *Alternaria* spp. for *Aphelenchoides* spp. and other aphelenchids, including *Bursaphelenchus xylophilus*.
- On carrot disks for *Pratylenchus* spp. and *Ditylenchus* spp.
- Other specific media, such as garlic, potato, and sweet potato for *D. destructor*, and cocoyam disks for *Radopholus similis*.

Guidelines are also given to store different nematodes for rather long times, including in vitro methods and in infected seeds, hay, and other plant parts. No information is given on how to prepare and store fixed materials.

Key words Nematode supply, Greenhouse and in vitro cultures, Nematode storage

1 Introduction

In Chap. 1 of this book, methods of sampling and techniques for extracting nematodes have been described and their usefulness in different situations highlighted. In many cases, the nematodes extracted from a single sample are not sufficient to conduct studies and run experiments. Moreover, under field conditions, nematodes may not be available all the year round, and, therefore, there is the need to rear them, to obtain the necessary amount, and store for

Sergio Molinari (ed.), *Plant-Nematode Interactions*, Methods in Molecular Biology, vol. 2756,
https://doi.org/10.1007/978-1-0716-3638-1_2,

use throughout the year. This can be done by collecting nematodes from fields, during the season favorable to reproduction, or rear nematodes in greenhouse or growth chamber, and when possible, under in vitro conditions. Usually, when nematodes are at optimum environmental conditions, they reproduce continuously, while under natural conditions, or anyhow under varying conditions, they may go through a dormancy stage. Thereafter, a certain period of time and chilling conditions may be necessary before they return active and infective. This is the case of *M. naasi* [1], *Globodera* spp. [2], cysts of *Heterodera carotae* [3] and *H. goettingiana* [Greco, unpublished, 1986, Bari, Italy], and probably also *H. ciceri*. Cysts of the *H. avenae* group species in temperate areas require an 8-week period at 7 °C for a significant amount of hatched eggs [4]. Under Mediterranean climate conditions, second-stage juveniles (J2s) *H. avenae* do not hatch during summer [3, 5], but they do hatch as soon as first rains occur and temperature drops to 15 °C in autumn [3]. Therefore, nematode rearing and storage must be planned well in advance to have them ready to inoculate plants by the time of their use in experiments. When experiments require a large number of nematodes, such as pathogenicity tests in micro-plots or screening of large numbers of seeds/plants from gene banks or breeding programs. Hereafter, we will explain the easiest ways to rear and store nematodes. Much more information can be found in EPPO standards and diagnostics to rear and store quarantine nematodes [6] and yearly updates on the EPPO website (https;//www.eppo.int). The same information should be available from other regional plant protection organizations.

2 Materials

2.1 Collecting Nematodes from Host Plant Crops Under Field Conditions

The most economic way to obtain large population densities of nematodes is making profit of infested field crops. Therefore, it is necessary to identify fields infested with the target nematode and have knifes to cut plants, hoes to dig the soil, a place to store the soil (such as bins or sheds), a lavatory to free the roots from the soil, and a nematology extraction laboratory.

2.2 Rearing Nematodes in Greenhouse

This is a good way to obtain large quantities of nematodes throughout the year and to maintain nematode in a collection, assuming that temperatures adequate to different nematodes are available.

A greenhouse is required in which soil temperature can be set at 18–22 °C for temperate nematodes and crops, or at 23–28 °C for warm season nematodes. The greenhouse should be provided with:

- An automated irrigation system.
- Benches made of inoxidizable material.

- Grating shelves to allow easy flow of water in case of overirrigation.
- Shields to separate pots of different nematode species, virulence groups, and populations.

Pots must be of the desired size, better made of clay to be filled with sterilized soil containing at least 80–90% sand. We suggest to avoid the use of organic material that would make cleaning the roots and extraction of the nematodes, especially cysts, more problematic. To improve fertility of the sandy soil, a solution of a compound fertilizer containing also microelements can be periodically (7–10 days) applied (*see* **Note 1**).

2.3 Obtaining Nematode Populations from a Single Female, Egg Mass, and Cyst

In the case of root-knot nematodes (RKNs) (*Meloidogyne* spp.), reniform nematodes, *Rotylenchulus* spp., and *Tylenchulus* spp., the materials needed include:

- Plants with roots parasitized with the desired nematode.
- Sterilized sandy soil.
- 10–12 cm diam. pots
- A stereoscope.
- Forceps, lancets, needles, and other dissecting items.
- Plants to infect at the 4–6-leaf stage if annuals or 10–15 cm tall if perennials.
- A granular NPK + microelement fertilizer.

2.3.1 Other Nematodes

For all other species of nematodes, just a single female must be used, and, therefore, the chance to establish on a host plant and reproduce is much less than using egg masses or cysts. This is the case of both endoparasitic migratory and ectoparasitic nematodes.

Therefore, you need a population of the nematodes to reproduce, seeds of the host plant, Petri dishes, plastic or Styrofoam coffee or teacups, 12–14 cm diam. pots, a growth cabinet, greenhouse, and fertilizing solution.

2.4 Rearing Nematodes Under In Vitro Conditions

The continued availability of nematodes or their reproduction under field or greenhouse conditions may require much space and time for maintenance. In several studies, sterile nematodes are necessary. This goal can be achieved by growing nematodes on monoxenic culture. This has been possible after several attempts and at least with species of the nematode genera *Ditylenchus*, *Pratylenchus*, *Radopholus*, and *Aphelenchoides*, by growing them on callus of host plants. In turn, this requires the use of proper media. Other nematodes, especially aphelenchids, are known to reproduce on fungal cultures.

2.4.1 Rearing Nematodes on Host Plant Callus Cultures

Common equipment and materials are laminar fume hood to work under sterile conditions, autoclave to sterilize media, sterile glass tubes or other container for media, and flasks of 0.5–2 L according to workload, ingredients to prepare media, precision scale to measure ingredients, distilled water, cabinet to grow callus and reproduce nematodes, chemicals to sterilize seeds [sulfuric acid (H_2SO_4) and chloride mercury ($HgCl_2$)], and nematodes [6–8]. The following media are suitable to growth callus of onion, alfalfa, and clovers, for reproduction of *D. dipsaci* and several species of *Pratylenchus*, and of peanut for *D. africanus*:

- White's medium as integrated by Riedel and Foster [9]: 1% agar; 2% sucrose; mg per liter of sodium sulfate (Na_2SO_4) 200, calcium nitrate [$Ca(NO_3)_2{\cdot}4H_2O$] 300, magnesium sulfate (-$MgSO_4{\cdot}7H_2O$) 750, potassium nitrate (KNO_3) 80, potassium chloride (KCl) 65, sodium phosphate ($NaH_2PO_4{\cdot}H_2O$) 19, manganese sulfate ($MnSO_4{\cdot}4H_2O$) 20, zinc sulfate (-$ZnSO_4{\cdot}7H_2O$) 12, boric acid (H_3BO_3) 6, potassium iodide (KI) 3, ferric tartrate [$Fe_2(C_4H_4O_6)_3{\cdot}4H_2O$] 2.5, cupric sulfate ($CuSO_4{\cdot}5H_2O$) 0.4, molybdenum trioxide (MoO_3) 0.004, glycine 3, thiamine HCl 0.1, 2,4-dichlorophenoxiacetic acid 2, nicotinic acid 0.5, pyridoxine 0.1, integrated with α-naphthaleneacetic acid 0.1.
- Krusberg's medium as integrated by Riedel and Foster [9]: 1% agar; 2% sucrose; mg per liter of sodium sulfate (Na_2SO_4) 800, calcium nitrate [$Ca(NO_3)_2{\cdot}4H_2O$] 400, magnesium sulfate ($MgSO_4{\cdot}7H_2O$) 180, potassium nitrate (KNO_3) 80, potassium chloride (KCl) 65, sodium phosphate ($NaH_2PO_4{\cdot}H_2O$) 33, manganese sulfate ($MnSO_4{\cdot}4H_2O$) 4.5, zinc sulfate (-$ZnSO_4{\cdot}7H_2O$) 6, boric acid (H_3BO_3) 0.375, potassium iodide (KI) 3, ferric tartrate [$Fe_2(C_4H_4O_6)_3{\cdot}4H_2O$] 40, glycine 3, thiamine HCl 0.1, 2,4-dichlorophenoxiacetic acid 2, integrated with α-naphthaleneacetic acid 0.1 and 150 ml of coconut milk/liter.
- Murashige and Skoog medium: Van Der Walt and De Waele [10], for reproduction of *D. destructor* isolated from peanut in South Africa (later described as the new species *D. africanus*), used callus of peanut grown on nutrient medium modified after Murashige and Skoog [11], containing the following ingredients expressed in mg/L, if not otherwise specified:

NH_4NO_3 1650, KNO_3 1900, $CaCl_2{\cdot}2H_2O$ 440, $MgSO_4{\cdot}7H_2O$ 370, KH_2PO_4 170, KI 0.83, H_3BO_3 6.3, $MnSO_4{\cdot}7H_2O$ 22.3, $Na_2MoO_4{\cdot}2H_2O$ 0.25, $CuSO_45H_2O$ 0.025, $CoSO_4{\cdot}6H_2O$ 0.025, Na_2ETDA 37.3, vitamins: inositol 100, nicotinic acid 1, pyridoxine· HCl 1, thiamine·HCl 10, sucrose (g/L) 30, caseinhydrolipate (g/L) 0.5, 2,4-D (mg/L) 1, agar (g/L) 8, distilled H_2O up to 1 L.

2.4.2 Rearing Nematodes on Fungus Cultures

It is reported the material to grow *Aphelenchoides besseyi* or other aphelenchids, *B. xylophilus* in particular, on cultures of the plant pathogenic fungus *Botrytis cinerea*. The needed materials consist of a monoxenic culture of the fungus *B. cinerea*, sterile specimens of *A. besseyi*, Petri dishes of about 9 cm diam., PDA (potato dextrose agar) medium, autoclave, and growth cabinet.

2.4.3 Rearing Nematodes on Carrot Disks

According to EPPO standards [6] and Coyne et al. [7], specific materials and equipment needed are:

- Laminar flow hood, Bunsen burner or equivalent, stereomicroscope, sterile pipette, picking needle, paring knife, wooden cutting board, counting dish cleaned with 70% ethanol, sterile Petri dishes 5 cm diam., parafilm, carrot plants with foliage (at least 4, with a minimum diameter of 2 cm), plastic containers with transparent lids, tissues, weighing paper, sterile Falcon tube (15 mL) (= plastic centrifuge tube with conical bottom), glass jar (100 mL), mistifier, sieve tray with filters for mistifier, 1% chlorine solution, ethanol 70%, sterilized tap water, streptomycin sulfate solution 0.1%, 50 mg/L carbenicillin, 50 mg/L kanamycin, Miracloth filter, small sieve with large mesh aperture (1–2 mm), beakers, bench centrifuge, Erlenmeyer flasks.

2.5 Other Methods of Rearing Nematodes

In several laboratories, according to specific needs, other methods for rearing and storing nematodes have been developed. Besides laboratory facilities, in general, other materials needed consist of the target nematodes (*D. destructor*, *R. similis*) and the host plant part (sweet potato tubers, garlic cloves, cocoyam corm) on which to reproduce the nematode.

2.6 Conservation of Nematodes

To use nematodes in periods of the year when live specimens are not available under natural field conditions, it is necessary that nematodes that are continuously reared in greenhouse or in vitro should be somehow stored. Materials to be used can vary according to nematode species. In this chapter, only suggestions on how to store leaving nematodes are given, while no protocol is described to prepare and store fixed materials (*see* **Note 2**).

3 Methods

3.1 Collecting Nematodes from Host Plant Crops under Field Conditions

The suggested procedure is as follows:

1. For cyst nematodes (*Globodera* spp. and *Heterodera* spp.), the best time is at the end of the crop cycle of annual plants when the density of cysts in the soil is at its maximum. Therefore, soon after harvest, perform the following:
   - Cut plants at soil level, discard the top 5 cm soil, and collect the soil up to 20 cm deep of the soil profile, just along the

plant rows or from each plant rhizosphere; this soil may contain 500–1000 eggs per cm^3 and can be stored under conditions similar to those the cyst would undergo in the field—this soil can be used directly, after estimating nematode density, or used for cyst extraction.

- In case of RKNs, or other nematodes forming egg masses, such as *Rotylenchulus* spp., it is preferable to collect roots or soil as soon as host plants approach senescence and roots become unsuitable for nematode feeding; afterward, nematode eggs may continue to hatch, but free-living stages, generally, start their natural decline that increases with time.

2. Therefore, uproot plants, cut the aerial part, and wash the roots free of soil.
3. Store roots at 5–8 °C for a few days and extract eggs.
4. Nematodes of the aerial plant parts can be stored in the infected organs (seeds, bulbs, stems, hay of alfalfa, and clover) or extracted soon from any infected plant part.
5. For the root-lesion nematodes *Pratylenchus* spp. and *R. similis*, it is preferable not to collect nematodes from the field but to use plants having rather thick roots, such as leguminous plants, walnut, and bananas.
6. Ectoparasitic nematodes can also be stored in the infested soil that should be collected along the crop row before plant senescence (*see* **Note 3**).

3.2 Rearing Nematodes in Greenhouse

3.2.1 Rearing Cyst Forming Nematodes (Globodera spp. and Heterodera spp.)

1. Determine the population density of the infested soil that will be used for reproduction.
2. Calculate the quantity of the sterilized potting soil you can mix with potting soil to have an initial population density of 10–20 nematode eggs per g.
3. Mix at the same time potting soil, granular fertilizer, and infested soil. This operation can be done by hand for a small amount of soil and with a concrete mixer, about 5 min each mixing lot.
4. Fill the pots (20–25 cm diam.) with the infested soil mix and arrange them on the benches (no need to leave space between pots infested with the same nematode species or virulent group; otherwise, keep them separated with plexiglass or other material shields).
5. To reproduce *G. rostochiensis* or *G. pallida*, plant in each pot of 20–25 cm diam. a potato tuber (about 40–50 mm diam.) or in small pots of 10–12 cm diam. a sprout with a piece of tuber of 15–20 g. If the nematode host plant is a legume, you may sow three to four seeds per pot.

6. Set illumination at 12–14 h per day; however, you may follow just natural day light—set temperature at 18–22 °C for potato cyst nematodes (*Globodera* spp.), *Heterodera avenae* group species, *H. carotae*, *H. ciceri*, *H. cruciferae*, *H. goettingiana*, and *H. trifolii*, or at 24–28 °C for *G. tabacum*, *H. oryzae*, *H. oryzicola*, *H. sacchari*, *H. schachtii*, *H. betae*, *H. daverti*, *H. glycines*, *H. cajani*, and *H. zea*.
7. Thereafter, irrigate as needed, fertilize every 7–10 days with a fertilizer solution at 2‰, max 3‰, and keep pest and disease under control; realize that the healthier are the plants, the more the nematode reproduce.
8. When plants start to senesce, stop irrigation and fertilization, cut plants at soil level, and a week later, uproot plants and mix thoroughly soil and roots to favor detachment of cysts from roots.
9. Now take subsample to determine the nematode soil density and use the soil for your need (*see* **Note 3**).

3.2.2 Rearing Root-Knot Nematodes, Meloidogyne *spp., and Other Egg Mass Forming Nematodes (*Rotylenchulus *spp.,* Tylenchulus *spp.)*

The procedure is basically the same as that of cyst forming nematodes with a few changes:

1. Collect nematode eggs and second-stage juveniles (J2s) preferably by the sodium hypochlorite method [12].
2. Mix at the same time potting soil and granular fertilizer. This operation can be done by hand in case of a small amount of soil and with a concrete mixer, about 5 min each mixing, in case of large quantities.
3. Fill the pots (20–25 cm diam.) with the soil mix and arrange them on benches (no need to leave space between pots infested with the same nematode species or virulent group); otherwise, keep them separated with plexiglass or other material shields; alternatively, instead of pots, trays of about 12 L capacity containing 10 dm^3 soil can be used.
4. Transplant in each pot a seedling of the host plant at the 4–6-true leaf stage. If the nematode host plant is a legume, you may sow two to four pre-germinated seeds per pot. One–two weeks after transplanting or when plants are at the 6–8-leaf stage, inoculate each plant with 3000–5000 eggs and J2s in three–four holes 2–3 cm deep around the stem, using a calibrated pipette—close the hole and add a few mL of water to allow the soil to adhere to eggs and J2s.
5. Set light at 12–14 h per day; however, you may follow just natural daylight—set temperature at 18–22 °C for temperate climate species RKNs (*M. artiellia*, *M. hapla*, *M. naasi*, *M. chitwoodi*, and *M. fallax*) and at 24–28 °C for the tropical species (*M. arenaria*, *M. incognita*, *M. javanica*, *M. enterolobii*,

M. ethiopica, *M. floridensis*, rice root-knot nematodes *M. graminicola* and *M. salasi*), *Rotylenchulus reniformis*, *Tylenchulus semipenetrans*, and *Nacobbus aberrans*.

6. Thereafter, irrigate, fertilize every 7–10 days with a fertilizer solution at 2‰, max 3‰, and keep pest and disease under control.
7. 60–80 days after inoculation, just before the host plants start to senesce, uproot plants and collect nematodes using both infested soil and infected roots. If you prefer to have entire egg masses, we suggest cleaning the roots free of adhering soil, cutting them in 0.5–1 cm long pieces, mixing in sterile sandy soil, determining its population density, and using it for further inoculation studies. This procedure is especially suited for yield loss assessment in microplots [13] (*see* **Note 4**).

3.2.3 Rearing Migratory Root-Lesion Nematodes *Pratylenchus* spp. and *Radopholus similis*

The most common *Pratylenchus* species involved in breeding programs and of which large populations are needed for screenings include *P. brachyurus* (soybean), *P. goodey* (bananas), *P. neglectus* and *P. thornei* (bread and durum wheat and chickpea), *P. penetrans* (potato), and *P. vulnus* (rootstocks of stone fruits).

Hereafter is the suggested procedure:

1. Throughout mix potting soil (rather sandy soil) with granular fertilizer and fill pots.
2. Transplant or sow in each pot one to three seedlings or geminated seeds of a good host plant.
3. When plants are about 10–15 cm tall and roots are already exploring half of the soil volume (check some pots), extract nematode specimens from any source available to you (infested soil, infected roots, callus culture, carrot disk culture).
4. Inoculate each pot by pouring in three to four holes 3–4 cm deep a nematode suspension to have an inoculum density of two to three nematodes/cm^3 soil, close the holes, and add a few mL of water to favor adhesion of soil to nematodes—set greenhouse temperature at the optimum for nematode reproduction and host plant growth.
5. Thereafter, irrigate as required, but every 7–10 days with a 2‰ solution of a NPK + microelement fertilizer; realize that plants with abundant and well-growing roots are instrumental for great reproduction of nematodes.
6. When the nematodes have completed two to three generations (about 70–90 days after inoculation, corresponding to flowering–early fruiting stage), nematode reproduction is expected to be at its maximum; then, uproot plants, collect all roots, wash them free of soil, and extract nematodes from soil and roots (*see* **Note 5**).

3.2.4 Rearing Migratory Nematodes Attacking Tubers, Bulbs, Stems, and Leaves

This subsection applies to *D. dipsaci* (parasitizing onion, garlic, potato, carrot, sugar beet, fodder crops, and many other plants), *D. gigas* (damaging broad bean), *D. destructor* (on potato tubers), *D. africanus* (on peanut) [14], and *D. angustus* (a severe parasite of rice). The procedure is the same as that of the previous nematode groups, at least for nematodes attacking belowground plant parts, as *D. dipsaci* and *D. destructor* on potato tubers and *D. africanus* on peanut. For these species, we suggest the following procedure:

1. Fill pots with sterilized and fertilized potting soil; sow the main host plant (potato or peanut) and when plants are very young (5–7 cm tall), extract nematodes from the sources available to you and inoculate 300–400 nematodes per plant in four holes 3–4 cm deep, close the holes, and add a few mL water just to let the soil adhere to nematodes.
2. Set temperature at the most suitable for the nematode (15–20 °C for *D. dipsaci* and *D. destructor* and 20–25 °C for *D. africanus*).
3. About 3 months after inoculation, when potato tubers and peanut pods are well formed and plants near to senescence, extract nematodes from tuber and soil pots or store plants at 8–10 °C, to extract nematodes before their use.

When the parts of the host plant attacked by the nematodes are bulbs, stems, leaves, and seeds, the inoculation procedure is a little different. We suggest the following:

1. Germinate seeds of the host plants in a Petri dish (10–12 cm diam.) at 23–25 °C in dark (switch light off or wrap with aluminum foil or paper).
2. Three–four days later, sow germinated seeds in 12–15 cm diam. pots, containing about 800–1000 cm^3 sandy soil, in holes 1–2 cm deep.
3. When seedlings are at cotyledon to early true leaf stage, extract nematode, concentrate them to have 50–100 nematode specimens per mL, add 1–2‰ carboxymethyl cellulose, and inoculate them by depositing on the central sprout or between the leaf axil and stem one to two drops of the nematode suspension.
4. Cover plants with a plastic transparent sheet or a plastic cage for 3–4 days.
5. Set temperature at 15 °C.
6. If necessary, a week later inoculated plants can be transferred to larger pots and temperature set at 18–20 °C.

However, we point out that this method is not so good as reproducing *Ditylenchus* spp. on carrot disk or callus of host plants.

3.2.5 Rearing Ectoparasitic Nematodes

Many ectoparasitic nematodes also cause damages to host plants and may also act as vectors of plant viruses. These include several longidorids (*Xiphinema* spp., *Longidorus* spp., *Paralongidorus* spp., *Trichodorus* spp., and *Paratrichodorus* spp.), among which are of particular importance *H. index*, vector of grape fan leaf virus (GFLV) and reproducing very well on both grapes and figs, and members of the *X. americanum* group and vectors of several plant viruses [15]. Other important ectoparasitic nematodes are *Mesocriconema xenoplax* on many plants including grape and peach; *Helicotylenchus multicinctus* especially on banana; *Hemicycliophora arenaria* on citrus in south California and *H. typical* on rice; *Hoplolaimus* spp. and in particular *H. columbus* on vegetables, cotton, and soybean; *Belonolaimus longicaudatus*, which is very polyphagous and widely distributed in the USA but of great importance on several vegetables in Florida; and *Dolichodorus heterocephalus* which is of importance in the USA both on annual and perennial plants.

The procedure is similar to the previous one, but nematodes must be extracted from the rhizosphere soil of the host plant or from non-cropped soil.

1. Fill pots with sandy-sterilized soil fertilized with a composite granular fertilizer.
2. Grow one or more host plants per pot depending on adult plant size, usually only a host plant (15–20 cm tall in case of perennial plants).
3. Adjust greenhouse temperature to that suitable for the target nematode.
4. When new and actively growing roots are exploring the soil, introduce in three to four holes 2–5 cm deep in the plant rhizosphere 100–300 nematode specimens of the target nematode; close the hole and add a few mL of water.
5. Thereafter, irrigate regularly and once in 7–10 days with a 2‰ NPK + microelement fertilizer, and keep pest and diseases under control.
6. Seventy–ninety days after inoculation, the nematode population should have increased; if not, use the soil of the first or few pots to infest more soil and repeat the same procedure.

In case of woody plants, it is suggested to wait at least 8--12 months after inoculation to check for the nematode population increase.

3.3 Obtaining Nematode Populations from a Single Female, Egg Mass, and Cyst

More genetically uniform populations of a nematode species may be required, for instance, in biochemical and molecular studies. For a better characterization of nematode virulence and plant resistance, availability of nematode populations derived from a single female is highly desired. This can be achieved by starting nematode rearing from a single egg mass, a single cyst, or a single female [16–18].

Nematode populations from a single egg mass:

1. Fill pots up to 2–3 cm of the edge with sterilized soil and fertilized with a 1–2 g/kg soil of the granular fertilizer.
2. Transplant the host plant and irrigate whenever needed and every 7–10 days with a 2‰ solution of a very soluble fertilizer.
3. Set greenhouse temperature at the optimum for plant to grow and nematode to infect and light on for 10–14 h per day.
4. When plants are about 10–15 cm tall and roots exploring at least half of the soil in the pot, inoculate them with the target nematode.
5. Forty–fifty days later, check the roots and make sure that on the roots of the infected plants, large egg masses have already developed.
6. Then take a piece of root, clean it under running water, put in a Petri dish and this on the stage of the stereoscope with incident light adjusted at 10–40× magnification, identify an egg mass, and with the help of lancets, detach it from the root and put each in a small container (2–3 cm diam. Petri dish, watch glass, or others) in water; collect about ten egg masses to be sure that at least one will successfully reproduce.
7. Now take the egg masses to the greenhouse or growth chamber; do a hole 2–3 cm deep below the plant stem, just in the plant rhizosphere, and with a forceps or a needle, take one from the container and introduce it in the hole; close the hole; add a few mL of water; and continue to grow the inoculated plants.
8. Keep plants inoculated with different egg masses (even if of the same species) separated from each other, if it is necessary, using shields.
9. Allow the nematode to develop a couple generations (60–80 days) and then check for nematode reproduction. For root-knot nematodes and cyst forming nematodes, checking can be done just taking out of the pot the plant with the entire root bowl and with the naked eye or with the help of a lens, check for egg masses (all nematodes inoculated), galls (root-knot nematodes), and white or yellow females or brown cysts (*Globodera* spp., *Heterodera* spp.).
10. The pots in which nematodes have successfully reproduced can be used to further increase the population or maintain it in collection in small amount (*see* **Note 6**).

Nematode populations from a single cyst:

The method is similar to that of egg mass forming nematodes with the following few differences. When the host plant is ready to be inoculated, collect from soil a well-formed cyst and with a forceps introduce it in a hole into the rhizosphere, and continue to grow the plants under optimal temperature until the nematodes are expected to have developed one or two generations (50–70 days after inoculation). Then check if white or yellow females or brown cysts are present on the root bowl. Pots showing positive reproduction can be left 10–15 days more, and then plants cut at soil level and irrigation discontinued. Seven days later, thoroughly mix the infested soil with other sterilized soil (three to four times volume), pour it in larger pots (20–25 cm diam.), and sow or transplant in it a susceptible host plant to further increase the nematode population (*see* **Note 7**).

3.3.1 Other Nematodes

The following procedure is suggested for successful reproduction.

In case of endoparasitic nematodes:

1. Germinate seeds of the host plant in a Petri dish having on the bottom a well-wetted filter paper, and keep in the dark at a suitable temperature (23–25 °C).
2. When seed radicle is about 1 mm long, transfer a seed in each of the eight to ten coffee or tea cups containing just fine sand (possibly quartz sand) and irrigate with a fertilizing solution such as Hoagland's solution.
3. Transfer the cups in a grow cabinet at the temperature suitable for the host plant growth and with a daylight of 14–16 h.
4. As soon as the roots of the new plant are exploring nearly all the soil, inoculate in its rhizosphere a young or adult female specimen of the nematode species to reproduce.
5. Four–five days later, when the nematode is supposed to have already penetrated into the roots, transfer the plant and quartz content in a larger (12–14 cm diam.) pot filled with sterilized sandy soil.
6. Allow the plants to grow for about 60–70 days in the same cabinet or in a greenhouse at the same temperature and light period.
7. Then check for nematode reproduction and use the successful pots for increasing the nematode population or to maintain it in a collection.

In the case of ectoparasitic nematodes, the procedure is simpler and we suggest the following:

1. Transplant a host plant in each of the eight to ten pots, filled with rather sandy soil, of 12–14 cm diam. For nematodes

having rather short life cycle (1–2 months) and 20–25 cm diam. for nematodes having rather long life cycle, as are longidorids.

2. When abundant and actively growing roots are exploring the potted soil, select a mature female of the nematode to reproduce and introduce it in the plant rhizosphere in a 4–5 cm deep hole, close the hole, and add few mL of water.
3. When it is expected that the nematode has already developed at least one to two generations, uproot the plants and check the roots for symptoms of the nematode attack and/or use a portion of the pot soil to extract nematodes (*see* **Note 8**).

3.4 Rearing Nematodes Under In Vitro Conditions

3.4.1 Rearing Nematodes on Host Plant Callus Cultures

On White's medium or Krusberg's medium:

1. Prepare the two media by dissolving all ingredients in distilled water in a flask.
2. Sterilize in an autoclave at 120 ° C for 20 min.
3. Leave to cool to about 70–80 °C (until the media are still very fluid).
4. Then, under a laminar fume hood, poor about 7 mL of each media in each 18 mm diam. test tubes to have slants.
5. Also, prepare medium composed of 10 g sucrose, 2 g yeast extract, 10 g agar, and 1000 mL distilled water, sterilize it as above, leave to cool, and then pour in Petri dishes (2–3 mm thick in each dish).
6. Sterilize seeds of onion, alfalfa, clover, or other host plants by soaking 15 min in concentrated sulfuric acid (H_2SO_4), rinsing four times in sterile distilled water, soaking for 15 min in mercury chloride ($HgCl_2$) 30% alcohol solution (1000 ppm), and finally rinsing four more times in sterile distilled water.
7. Sow sterile seeds in the Petri dishes containing the agar medium and keep them at 23–25 °C.
8. When seedlings are 7 days (onion) or 4 days (alfalfa) old, transfer seedling on the slant of each tube.
9. Two–three weeks later, when each tube contains a callused tissue, inoculate with a small piece of callus from stock culture or sterilized nematodes.
10. Maintain inoculated tubes at 23 °C in the dark for at least 8 weeks.
11. Now nematodes can be extracted from the callus by Baermann's funnels or one of its modifications.

On Murashige and Skoog medium:

1. Prepare the medium, sterilize as before, leave to cool to about 70–80 °C, and then, under a laminar fume hood, pour 25 ml of it in 9 cm diam. Petri dishes and leave to cool completely.

2. Sterilize leaves of 4-week-old groundnut cv. Sellie in 70% ethanol for 30 s and then in 0.5% NaOCl containing Tween 20 (for complete wetting of the leaves) for 15 min and rinse four times in sterile distilled water.
3. Under a laminar flow cabinet, transfer leaf pieces of about 1 cm^2 in each of the Petri dish containing the medium.
4. Transfer the Petri dishes in the dark at 25 °C for 4 weeks.
5. Then callus with increased volume can be transferred to fresh Petri dishes for inoculation of nematodes.
6. Therefore, extract nematodes from an available source, and sterilize them according to Meyer [21] under a laminar flow cabinet as follows:
7. Transfer 50 nematode specimens of mixed stages in a Syracuse watch glass containing sterile water and then, using a sterilized Pasteur pipette, into a sterilized centrifuge tube containing 2 ml of 0.1% mercury chloride ($HgCl_2$) and at its bottom a piece of milk filter or a small piece of cotton wool.
8. Centrifuge at 1750 *g* for 2 min.
9. Discard the supernatant and rinse three times in 2 mL sterile water the tube with the nematodes trapped in the milk filter or cotton wool.
10. Using a sterilized tweezer, transfer the milk filter in the Petri dish next to the callus, close with parafilm the dish, and incubate in the dark at 25 °C.
11. Four–five weeks later, nematodes can be extracted by transferring the callus culture in a Petri dish containing 25 mL water and sectioning the callus with a scalpel or lancet. Alternatively, the nematodes can be extracted by putting the pieces of callus on Baermann's funnels and checking and counting nematodes for a couple consecutive days (Figs. 1–2) (*see* **Note 9**).

3.4.2 Rearing Nematodes on Fungus Cultures

Many plant parasitic nematodes, such as some species of *Ditylenchus* and *Aphelenchoides*, can reproduce on fungi. Several of them (*D. myceliophagus*, *A. agarici*, *A. composticola*, *A. swarupi*, *Aphelenchus avenae*, and a few others) have fungal mycelium as the primary host. They are known to damage the mycelium of edible mushrooms, which is being a severe problem in mushroom cultivation. This alternative way of reproduction can be exploited for rearing nematodes in vitro and thus in little space. The most suitable fungus species are *Botrytis cinerea*, *Alternaria* sp., and *Fusarium* spp. These fungi can be easily grown on artificial media and then inoculated with sterilized nematodes.

Hereafter is a procedure that can be followed to grow *A. besseyi* or other aphelenchids, *B. xylophilus* in particular, on cultures of the plant pathogenic fungus *Botrytis cinerea* (Fig. 3):

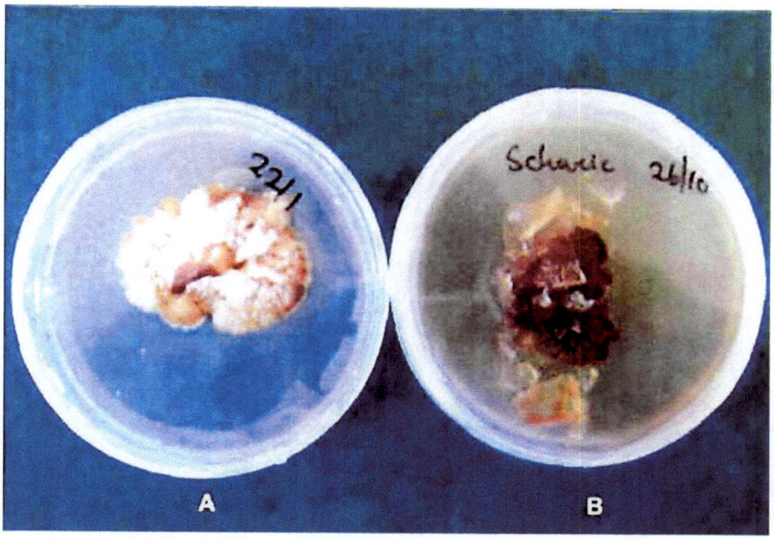

Fig. 1 Callus tissue of *Arachis hypogaea* for rearing *Ditylenchus africanus*. (**a**) *Ditylenchus africanus*-infested groundnut callus; (**b**) *Ditylenchus africanus*-infested groundnut callus tissue, culture ready for renewal. (From Ref. 14, no permission required)

1. Prepare the PDA medium, sterilize in an autoclave at 120 °C for 20 min, leave it to cool to 70–80 °C and then, under a laminar flow fume hood, pour it in the Petri dishes (to form a layer of 2–3 mm thick), and leave it to cool until it solidifies.
2. Under the laminar flow, take a plug of the fungal culture and transfer on the PDA medium in Petri dishes and maintain dishes in the dark at 24 °C for 7–10 days.
3. When the fungus covers nearly the entire dishes, extract nematodes from the source available to you, select nematode stages, better adults of both sexes or gravid females, sterilize them (see previous paragraphs), and then pipette 100 μl of the nematode suspension, containing about 80–100 specimens of the nematode species, on the fungus culture in the plates.
4. Cover the plates with the lid, seal with parafilm, and incubate at 25–28 °C for 45 days, to allow the nematode to develop two to three generations.
5. Now you can extract nematodes just by cutting pieces of the fungal mat and putting them in water in a Petri dish or on Baermann's funnel overnight. The first successful culture can be used to inoculate, under sterile conditions, other fungal plates to increase the nematode population or to maintain nematodes in a collection (*see* **Note 10**).

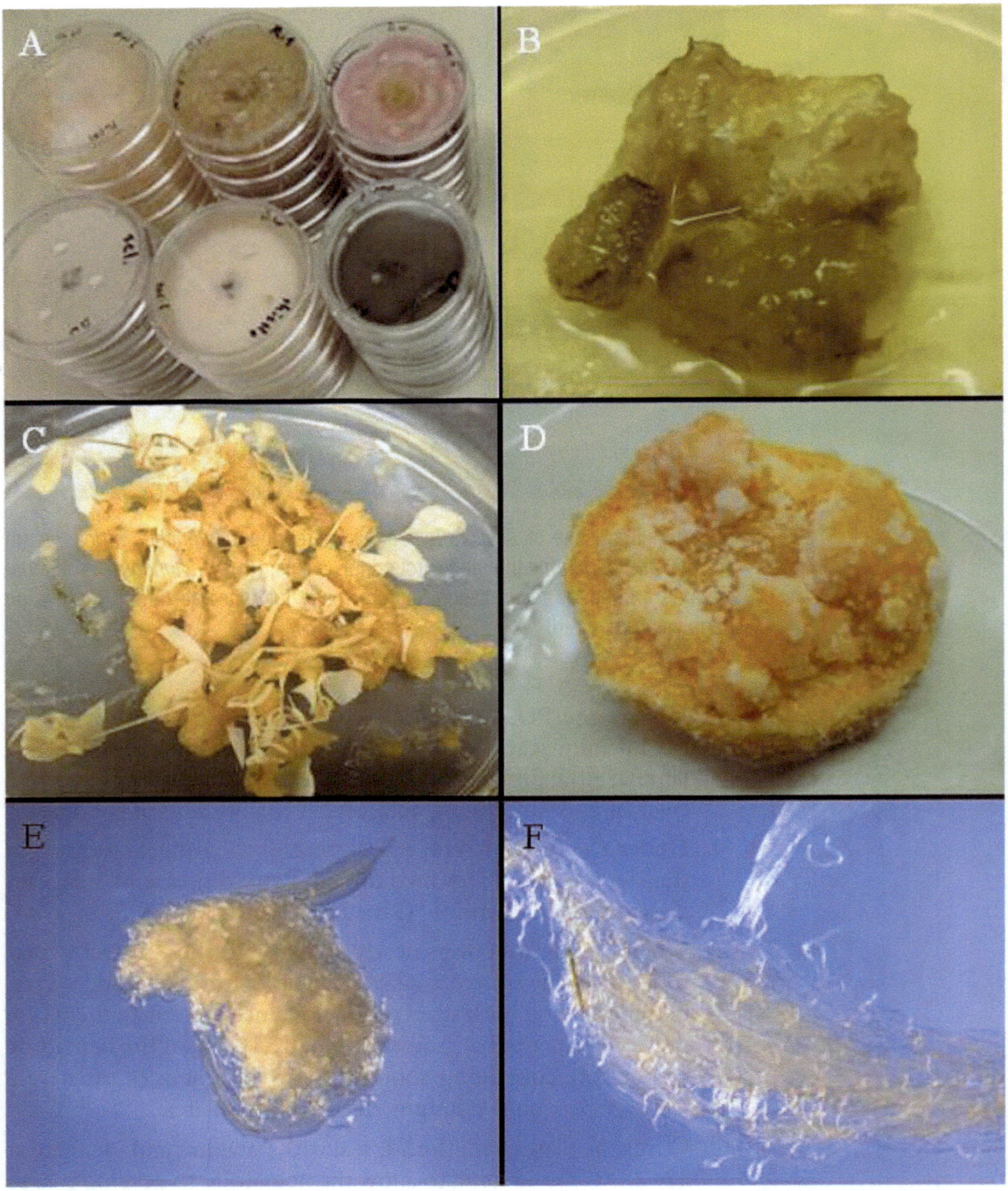

Fig. 2 Fungal cultures (**a**), callus tissues of creeping thistle (**b**), and alfalfa (**c**), and callused carrot disks (**d**) added with *Ditylenchus weischeri* or *D. dipsaci*. Aggregation of *D. weischeri* (**e**) and *D. dipsaci* (**f**) on the surface of inside lids of callused carrot disk cultures (90× magnification). (From Ref. 19, permission from the authors)

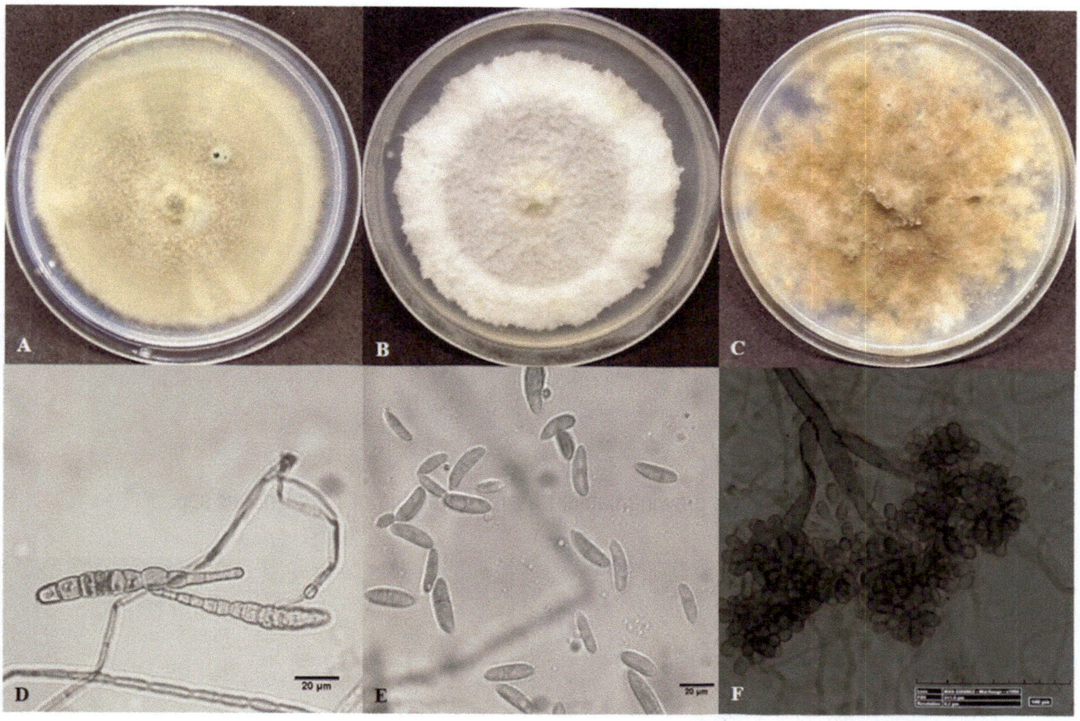

Fig. 3 Fungal cultures and microscopic appearances of species used as rearing media of *Aphelenchoides besseyi*. *Alternaria padwickii* (**a**, **d**), *Fusarium semitectum* (**b**, **e**), *Botrytis cinerea* (**c**, **f**). (From ref. 31, with kind permission of the Journal Perlindungan Tanaman Indonesia)

3.4.3 Rearing Nematodes on Carrot Disks

In 1968, O'Bannon and Taylor [22] reported a method to rear root-lesion nematodes on carrot disk. The method was later refined by Moody et al. [23] and then used in many laboratories to obtain large populations of these nematodes or have them in a collection [8, 19, 24–26]. All these authors and others adapted and improved the method according to specific needs. Protocols are reported by Coyne et al. [7] and more recently by EPPO standards [6], where detailed description and many figures illustrating the method are reported which are useful also for those not familiar with handling of nematodes. The procedure reported here (Figs. 4–5) is abstracted from Castillo et al. [8], Ajihassani et al. [19], Coyne et al. [7], and EPPO standards [6]:

Nematodes extracted from infested plant material or from soil must always be surface-sterilized.

1. Select carrots that are as fresh as possible (better directly from a crop) and apparently healthy. The most suitable are tap roots cylindrical with a large diameter (max. 4 cm), not rich in water (better of Nantes-type cultivar); remove the foliage and scrub the carrots with water to remove soil and debris, and dry them with paper; a minimum of four carrot tap roots should be selected so that slices of four different carrots can be used per

Fig. 4 A number of carrot disks in a growing cabinet for mass rearing of *Pratylenchus* spp.

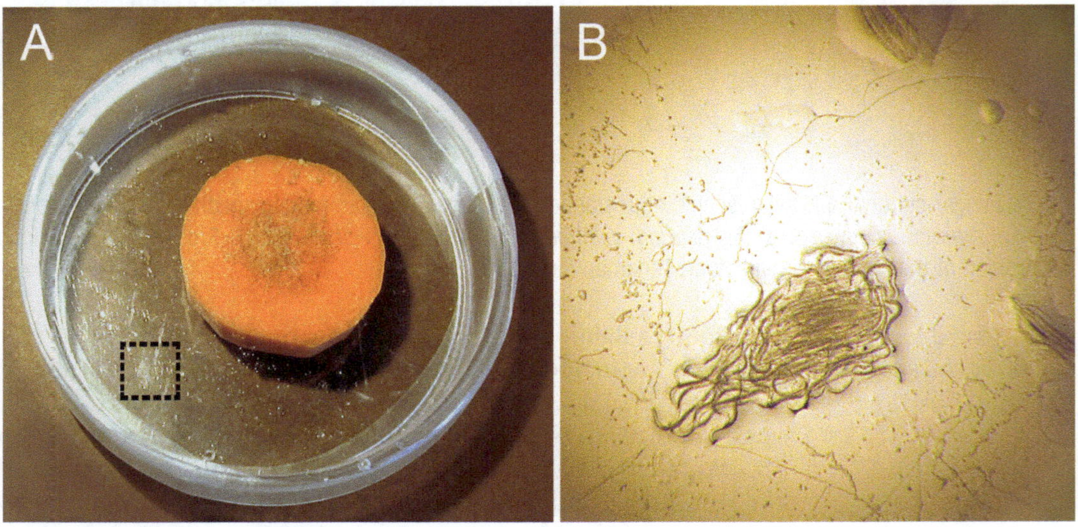

Fig. 5 (**a**) Rearing of *Pratylenchus* sp. on carrot disks. (**b**) Several nematode specimens gathered in the water drop highlighted in A. (From Ref. 29, with kind permission of SIN (Italian Society of Nematology))

nematode isolate, to reduce the risk of isolate loss—an extra carrot can be used, if needed, to ensure adequate number of slices. The carrot disks need to be at least 2 cm in diameter and 0.5 cm thick.

2. Extract nematodes from soil or infected plant material using the most appropriate extraction method available.
3. Surface-sterilize nematodes: a streptomycin sulfate solution (0.1%) can be used as disinfectant, higher concentrations (up to twice) of streptomycin sulfate solution can be used if necessary; pour the nematode suspension into a sterile Falcon tube (15 mL); allow the suspension to settle for at least 30 min

until a pellet with nematodes is formed at the bottom of the tube; use a sterile pipette to discard the water, and stop just above the pellet; add 10 mL of 0.1% streptomycin sulfate solution, shake the tube to homogenize the suspension, and leave the suspension overnight at room temperature; nematodes must then be washed with sterilized tap water to remove the streptomycin sulfate solution by using a sterile pipette to discard the solution and stopping just above the pellet; add 1 mL of sterilized tap water and shake to homogenize the suspension; and wait at least 15 min to allow the nematodes to form a pellet at the bottom of the tube. Use a sterile pipette to discard 1 mL of the solution from the tube; check the extracted liquid suspension for the presence of nematodes; and repeat this procedure three times.

4. Inoculation: preferably work in a laminar flow hood, but if not possible, work in a clean room as possible and open the Petri dishes on a little space; clean a counting dish and a picking needle with 70% ethanol, pour nematode suspension into the counting dish, prepare at least three Petri dishes per nematode isolate; pipette with a sterile pipette a small droplet of sterilized tap water onto the middle of each Petri dish; and inoculate at least three carrot disks per isolate, derived from different carrots. Use one carrot disk, per used carrot, as a negative control, by adding only a small droplet of sterilized tap water onto the bottom of a sterile Petri dish, using a sterile pipette; work under a laminar flow hood in sterile conditions and wear sterile gloves; flame the surface of a wooden cutting board and a weighing paper to sterilize the surface; put the paper on top of the cutting board; and flame shortly the surface of two more weighing papers to use one as a clean work spot in the laminar flow hood and the other one to dry the carrot later on. Sterilize the paring knife and the carrot by soaking them in a 1% chlorine solution for 1 min or by spraying them with alcohol (70%) and flaming and use a surface-sterilized weighing paper to dry the carrot and the paring knife. Peel the carrot (peel thickness a few millimeters) from the top downward, with the growing point of the foliage at the top and drop the peels on top of a weighing paper to keep the laminar flow hood as clean as possible. Put the peeled carrot on top of the weighing paper on the cutting board; cut and discard the top of the carrot with the growth point of the foliage; cut the carrot into disks with a thickness of 0.5–1 cm, depending on the height of the Petri dishes used, and 2 cm or more in diameter; now put the carrot disk with the bottom side down on the droplet with nematodes in the Petri dish; alternatively, a droplet containing nematodes (about 20–40/drop) can be placed directly onto the margin of the carrot disk (not in the center), close the Petri dish, and seal the

sides with parafilm to prevent contamination and dehydration; repeat these steps for the remaining carrots, put the Petri dishes inoculated with the same nematode isolate in a plastic container (each box with the same isolate) with a transparent lid or pile them up, with each pile containing the same nematode species, and put them into a larger plastic container; write on the lid the date of inoculation, isolate number, and name of the species; and keep the containers in the dark at ±22 °C.

5. For maintenance, after 2 weeks, check for living nematodes in the Petri dishes using a stereomicroscope; nematodes can be incubated at ±22 °C for 6–8 weeks; and when the carrot disks start to deteriorate, the nematodes should be transferred to new carrot disks, but this time using already sterile nematode.

Therefore:

Extract nematodes from existing carrot disk culture. Several options are possible, depending on the facilities in the laboratory. Anyhow, all steps must be performed under sterile conditions. This is particularly important if you decide not to disinfect the nematodes again before inoculating them on new carrot disks; otherwise, the sterilization steps listed before must be followed.

Nematodes must be extracted using sterile materials and can be done in different ways. One of them is direct removal by washing. Therefore, rinse the lid of the Petri dish with sterilized tap water to collect the nematodes from the lid and/or add a few droplets of sterilized tap water onto the previous carrot disk culture. Pour or pipette the suspension with nematodes into a clean glass jar or counting dish.

Nematodes can also be extracted by cutting each carrot disk in several pieces with a sterile knife and then using the modified Baermann's funnel method (sieve in a Petri dish) or the enzymatic method as suggested by Kaplan and Davis [27]. The centrifuge method is also suitable, but in this case, sterile water and small capped sterile tubes must be used (*see* **Note 11**).

3.5 Other Methods

3.5.1 Sweet Potato Tuber Hole Method

In China, where *D. destructor* is very severe on sweet potato, Wang et al. [28] developed a method to reproduce nematodes on sweet potato tubers, called the tuber hole method (Fig. 6):

1. Select a fresh and good sweet potato tuber, clean it and surface-sterilize in 75% ethanol for 5 min, and cut it longitudinally to have a 1/3 of the top part and 2/3 of the base part.
2. On the basal section, do a 5 mm diam. and 1 cm deep holes at 3 cm distance, pour in each hole 20 μL sterile water, and then inoculate nematodes (gravid females).
3. After inoculation, put the top section of the tuber on the basal section and seal the edges with paraffin wax.

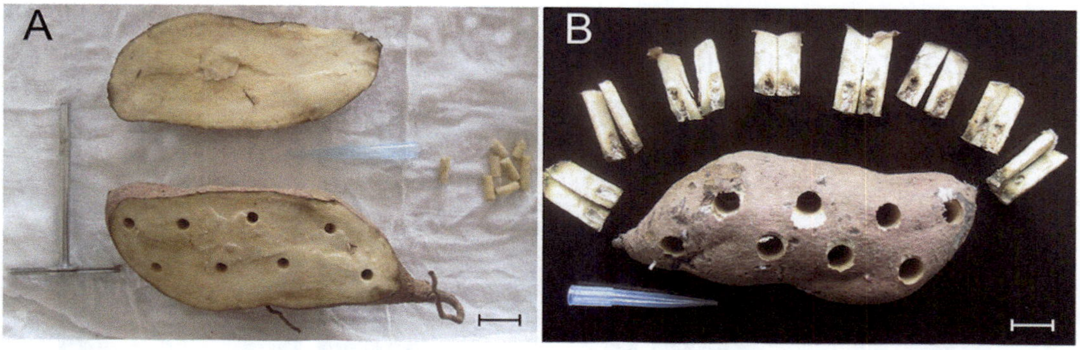

Fig. 6 Pure culture of *Ditylenchus destructor* on sweet potato. (**a**) Longitudinal cut showing the holes made on the base section of the tuber for nematode inoculation; (**b**) Lesions on tissue chips caused by nematode infestation. (Scale bars = 2 cm). (From Ref. 28, with kind permission of Nematology)

4. Put the tubers in a plastic box and a small amount of sterile water sprayed in it.
5. Now incubate tubers at 25 °C for at least 30 days before checking for nematode reproduction.

3.5.2 Garlic Clove Method

In Japan, severe rot of garlic is caused by *D. destructor*. Therefore, Yoshiga [30] developed a method to reproduce the nematodes on garlic. It is summarized hereafter:

1. Collect garlic cloves from dried garlic bulbs.
2. Remove the protective leaves and check at the base of the clove for absence of nematode infection as here early symptoms of infection appear.
3. Surface-sterilize cloves by soaking in 10% sodium hypochlorite for 5 min and then rinse three to four times in sterile deionized water.
4. Remove the base of the cloves with a sterile knife.
5. Do a slit on the clove to hold the inoculated nematodes.
6. Place the clove on 10 mL of a 1% sterilized agar medium, in a 100 mL medium bottle, (Fig. 7) with the cut side up.
7. Extract nematodes from a source available to you and sterilize by soaking overnight in an antibiotic mixture made of 50 μg/mL of penicillin, 50 μg/mL of streptomycin, and 100 μg/mL of neomycin.
8. From a sterile and concentrate nematode suspension, take 20 mL containing around 200 mixed nematode stages and inoculate the clove slit.
9. Now incubate the bottles at 25 °C in the dark.
10. Seven weeks later, when garlic cloves become partially yellow, nematodes accumulate in the agar below the cloves and can be extracted from the culture bottles.

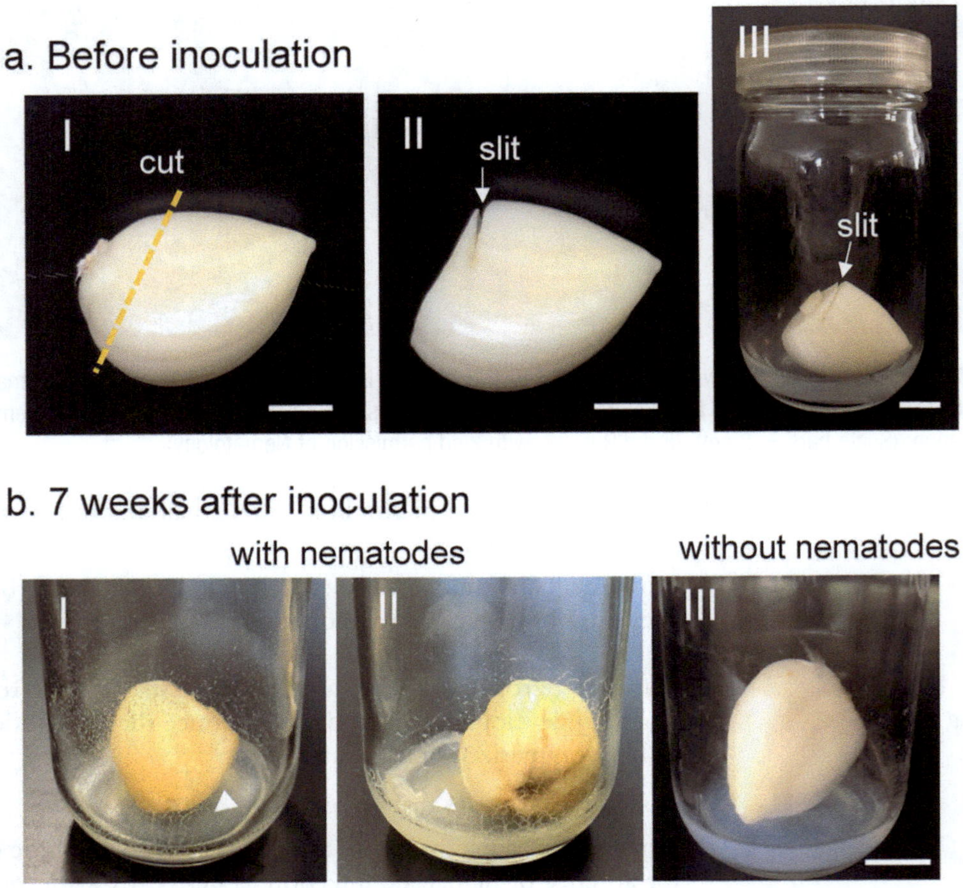

Fig. 7 Rearing of *Ditylenchus destructor* on garlic storage leaves. (**a**) Preparation of the storage leaf. Before nematode inoculation, the base of the storage leaf is aseptically removed with a knife (I), a slit is made for nematode inoculation (II), and the storage leaf is placed in a 100 ml culture bottle containing 1% agar (III). (**b**) Appearance of the storage leaves 7 weeks after inoculation. The nematode-inoculated storage leaves have become partially yellowish and nematodes emerged from them (I, II), whereas the control leaves (without nematodes) show no change in appearance (III). Arrows indicate emerged nematodes. Bars, 1 cm. (From Ref. 30, copyright Springer Nature)

11. Therefore, cut the clove in slices less than 5 mm thick with a sterilized knife, remove the agar by cutting it in pieces, and transfer them on Baerman's funnels overnight. Also, rinse the bottles with sterile water to collect nematodes on the inner surface.

We believe that this method can be used also to reproduce *D. dipsaci* and *D. destructor* on potato tubers, just by doing a slit in tubers, inoculating sterile specimens and then closing the slit with a tape.

Fig. 8 Fresh cocoyam disk inoculated with 100 *Radopholus similis* showing nematodes accumulating at the edge of the 45 mm diam. Sterile glass Petri dish, 3 weeks post-inoculation. (From Ref. 32, with kind permission of Nematology)

3.5.3 Cocoyam Disk Method

At the International Institute of Tropical Agriculture (IITA), Nairobi, Kenya, Oudori et al. [32] have developed a method, for mass culturing of *R. similis*, similar to the carrot disk methods but using, instead, disks of cocoyam corm (Fig. 8), which seems to last longer. The following procedure is suggested:

1. Wash corms of cocoyam under running water, dry them with a tissue paper, and put under a laminar flow cabinet.
2. Cut cocoyam corms in four vertical quarters using an ethanol flame-sterilized knife.
3. Surface-sterilize each quarter by spaying absolute ethanol and flaming.
4. Peel the pieces at least three to four times, taking care to sterilize the peeler between peelings, until obtaining cylinders of about 40 mm diam.
5. Cut cylinders in disks of 8 mm thick and put them in glass Petri dishes of the required size (at least 45 mm diam. For 40 mm diam. disks).
6. Inoculate disks with about 100 nematode specimens from a population of *R. similis* available to you, after sterilizing in streptomycin sulfate at concentration of 0.06 g/10 mL.
7. Seal Petri dishes with parafilm and then incubate in the dark at 27 °C for at least 7 weeks.

Disks of cocoyam deteriorated at 16 weeks, while similar disks of carrot did it at 10 weeks.

3.6 Conservation of Nematode Populations

First of all, we wish to emphasize that all the previous methods reported to provide and reproduce nematodes are at the same time also methods to store nematodes. However, hereafter, we give more suggestions also for other nematodes and methods. *See* also EPPO standards and diagnostics [6] and EPPO web site (https://www.eppo.int).

3.6.1 Conservation in Soil

All soil samples that have been shown to contain interesting nematodes can be stored at room temperature for a few days and for some weeks at 5–7 °C. Cysts extracted from soil and concentrated in soil debris, still containing some moisture (dried at room temperature), can also be stored in a cool room at 4–7 °C for 1–2 years. However, if this inoculum is used after more than a year, we suggest to assess again its viable egg content. The same can be done with cyst nematodes reproduced in pots in greenhouses; the entire pot or only the soil can be closed in a plastic bag and put on a shelf of a cool room. However, despite being stored at low temperature, the nematodes will decline with time, and, therefore, we suggest that time by time (every 2 years), the soil is used to renew the nematode population by planting in it a host plant for an entire life cycle and then the nematode population density assessed again.

3.6.2 Conservation in Infected Plant Parts (Bulbs, Stems, Seeds, Hay, Tap Roots, Tubers)

Many nematodes are known to attack and survive in plant parts. Therefore, *D. dipsaci* can be stored in infected bulbs (onion, garlic, tulips, tap roots of carrot and sugar beet, and potato tubers) for months, until they become exhausted. This nematode survives also in hay of alfalfa and clover from which thousands of nematodes per g can be extracted by putting infected hay on Baermann's funnels overnight. Infected tubers are also useful to store *Pratylenchus* spp. (especially *P. penetrans*) and root-knot nematodes, mainly *M. hapla*, *M. chitwoodi*, and *M. fallax*, but also the tropical species, and *Nacobbus* spp.

D. dipsaci can be stored for years in seeds of onion, alfalfa, and clovers, *D. gigas* in seeds and stems of broad bean, *Aphelenchus besseyi* in seeds of rice, and *Anguina tritici* even for a decade in infected wheat kernels of which a single kernel may contain thousands of motile second-stage juveniles. These can be extracted simply by crushing kernels and incubating them in water overnight.

3.6.3 Conservation in Monoxenic Cultures

Monoxenic cultures on carrot disk and on callus or fungal cultures are also useful to store and perpetuate nematode populations if these media are renewed before becoming deteriorated.

3.6.4 Conservation in Water Suspension

After extraction, nematodes can be stored in water for a few weeks, before counting or inoculating. However, with time, especially juvenile stages, their energy reserves are reduced and they may float in water. This may cause problems while counting especially with root-knot nematodes. Nematodes floating in the counting dish, while counting under a stereoscope, may be disregarded if the focus is not chanced continuously to observe both specimens floating or decanted on the bottom of the counting dishes. This problem can be avoided using Peter's 1 mL counting slides.

The first author found that eggs and juveniles of *M. artiellia*, extracted with the sodium hypochlorite method [12] and stored at 5–7 °C, were 100% very actively moving and full of energy reserves after a year. Only a small proportion did not move after a year and a half, but juveniles had reduced energy reserves.

3.6.5 Conservation on Filter Paper

A large number of *D. dipsaci* extracted from plant tissues, especially bulbs and stems, can be stored on filter paper (Seinhorst JW, personal communication to the first author, December 1975, Wageningen, the Netherlands), using the following procedure:

1. Extract nematodes from bulbs, tap roots, or stems, using Baermann's funnel method overnight and collect them in a beaker. It is preferable that bulbs and stems are collected just before plants approaches senescence, when the great majority of the specimens are of the third- and fourth-stage juveniles, i.e., those mostly undergoing quiescence.
2. Take a Whatman filter or a disk of common laboratory filter, adapt it to the internal bottom base of a Buchner funnel, and add a few mL of water so that it sticks perfectly to the funnel (Fig. 9).
3. Insert the Buchner funnel on top of a flask, pour in it the nematode suspension, and exert a small suction to favor filtering. In case of large quantity of nematode suspension, repeat this procedure using a new filter.
4. Remove the filter from the funnel and put it in a larger Petri dish without lid.
5. Leave the Petri dish at 15–20 °C, until the excess water has evaporated, and then in a refrigerator for a few days until the filter paper appears rather dry. The drying of the filter must occur slowly and may take a few days.
6. Then store each filter separate from the other, in a paper envelope, write on it notes to identify the nematode population, and store envelopes in a refrigerator at 4–6 °C Nematodes will remain active on the filter for years.
7. When you need nematodes, just cut a piece of filter, put it in a Petri dish of 6–10 cm diam., add sterile water, and leave overnight after which most of the third- and fourth-stage juveniles will return active and ready for use.

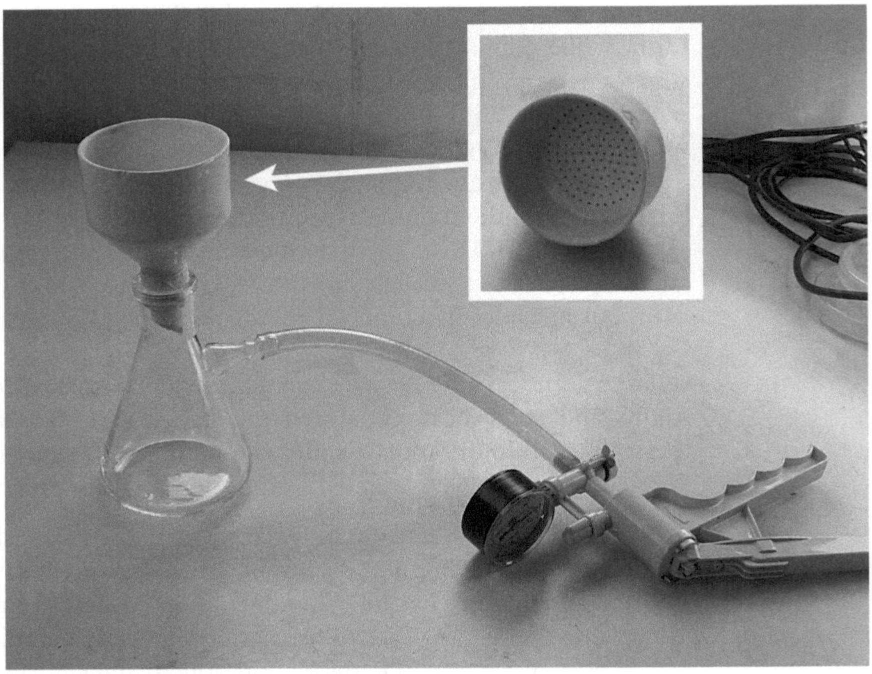

Fig. 9 Ceramic Buchner funnel for filtration under suction conditions. The funnel is inserted in the large hole of a rubber stopper and all inserted on top of a flask that has an outlet inserted in a rubber tube connected to a vacuum maker devise. This is manually operated after pouring the nematode suspension (rather dense when nematodes have been extracted from plant tissues) in the funnel to facilitate filtering through the paper filter disk adapted in the funnel. (From Ref. 29, with kind permission of SIN (Italian Society of Nematology))

3.7 Final Considerations

It is impossible to describe protocols particularly suited for extraction, reproduction, and storage of all nematode species. We believe that those reported in this chapter can be of great help to be used as described or after adjustments to different situations. For more information, readers are suggested to consult EPPO standards [6] where several protocols are presented to rear and store quarantine nematodes as reference materials.

Some protocols have been developed to grow in vitro excised roots or host plants to be inoculated with nematodes [16], especially to work with some cyst nematodes, such as *Globodera* spp. and *Heterodera* spp. However, such methods rather than to obtain large numbers of nematodes are designated to conduct screening or biological studies.

4 Notes

1. However, if greenhouse temperatures are constantly optimal for nematode and plant development until plant senescence, nematodes may not be induced to enter in dormancy, and the physiological state of nematodes reproduced in greenhouse

may not be the same of those obtained under field conditions. In general, if nematodes are collected long before plant senescence, they are ready to infect, and it does not matter how they behave normally under field conditions. Failure has been observed of *M. incognita* to damage sown melon and *M. artiellia* to wheat and chickpea in microplots infested with nematode reproduced in greenhouse. Nevertheless, nematodes reproduced in greenhouse, especially *Meloidogyne* spp., can be always useful in greenhouse, while others may require a chilling period. How large the nematode population must be increased depends on the needs. Several hundred million individuals can be necessary when screening gene bank germplasm or progenies from breeding programs. When eggs of root-knot nematodes (RKNs) are used, it should be realized that single tomato plants that are well grown in a 20–25 cm diam. pots, 70–90 days after inoculation, may contain at least 100,000 J2s and eggs. When *Meloidogyne*, *Globodera*, *Heterodera*, *Nacobbus* spp., *Rotylenchulus reniformis*, endoparasitic migratory nematodes (*Pratylenchus* spp., *R. similis*, *Ditylenchus* spp., *A. besseyi*), and some ectoparasitic nematodes, such as *X. index* and *X. americanum* sensu lato, and criconematids are the target nematodes, often large population densities are required. It is not convenient to store nematode attacking leaves for a long time as leaves easily rot. Also realize that with the use of field-infested soil, other microorganisms including plant pathogens may be brought about. When dealing with cyst nematodes, extraction of cysts would eliminate most of the other organisms through the rather large sieves used. In case of nematodes forming egg masses, these microorganisms would be killed by the sodium hypochlorite solution if this is used to extract eggs. Whether it is more convenient to collect nematodes from a limited area or different parts of the same or more fields depends on the use of nematodes. Collecting nematodes from a single plant or several close plants may guaranty the chance to deal with a rather genetically uniform population. Some investigators prefer to use mixed populations for screening. Of course, nematodes can be maintained on host plants in large pots (50–100 cm diam.), outside, or in microplots all year, especially in tropical areas and when dealing with nematodes of perennials.

2. However, the reader may wish to consult EPPO standards and diagnostics [6], which is a huge source of information and protocols for nematodes listed as quarantine organisms in the EPPO region (Europe and most Mediterranean countries), containing beautiful explicative photos.

3. Realize that in amphimictic nematodes, a single female can be fertilized by more males. However, for more genetically uniform nematode populations, it is suggested that mating

may occur under in vitro conditions, in which only a male is allowed to fertilize the selected female. To reproduce species of the *H. avenae* group, it is suggested to use sandy soil (95–98% sand) as they suffer from lack of oxygen. For nematodes whose eggs hatch in water in normal (*H. avenae*) or large proportion (*H. schachtii*), collect the soil as soon as the host plant approaches senescence. The risk is that nematodes do not reproduce anymore and start their natural decline so that cysts end up containing few or no eggs. Avoid that the soil in the pots become too dry for storing. If at beginning of the rearing only a few nematodes are available, first start with a few pots; thereafter, use the infested soil of these previous pots to infest more pots.

4. Do not wait long time after the start of the senescence to collect nematodes. If no sufficient food is available, nematodes will start their natural decline; the ones extracted become much less and exhausted. However, in case of *T. semipenetrans* that must be reared on citrus plants (woody), and that are less vigorous, it may take 6–10 months before collecting eggs and J2s. To avoid such a long time, it is suggested to collect nematodes from the roots of a citrus grove or from a few infected citrus plants grown in large containers in greenhouse.
5. A source of inoculum for reproduction can also be used infested field soil or roots. However, this may also carry over other soil microorganisms, including plant parasites. We emphasize that this is not the most convenient way to reproduce root-lesion nematodes; instead, the carrot disk culture is the most suitable.
6. With the exception of *M. naasi*, whose eggs require a vernalization period (about 2 months at 7–8 °C) before hatching, eggs in other egg masses will hatch as soon as they have completed their embryogenesis. Several species of cyst nematodes, besides retaining eggs in cysts, lay eggs also in egg masses (*H. carotae*, *H. cruciferae*, *H. glycines*, *H. cajani*). Therefore, egg masses of these species can also be used to obtain populations deriving from a single female.
7. Be sure that at inoculation time, the eggs in cysts have already fulfilled their vernalization or diapause waiting period.
8. In the case of nematodes reproducing parthenogenetically, any female stage can be used. However, we suggest to prefer pre-adult or adult females. Instead, of the amphimictic nematodes, it is mandatory to inoculate just gravid females, after observation under a stereoscope for the presence of eggs inside the gonads or sperms in the spermatheca. Maintain plants under good growth conditions (fertilizing and keeping pest and diseases under control) and avoid excess irrigation; water

should not leak from the pots. Moreover, for nematodes reproducing on different in vitro media (callused plant tissues, fungal culture), obtaining single female nematode populations by inoculating these media is highly recommended.

9. Although both *D. dipsaci* and *P. penetrans* reproduced on both media, greater reproduction can be achieved of *P. penetrans* on callus of alfalfa on White's medium supplemented with 2,4-D and αNAA and of *D. dipsaci* on callus of onion on Krusberg's medium supplemented with 150 ml/L of coconut milk. More details on the rearing of *D. dipsaci* can be found in Plowright et al. [20]. *D. destructor* can also be increased on callus tissues of carrot, clover, tobacco, and potato obtained on the Murashige and Skoog [11] medium. Nematodes on callus can be perpetuated as long as you wish and callus can be grown also on different media put in small (50–200 ml capacity) flasks, whose size depends on the aim: small size to maintain nematodes in collection and larger ones for obtaining large nematode populations. The mentioned media can be suitable to grow also callus of different host plants and nematode species other than those listed above could be reared on them.

   When preparing and handling in vitro culture, especially when sterilizing nematodes, seeds, or plant parts, you may be using toxic substances; therefore, always wear gloves and work under a laminar flow hood.

10. The procedure described can be followed with other fungus–nematode combinations. However, not all fungus culture will support the reproduction of any nematode species. Therefore, before undertaking large-scale nematode reproduction on different fungal cultures, we suggest to do first a preliminary test. Also, to obtain the axenic culture of the desired fungus species, the specific expertise of mycologists is needed. Starting from a monoxenic culture of the fungus will save time and guaranty success.

11. Today, the carrot disk method is one of the best and widely used to have sterile nematodes in a collection and reproduce large population densities for different uses, including pathogenicity tests and screening of genetic planting material materials. Sterilization of the external cuticle of nematodes is necessary whenever bacteria or fungi are suspected to be present in the nematode suspension or on the nematodes themselves, as generally occurring when extracting nematodes from soil or plant material. The method is suitable to rear root lesion (*Pratylenchus* spp., *Radopholus* spp.), several *Ditylenchus* spp., and probably other nematode species. It is rather simple and requires little space and only common materials and equipment generally available in every laboratory. Always select females for inoculation. If dealing with amphimictic species, prefer gravid

females; otherwise, both female and male specimens must be inoculated on the same disk. Depending on the diameter of the carrot disk and the aim of the culture, the number of nematodes to inoculate may vary greatly. When the aim is to obtain a large number of nematodes for screening purposes or pathogenicity tests, the use of a relatively large number of nematodes is preferable, while less nematodes are suggested to maintain nematodes in a collection. Temperature is also important. Therefore, carrot disk must be incubated at the most suitable temperature for the nematode species and geographical origin and at a lower temperature to maintain nematode in collection. Inoculating each carrot disk 10 mm thick and 30–40 mm diameter with 25 specimens of *P. thornei* at 25 °C resulted in nematode numbers of 31,370 and 90,480 and reproduction rate of 1255-fold and 3619-fold at 75 and 100 days, respectively [8]. If, during checks, carrot disks appear contaminated or anyhow not in good conditions, they must be discarded; keeping in collection noncontaminated nematode will save time in the future. The temperature of incubation usually depends on the species: more rapid nematode multiplication may occur at 28 °C, but lower temperatures allow the isolate and the disk to be maintained for a longer period.

Acknowledgments

We wish to thank all of the journals and authors that granted us permissions to reuse their figures.

References

1. Gooris J, D'Herde CJ (1972) Le cycle de développement de *Meloidogyne naasi* Franklin sur cereals de printemps et d'hiver et sur betteraves. Rev Agric 25:651–657
2. Hominick WM, Forrest JMS, Evans AAF (1985) Diapause in *Globodera rostochiensis* and variability in hatching trials. Nematologica 31: 159–170
3. Greco N (1981) Hatching of *Heterodera carotae* and *H. avenae*. Nematologica 27:366–371
4. Fustey SG, Johnson PW (1966) The biology of the oat cyst nematode *Heterodera avenae* in Canada. I. The effect of temperature on the hatchability of cysts and emergence of larvae. Nematologica 12:313–320
5. Banyier RJ, Fisher JN (1971) Effect of temperature on hatching of eggs of *Heterodera avenae*. Nematologica 13:273–278
6. Anonymous (2021) PM 7/148 (1) guidelines for the management of the nematode collection used for the reproduction and management of reference materials. EPPO Bull 51: 507–548. https://doi.org/10.1111/epp.12798
7. Coyne DL, Adewuyi O, Mbiru E (2014) Protocol for in vitro culturing of lesion nematodes: *Radopholus similis* and *Pratylenchus* spp. on carrot discs. International Institute of Tropical Agriculture (IITA), Ibadan, Nigeria. 15 pp
8. Castillo P, Jl T-C, Jimenez-Diaz RM (1995) Effect of time, temperature, and inoculum density on reproduction of *Pratylenchus thornei* in carrot disk cultures. J Nematol 27:120–124
9. Riedel RM, Foster JC (1970) Monoxenic culture of *Ditylenchus dipsaci* and *Pratylenchus penetrans* with modified Krusberg's and White's media. Pl Dis Rep 54:251–254
10. Van Der Walt PCW, De Waele D (1989) Mass culture of the potato rot nematode *Ditylenchus*

destructor on groundnut callus tissue. Phytophylactica 21:79–80

11. Murashige T, Skoog F (1962) A revised medium for rapid growth and bioassay with tobacco tissue cultures. Physiol Plant 15:473–497
12. Hussey RS, Barker KR (1973) A comparison of methods of collecting inocula of *Meloidogyne* spp., including a new technique. Pl Dis Rep 57: 1025–1028
13. Greco N, Di Vito M (2009) Population dynamics and damage levels. In: Perry RN, Moens M, Starr JL (eds) Root-Knot Nematodes. CABI, Wallingford, pp 246–274
14. Steenkamp S (2008) Host plant resistance as a management tool for *Ditylenchus africanus* (Nematoda: Tylenchidae) on groundnut (*Arachis hypogaea*). PhD Thesis, Potchefstroom Campus of North-West University, South Africa
15. Brown DJF, Zheng J, Zhou X (2004) In: Chen ZX, Chen SH, Dickson DW (eds) Virus vectors. Nematology: advance and perspectives, vol 2. CABI Publishing, Wallingford, pp 717–770
16. Mugniery D, Person F (1976) Méthode d'élevage de quelques nematodes a kistes du genre Heterodera. Sciences Agronomique Rennes 1976:217–220
17. van Bezooijen J (2006) Methods and techniques for nematology. Revised version 2006. Wageningen, p 118
18. Mwangi JM, Niere B, Daub M, Finchk MR, Kiewnick S (2019) Reproduction of *Globodera pallida* on tissue culture-derived potato plants and their potentials use in resistance screening process. Nematology 21:613–623
19. Ajihassani A, Tenuta M, Gulden RH (2017) Monoxenic rearing of *Ditylenchus weischeri* and *D. dipsaci* and microplot examination of the host suitability of yellow pea to *D. weischeri*. Plant Protec Sci 53:254–264. https://doi.org/10.17221/158/2016-PPS
20. Plowright RA, Caubel G, Mizen KA (2002) Ditylenchus species. In: Starr JL, Cook R, Bridge J (eds) Plant resistance to parasitic nematodes. CAB International, Wallingford, pp 107–139
21. Meyer AJ (1984) Die in vitro-aseptiese massateling van *Pratylenchus zeae* (Nematoda: Pratylenchinae). Phytophylactica 16:259–261
22. O'Bannon JH, Taylor AL (1968) Migratory endoparasitic nematodes reared on carrot discs. Phytopathology 58:385
23. Moody EH, Lownsbery BF, Ahmed JM (1973) Culture of the root-lesion nematode *Pratylenchus vulnus*, on carrot disks. J Nematol 5: 225–226
24. Verdejo Lucas S, Pinochet J (1992) Population densities of five migratory endoparasitic nematodes in carrot disk cultures. J Nematol 24:96–98
25. Di Vito M, Catalano F, Zaccheo G (2002) Reproduction of six populations of *Pratylenchus* spp. from the Mediterranean region on selected plant species. Nematol Medit 30: 103–105
26. Mokrini F, Viaene N, Waeyenberg L, Dababat AA, Moens M (2018) Investigation of resistance to *Pratylenchus penetrans* and *P. thornei* in international wheat lines and its durability when inoculated together with the cereal cyst nematode *Heterodera avenae*, using qPCR for nematode quantification. Eur J Plant Pathol 151:875. https://doi.org/10.1007/s10658-018-1420-0
27. Kaplan DT, Davis EL (1990) Improved nematode extraction from carrot disk culture. J Nematol 22:399–406
28. Wang H, Mao J, Li R, Luo D, Zhao G, Li H (2016) A culture technique for *Ditylenchus destructor* on sweet potato (*Ipomoea batatas*). Nematology 24. https://doi.org/10.1163/15685411-00002969
29. Greco N, Carletti B (2014) Estrazione, allevamento e conservazione dei nematodi. In: Ambrogioni L, d'Errico FP, Greco N et al (eds) Nematologia Agraria Generale e Applicata. Società Italiana di Nematologia (SIN), Firenze, pp 115–153
30. Yoshiga T (2020) In vitro culture method of *Ditylenchus destructor* (Tylenchida: Anguinidae) using garlic. Appl Entomol Zool 55: 435–437. https://doi.org/10.1007/s13355-020-00701-x
31. Imamah AN, Supramana DTA (2020) *In vitro* cultivation of *Aphelenchoides besseyi* Christie on fungal cultures. J Perlindungan Tanaman Indonesia 24:43–47
32. Oduori CA, Atandi J, Kysaakye J, Coyne D (2021) Cocoyam (*Colocasia esculenta*) provides an effective monoxenic culture media for *Radopholus similis*. Nematologica 23:597–599. https://doi.org/10.1163/15685411-bja10088

Chapter 3

Detection of Resistance, Susceptibility, Tolerance, and Virulence in Plant–Nematode Interactions: Part I—Sedentary Endoparasitic Nematodes

Nicola Greco and Renato N. Inserra

Abstract

The use of nonhost, tolerant, or resistant plants, to manage plant parasitic nematodes (PPNs), is an appealing, economic, and environmentally friendly agronomic practice, which is effective when precise information on the identification of PPN species and their virulence to target host crops is available. This chapter describes suggested protocols to evaluate the reaction of the most important crops and fruit trees to infestation by the most damaging PPN with sedentary endoparasitic habits, with the aim of assessing resistance and tolerance traits, sources of resistance in progenies from breeding programs, the reaction to nematodes of newly released cultivars, and the virulence of the most noxious PPNs. These protocols consist of classical screening techniques not involving biochemical and molecular analyses. PPN species and genera considered in this chapter include (i) the most important species of root-knot nematodes *Meloidogyne* spp., including also *M. chitwoodi*, *M. enterolobii*, and *M. graminicola*, and (ii) the cyst-forming nematodes of the genera *Globodera* and *Heterodera*, such as the potato cyst nematodes (PCNs) *Globodera rostochiensis* and *G. pallida*, and also *Heterodera avenae* group, *H. ciceri*, *H. glycines*, and *H. schachtii*. Schemes are given to identify virulence groups for most of these nematodes.

Key words Cyst-forming nematodes, Root-knot nematodes, *Globodera* spp., *Heterodera* spp., *Meloidogyne* spp., Screening for resistance, Tolerance, Virulence

1 Introduction

Control of plant parasitic nematodes (PPNs) is instrumental in avoiding or reducing crop loss and making the crop profitable for the farmer. In the past decades, control of nematodes was mainly based on the use of synthetic nematicides [1]. The use of such chemicals may cause hazards for the applicator and bystanders and is badly affecting the quality of the environment. Safety concerns have forced environmental protection agencies to limit the use of

Author Nicola Greco is now retired.

Sergio Molinari (ed.), *Plant-Nematode Interactions*, Methods in Molecular Biology, vol. 2756,
https://doi.org/10.1007/978-1-0716-3638-1_3,

some agrochemicals, replaced, when feasible, by non-chemical control methods within the framework of an integrated pest management system. Research on alternatives to chemical management has been prioritized with encouraging results. Agronomic means of control are among the safest and most environmentally friendly approaches to limit damage caused by PPNs. Among these practices, crop rotations with nonhost plants and the use of resistant or tolerant cultivars [2–4] are very effective when used within a system of an integrated nematode management [5]. The implementation of these practices requires determining reaction of different plant species and cultivars to nematode infestation, including wild species and weeds growing in different ecosystems. Therefore, at least for the most important PPNs, it is necessary to assess the host reaction they induce on thousands of plant species occurring in genetic resource units. In addition, when a resistant line/cultivar is identified and the gene governing that resistance is characterized and used in breeding programs, the large numbers of offsprings must also be screened. The selected resistant plants must be grown to produce seeds for use in further crosses. In some cases, resistance genes can also be introgressed in plants, which should be tested for their reaction to nematodes after the introgression. Hereafter, suggestions are given about the procedures to screen the most important cultivated plant species to the most damaging PPNs. We are aware that it is impossible to provide suggestions for each plant–nematode interaction. Readers should refer to the most similar plant–nematode combination presented here to design screenings for their needs.

Breeding programs for PPNs should be carried out with nematode species present in each given country. The use of exotic nematodes requires permits from local official authorities and should be conducted in quarantine greenhouses. As a common practice, at the beginning of the test, plant and soil used for screening must be free from PPNs. At the end of the test, nematode-inoculated plants, soil, and containers should be steam-sterilized to avoid the spread of the PPNs in the environment. Finally, before selecting a plant/cultivar/line to be used as gene donors in breeding programs, it is imperative that their resistance had been carefully ascertained; a wrong selection would jeopardize the entire breeding program causing loss of time and funds.

Before starting a screening program, the reader should be familiar with the basic concepts and principles established for the assessment of plant reactions to PPN infestation [6–8]. The classification of the plant reaction to a PPN is based on the extent of the damage suffered from nematode infestation and on the reproduction factor (Rf) of the nematode on that plant, expressed as the ratio $Rf = Pf/Pi$. In this ratio, the symbol Pf represents the number of nematodes assessed at plant harvest (final population density) and Pi the number of nematodes at planting (initial population density). Therefore, a plant is classified as susceptible or non-susceptible depending on whether or not it suffers from nematode infestation and allows its reproduction. According to the severity of damage and reproduction degree, a plant may show different levels of susceptibility from low to very high. Based on nematode reproduction, a plant can be classified as host or nonhost, with its host status varying from poor to very good. When in contact with nematodes, a plant species may show immunity, resistance, tolerance, or intolerance. Immune is used to indicate a plant that does not attract or repel PPNs. Resistant is defined as a plant in which PPNs initiate infestation, but its development and reproduction is arrested. Tolerance is defined as the reaction of a plant that allows infestation and good nematode reproduction but nevertheless still yields satisfactorily. However, a plant can be tolerant up to a certain level of Pi and becomes non-tolerant at larger Pi. Finally, a plant is considered intolerant if it is damaged from nematode colonization, even at low nematode densities, and it may die prematurely preventing reproduction. To be useful for agronomic purposes, a resistant plant must also be tolerant (not suffering much from nematode infestation). On the other hand, if a resistant plant is highly intolerant (severe damage coupled with no or poor nematode reproduction), it is not useful at all.

When screening for resistance, there is a general agreement to consider a plant resistant to a PPN when the value of Pf/Pi ratio is <1 and partially resistant when Rf is slightly >1 (Rf in the range 2–3). Also, a plant is considered resistant if the number of PPN eggs and juveniles (J2) extracted from its roots and the surrounding soil at harvest is less than 10% of that of the susceptible control plant. This standard is also called reproduction index (RI) or reproduction rate (Rr), and if it refers to numbers of females on the roots, such as for cyst nematodes (*Heterodera* spp., *Globodera* spp.), it is called female index (FI). The reproduction and female indices (RI or Rr and FI) are expressed by the ratio RI or FI = [(Pf of the test plant/Pf of the susceptible control plant) × 100]. In this ratio, Pf represents the number of eggs and J2 or females only, determined at plant harvest (final population density) for the test and susceptible control plants. Other indices are root galling (RGI = root gall index) and egg mass index (EMI) for root-knot nematodes, *Meloidogyne* spp. Damage index is based on the extent of the most obvious symptoms caused by the nematode, such as the number of galls.

2 Materials

2.1 General Screening of PPNs

For a vigorous plant growth, PPN reproduction, and speedy soil and plant tissue processing, using a steam-sterilized sandy soil (85–95% sand) without clay or organic matter is recommended.

The fertility of this soil should be enriched by applications of a water solution of a fertilizer containing N, P, and K, in equal proportions plus microelements. When leguminous plant species are to be screened, treating seeds with the most appropriate root nodulation bacterium [9] is recommended, especially if tests are conducted in large pots or microplots containing sterilized soil. For all tests, the following is required:

- Greenhouse equipped to set temperature in the range 15–30 °C and light for 14–16 h per day.
- Screenhouse especially useful in tropical areas.
- Growth chambers of specific dimensions according to plant size and number
- Trays, pots, and sterilizing steamers.
- A nematology laboratory for processing soil, plants, and extracting nematodes.
- Well-trained personnel to process soil and determine PPN population levels.
- Infested field or microplots for final assessment of plant reaction (*see* **Note 1**).

2.2 Screening for Resistance to Root-Knot Nematodes (RKNs), Meloidogyne spp., in Annual Plants

The materials for screening resistance to RKNs are listed below and can be used for most of the species in the genus *Meloidogyne*.

Before starting, you need seeds of the cultivar/line/accession and races to screen, i.e., seeds of lines of the same plant species known to be susceptible, resistant, and tolerant to serve as controls. If no information is available on the reaction to RKNs, a cultivar of tomato or another solanaceous plant known to be very susceptible should be used as susceptible control. Used materials are the following:

- Sterilized sandy soil.
- A composite soluble fertilizer, such as 20-20-20 NPK plus microelements.
- Different kinds of pots to grow test plants for about 2 months. In general, pots containing 800–1000 cm^3 soil or 12–14 cm diameter are sufficient. The best solution would be having transparent plastic pots inserted in black plastic pots of the same size or wrapping the transparent pots in black plastic sheet or aluminum sheet, to avoid growth of algae that would make difficult the observation of the roots adhering to the walls of the transparent pots.

2.3 Screening for Resistance to RKNs in Perennial Plants

Unlike annual plants, perennials, if originated from seeds, do not generally produce good quality fruits as those of the parents and, therefore, the cultivars having good agronomic qualities are grafted on rootstocks. Therefore, most of the screenings deal with rootstocks the progenies of which come from crosses of wild relatives that are made to improve their resistance to different soilborne parasites and pathogens [10]. General rules are the same as those for annual plants except that the containers must be larger and the evaluation period after inoculation may be longer.

2.4 Screening for Resistance to Cyst-Forming Nematode (CNs) Species

The materials used for screening resistance to some selected species of economically important CNs are listed below:

- ***Heterodera avenae*** **group**: needed materials are the following:
  - Nematode populations of known species and pathotype; equipment for hatching test to obtain infective J2.
  - Seeds of genotypes to evaluate.
  - Seeds of known susceptible and resistant cultivar/line; a vernalization period before germinating may be required for seeds of spring cultivars.
  - Special tubes about 2.5–3 cm diam and 20 cm high filled with 200 cm^3 potting sterile soil made of quartz/river sand (70%) field soil (29%) and only a small proportion of peat (1%).
  - Fertilizers.
- ***H. glycines***: required materials are seeds of the entry genotypes, seeds of the suggested susceptible control cultivars, seeds to define the population type of HG. Three populations of the same HG type. Containers of the selected type and volume. Sterilized sandy soil.
- ***H. schachtii***: to proceed, you need populations of the nematode, seeds of the crop plants to be tested for their reaction to the nematode, pots or small tubes, and sterilized sandy soil.
- ***H. ciceri***: the screening tests for this nematode were designed to accommodate at first the screening of more than 10,000 lines from the ICARDA genetic resource unit (GRU) and then that of other thousands of progenies from breeding programs. Required materials include the availability of a large amount of nematode per g soil for the inoculum and adequate equipment to extract nematodes. A large amount (2–3 m^3) of sterilized soil (about 70% river sand + 30% field soil, containing around 33% clay) is also needed. Numerous benches, 3–5 dm^3 pots (buckets), and thousands of special 6 cm diam peat moss pots (Jiffy pots). Other materials are seeds of the chickpea lines to screen along with those of a susceptible cv. (Ghab 1 = ILWC 292), or a similar one, and fertilizers.
- ***Globodera rostochiensis*** **and** ***G. pallida*** **(PCNs)**: required materials include seeds, in vitro culture, seedlings, tubers, or rooted cuttings of the genotypes to screen susceptible, partially resistant, and resistant cultivars to serve as controls for comparison.

Small pots of about (5 × 5 cm) to transfer in vitro cultures or seedlings after germination. Larger pots (10–14 cm diameter), containing 500–1000 cm^3 soil, for the screening. For an easier observation of females on the root balls, it is recommended to use plastic transparent pots to be inserted in opaque pots or in containerized sand on benches in greenhouses. Steam-sterilized sandy soil (at least 85–95% sand) or a mixture of field soil with river sand, avoiding adding peat or other organic material that may cause problems during the extraction and counting of nematodes. PCN populations selected for the screening test. In breeding centers, it is suggested to have a collection of all pathotypes of both PCN species. They will be useful especially when screening wild genotypes. Progenies from parents having known resistance to certain pathotypes should be screened toward these pathotypes. In temperate areas where the potato cycle ends in autumn, eggs in cysts of both PCN species undergo a dormancy period. Therefore, it is suggested that cysts are to be extracted and concentrated in the remnant soil debris collected after sieving or in unprocessed heavily infested soil and kept at 4–5 °C for 4–6 months before use. After this vernalization period, J2 in the embryonated eggs inside the cysts will react promptly to hatching stimulus of root diffusates or other chemicals and are ready to colonize the plants soon after inoculation. Woody or plastic sticks to serve as tutors for the plants as they grow. Toothpicks or barbecue woody sticks are good to tutor short (5–7 cm tall) potato seedlings. Soluble fertilizer such as a 15-15-15 NPK + microelements to be used in solutions at concentration of 2‰.

2.5 Assessment of Plant Tolerance

As already discussed, the reaction of a plant to a nematode may vary greatly from highly resistant to very susceptible. Moreover, some cultivars can allow the reproduction of the nematode, or even more than a susceptible cultivar, and still yield satisfactorily. These cultivars are considered tolerant [6–8, 11] and may play an important role in crop rotation system. A tolerant cultivar has the advantage of not exerting any genetic pressure on the nematode population that is the major factor in the development of virulent individuals that after a few generations may displace the less virulent populations. However, at the end of the crop cycle, tolerant plants favor a rapid increase of nematode populations hampering the growth of any following susceptible crop used in the rotation system. To avoid selection of virulent nematode populations, it is suggested to include a resistant cultivar once in a 4–5-year rotation cycle that should be followed by a susceptible or tolerant one in the next rotation cycle. However, especially among farmers, there is uncertainty in considering a cultivar tolerant; sometimes, a cultivar considered tolerant resulted as either susceptible, moderately resistant, or resistant [12]. Therefore, before defining a cultivar as tolerant, it

Table 1
Approximate suggested inoculum densities, above the tolerance limit, to assess crop tolerance to nematodes for some plant–nematode combinations, expected to cause yield losses of 10, 30, and 50% of a susceptible crop, derived from the international literature

		Population density expected to cause yield losses of		
Crop Plant	**Tolerance limit**	**10%**	**30%**	**50%**
	Meloidogyne incognita eggs and second stage juveniles (J2) per cm^3 soil			
Tomato	0.55	1.5	4	7
Pepper	0.16	0.7	1.5	3
Melon	0.19	1	1.5	3
	Meloidogyne chitwoodi eggs and J2 per g dried soil			
Carrot	0.012	0.1	0.2	0.5
	Meloidogyne arenaria eggs and J2 per g dried soil			
Carrot	0.25	1	2	6
	Heterodera avenae eggs and J2 per g dried soil			
Wheat	1.3	5	14	22
	Heterodera ciceri eggs and J2 per g dried soil			
Chickpea	1	5	12	20
	Heterodera schactii eggs and J2 per g dried soil			
Sugar beet	1.8	5	15	30
	Heterodera goettingiana eggs and J2 per g dried soil			
Pea	0.5	2	5	10
Broad Bean	0.8	3	8	15
	Heterodera carotae eggs and J2 per g dried soil			
Carrot	0.8	2	7	20
	Globodera rostochiensis and *G. pallida* eggs and J2 per g dried soil			
Potato	1.9	7	15	32

is necessary that a proper test is conducted. In such tests, it is important to select a population density of the nematodes that may cause a damage varying from low (max 10–20%) to medium (50–60%) to a susceptible crop. A very low nematode population may fail to discriminate statistically small yield differences, while very large populations may cause severe damage to both susceptible and tolerant cultivars. Therefore, we suggest that tolerance tests are designed by using three different inoculum densities that should not cause yield losses greater than 50–60%. To meet these needs, in Table 1, the suggested inoculum densities for some nematode–plant combinations are reported; other indications can be obtained from local personal experience.

Fig. 1 Microplots useful to assess the effect of nematodes on growth and yield of several crop plant species. (**a**) Microplots infested with the sugar beet cyst nematode, *Heterodera schachtii*, and planted with sugar beet; (**b**) Microplots inoculated with the chickpea cyst nematode, *H. ciceri*, and just sown with chickpea

2.5.1 Assessment of Plant Tolerance in Greenhouse

The necessary materials include propagative plant organs of the cultivars to be tested. Susceptible, resistant, and if available, tolerant varieties are to be used for comparison:

- Large pots (3–5 dm^3 volume) to allow normal plant development and yield; soil mix consisting of sandy soil (at least 80% sand) sterilized 3–4 months before use.
- Nematode population, possibly reproduced under natural conditions from a susceptible crop grown in large pots or microplots sunk into the soil.
- Fertilizer, pesticides, and, if necessary, fungicides.
- Air-conditioned greenhouse, possibly equipped with an automatic irrigation system.

2.5.2 Assessment of Plant Tolerance Under Field Conditions

Microplots are used commonly to assess the tolerance of a crop to a nematode, because they mimic well-field conditions. Therefore, prepare microplots of about 1 m^2 large, to accommodate several plants, or smaller, about 30 cm diameter or 30 cm × 30 cm (Fig. 1), to be planted with a single plant. Microplots must be filled with about 30–40 dm^3 infested soil up to 5 cm below the ridge.

Other materials are as reported for the greenhouse test.

2.6 Assessment of Nematode Virulent Populations on Plants Carrying Resistance Genes

Populations of nematode species from different geographical areas, at region, state, national, and international levels, are needed. Precautionary measures should be adopted for quarantine exotic PPNs.

Plant species, cultivars, and plant introductions listed in the scheme to be used for the determination of virulent strains.

3 Methods

As a precautionary practice, the examination of the root system of susceptible controls should be carried out before that of the screened genotypes because it provides evidence of the effectiveness of the greenhouse growing conditions in allowing the penetration of the nematode J2s into the roots and their development into numerous healthy females. Presence of only a few small females in the roots of the susceptible control suggests insufficient colonization and development of the nematode, making the results of the test unreliable.

In searching for a genetic source of resistance to PPNs, the first step is checking the reaction of lines/cultivars available in gene banks or genetic resource units to these parasites. This may require the test of several thousand plants, each line replicated four to five times. This reaction must be compared with that of well-known susceptible, resistant, and partially resistant cultivars/lines. Moreover, each test should include pots inoculated and kept without plants. The number of nematodes extracted from these pots without plants at the end of the test (Pf) represents the portion of the initial inoculum (Pi) that survived during the test. By subtracting the surviving nematodes from the number of nematodes that are extracted from the inoculated pots with plants, it is possible to obtain a more precise measure of the nematodes (Pf) produced by the initial inoculum (Pi) in the presence of the plants. The assessment of surviving nematodes is very useful and facilitates the determination of the number of the inoculated nematodes that may have died soon after their placement in the soil, thus failing to infest and reproduce on the tested plants. When the mortality of the initial nematode inoculum (Pi) is high, the reproduction factor ($Rf = Pf/Pi$) calculated by taking into account this mortality becomes greater because the number of nematodes in the initial inoculum (Pi) that survived and reproduced is smaller than that of the nematodes that were inoculated, increasing the ratio Pf/Pi.

When designing host suitability tests, the genetics of different materials must be known. If the entries of the test are not homogeneous, such as populations of wild species, collected from different areas, or offsprings from crosses of a breeding program, the putative resistant plants should be saved and transplanted for immediate crossing or production of seeds for further uses. This can be done by using nondestructive indices for rating plants, such as RGI and EMI for RKNs, and FI for CNs. Moreover, sowing/planting must be done in containers to ensure that roots are not damaged when processed for rating. This can be done using a transparent plastic pot inserted in a black plastic pot of the same size to be removed at rating time. The transparent pots could also be wrapped in black plastic or aluminum foil and unwrapped at rating. The rating would be done only on the most external roots of the root ball without disturbing most of them to allow optimal growth and seed production after plants are transplanted into larger containers.

3.1 Screening for Resistance to RKNs

This genus includes more than 100 species of endoparasitic sedentary nematodes parasitizing belowground parts of plants. It is the most damaging group of PPNs worldwide [13]. The species of this genus cause evident symptoms in the roots and other plant organs, which become deformed, swollen, and transformed in galls of different sizes (Fig. 2). They have wide host ranges, which include many crops, fruit trees, and weeds. Some of the most common warm climate species are represented by *Meloidogyne arenaria*, *M. enterolobii*, *M. ethiopica*, *M. exigua*, *M. incognita*, *M. graminicola*, *M. javanica*, and *M. salasi*. Other selected cool climate species include *M. artiellia*, *M. chitwoodi*, *M. fallax*, *M. hapla*, and *M. naasi*. These species have selected populations

Fig. 2 Large galls caused by root-knot nematodes, *Meloidogyne* spp. in roots of (**a**) zucchini; (**b**) carrot; (**c**) tomato; (**d**) soybean

that show different degrees of reproduction and virulence toward several differential host plants. They have also different host preferences and are indicated as races. One or more genes conferring resistance to one or more RKN species have been identified in many crop plant species. Among these plants, the wild tomato *Solanum peruvianum* provided the first source of resistance [14], because it contains the resistance gene *Mi1.2*, which has been transferred in subsequent breeding programs to many tomato cultivars and F1. Other resistance genes have been identified and already introgressed in many other crop plants, including pepper, potato, soybean, and fruit tree rootstocks. The methods for screening for resistance to RKNs can be used for most of the species in the genus *Meloidogyne*. For convenience to the reader, these methods have been divided in methods for annual and perennial plants.

3.1.1 Screening of Annual Plants

Then proceed as follows:

1. Germinate seeds in Petri dish containing wetted filter paper disk in a growth cabinet at 23–25 °C.
2. Soon after germination (Fig. 3), transfer seedlings in trays (50 cm long × 25 cm wide × 15 cm high) filled by two-thirds with a mixture of 1/3 peat and 2/3 of sterilized river sand, or in small pots (8–12 cm diam.) such as plastic or Styrofoam polypots containing the same soil mix, and keep them in a greenhouse at 25 °C. Production of seedlings can also be done by specialized and reliable nurseries.

Fig. 3 A Petri dish lined with moisted filter paper and Styrofoam polypots useful to obtain seedlings. (**a**) Germinated tomato seeds in a Petri dish; (**b**) Empty Styrofoam pots; (**c**) Styrofoam pots with potted tomato seedlings

3. Water and fertilize as needed.
4. When seedlings reach the two- to four-true leaf stage, transplant them in the transparent pots of the size specified above and filled with a sterile sandy soil (90% sand). Avoid the use of excessive organic matter because it may cause problems when observing the roots for evaluation of the reaction to nematodes and extract nematodes from the soil.
5. Insert fertilized and watered pots into the black ones and arrange them in a randomized block design with six to ten replicates, on benches in a greenhouse set at 25 °C and 14–16 h light per day.
6. Ten to fifteen days later, when plants sprout two more leaves and new roots, inoculate them with extracted RKN eggs and second-stage juveniles (J2) at the rate of two to three eggs and J2 per cm^3 (initial population, Pi) soil by making four holes 3–4 cm deep and pouring in each 1/4 of the inoculum per pot. Close holes by pouring in each a few mL of water to allow soil to adhere to eggs and J2.
7. To check the right time for inoculation, we suggest having extra susceptible control plants to be used to observe development of new roots and their uniformity in the pots.
8. To obtain the inoculum, we suggest using the sodium hypochlorite (NaOCl) method of Hussey and Barker [15]. However, roots must be shaken for a shorter time (3 min) in a solution containing only 0.5% of NaOCl; wash away the excess hypochlorite by sieving through a 74 μm pore (200 meshes) sieve nested on a 10–20 μm pore (500 meshes) sieve, by spraying tap water on the upper sieve; and collect eggs and J2 that remained on the 10–20 μm pore sieve into a beaker. Finally, adjust nematode concentration to 500–1000 eggs and J2 per mL. This method is generally good for every *Meloidogyne* species, as it sterilizes nematode surface, but exposure of eggs and J2 to high hypochlorite concentration reduces by 80% nematode viability.
9. For some research, such as investigating the dynamics of the response of the resistance gene when challenged by a nematode, it may be preferable the use of J2 rather than eggs as J2 parasitize roots much faster.
10. Set greenhouse temperature and light at 25–28 °C and at 14–16 h per day, respectively.
11. Watering is a critical step: much water may cause lack of oxygen and may wash away nematodes. Different pots may require different amounts of water.
12. Therefore, well-trained personnel should water manually the plants. An automated and precise drip irrigation system could be helpful.

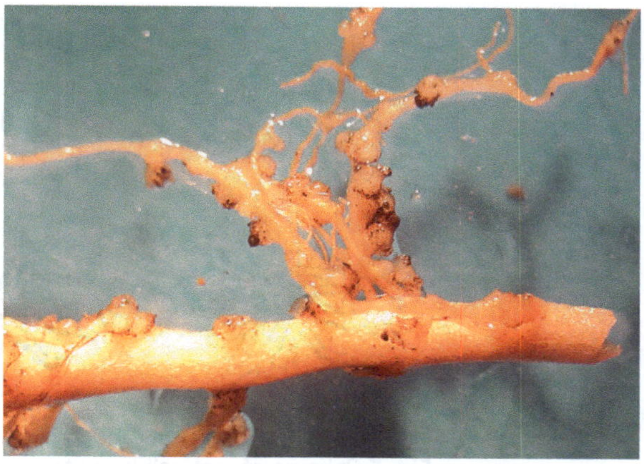

Fig. 4 Discrete galls and egg masses induced by *Meloidogyne* spp. in tomato roots

13. If necessary, insert a woody stick in the pot to serve as a tutor for the plant.
14. Fifty to sixty days after inoculation, check the roots of a few extra susceptible control plants to verify presence of galls of at least 3–4 mm in diameter and prominent egg masses. However, size of galls and egg masses produced by RKN may vary with each species and host plants.
15. If both galls and egg masses are discrete (Fig. 4), evaluate the plant reaction to the nematode.
16. Therefore, cut aboveground plant part and discard it if its growth does not differ from that of non-inoculated control; if differences are noted, record its weight and compare it to that of the control plants.
17. Evaluate the plant reaction to RKN using the root gall index (RGI) and egg mass index (EMI) (Fig. 5), or the reproduction factor (Rf) alone or in combination. When screening thousands of plants deriving from breeding programs, evaluation based on the visual observations of galls and egg masses is sufficient as a preliminary screening, while a more complete evaluation can be done later, if necessary.
18. The determination of the RGI and EMI can be done by inspecting visually the plant roots that are in contact with the wall of the transparent pots. If no transparent pots are used, at evaluation time, turn the pot upside down and unpot plants without damaging their top and root system. The most common rating scales are based on the percentage and severity of total root system galled and are from 0 to 5 (Fig. 5) or 0 to 10 (Fig. 6) with 0 for the entire root system without galls or egg masses and 5 and 10 for the entire root system galled or infested by egg masses. Other scales are based on the number of galls present in the entire root system. According to these

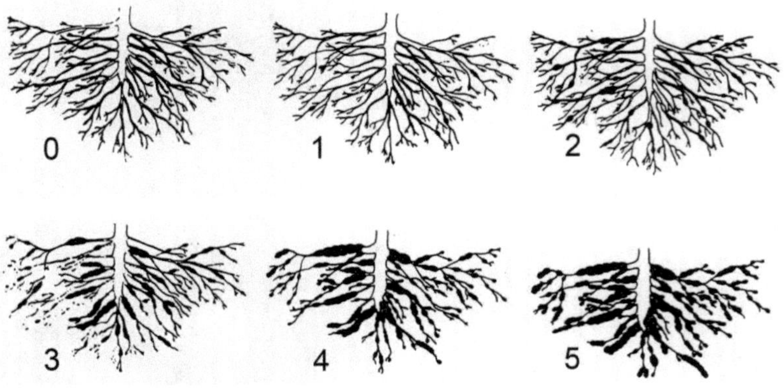

Fig. 5 Scheme of root gall index to score roots for degree of galling induced by *Meloidogyne* spp. on a 0–5 rating scale. (Permission by the Italian Society of Nematology, SIN)

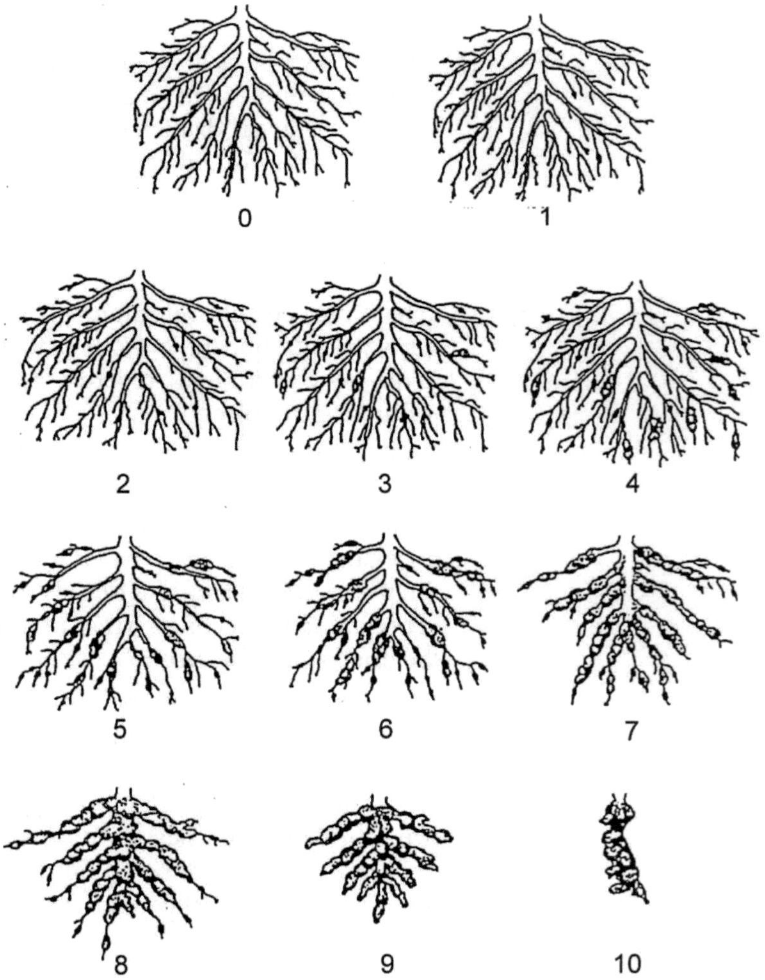

Fig. 6 Scheme of root gall index to score roots for degree of galling induced by *Meloidogyne* spp. on a 0–10 rating scale. (Permission by SIN)

scales, plants evaluated 0–2 in the 0–5 scale or 0–3 in the 0–10 scale for both indices can be considered resistant and must be saved to be transplanted in larger pots (20–25 cm diam containing 2–3 kg soil) to produce seeds for further breeding or studies. To discern egg masses clearly, the entire root apparatus should be freed from the adhering soil using a stream of tap water and then stained red in a water solution of 0.15% phloxine B for 15 min or other methods. Excess stain can be removed by gently washing the root system with running tap water.

19. To calculate the Rf, uproot plants from all pots separately and free gently each root system from the adhering soil using a stream of tap water. Excess water is removed by drying the roots between two sheets of laboratory paper. Then, weigh and soon after cut the roots in about 0.5 cm long pieces to be placed and mixed well in a bowl with water. An aliquot of 10 g of roots per pot or all the entire root system for smaller plants should be used to extract the eggs according to the Hussey and Barker [15] method; eggs are counted and referred to the total root weight per pot.
20. Mix the soil of each pot thoroughly in a tray and remove a subsample of 500 cm^3 soil or less for nematode extraction. We suggest the centrifugal flotation methods of Coolen [16] or Jenkins [17].
21. Count extracted eggs and J2 and refer them to the total amount of soil per pot. Add number of eggs and J2 from the soil to that from the roots to obtain their total number per pot at the end of the test (Pf).
22. Based on the value of Rf (see Definitions), potted plants with Rf < 1 can be classified as resistant and those with Rf ≥ 1 can be considered susceptible.
23. Finally, calculate the value of RI or Rr (*see* Sect. 1 "Introduction"). If RI or Rr is <10% of the susceptible control, plants can be considered resistant, if not susceptible. Similar comparisons can also be made with resistant and tolerant cultivars if included in the test.

It is not convenient to evaluate plant reaction late in the season when large galls are formed in the roots because they may contain females with eggs embedded inside the roots and not extractable by the Hussey and Barker [15] method. In this case, the roots must be chopped and then ground in a blender and the eggs extracted with Coolen's [16] centrifugation method.

3.1.2 Screening of Rice

The screening of rice to *Meloidogyne* spp. may require modifications of the protocols described before because of the environmental conditions required by this cereal crop. The following procedures are abstracted from several published articles. However, always include in the test well-known resistant and susceptible cultivars/lines to serve for comparison and ensure that the nematode has reproduced well.

At IRRI (International Rice Research Institute), Manila (Philippines), Cabasan et al. [18] have developed a hydroponic method to test the reaction of rice to *M. graminicola* or other species, such as *M. oryzae* and *M. salasi*. This method is suitable for large numbers of rice genotypes. It has the advantage of requiring limited space, number of nematodes, labor, and time. It is reported hereafter.

1. The population of *M. graminicola* was isolated from an Asian rice from the field and reared in greenhouse on Asian rice genotype UPLRi-5. To obtain J2 inoculum for the experiment, galled roots were washed free of soil, cut in 0.5 cm long pieces, incubated in a mist chamber, and J2 collected and counted after 48 h.
2. A modified Hoagland's solution by Hothem et al. [19, 20] was used as plant nutrient. The following stock solutions in distilled water should be prepared: potassium nitrate 10.1 g/100 mL, potassium dihydrogen phosphate 13.6 g/100 mL, calcium nitrate ($Ca(NO_3)_2 \cdot 4H_2O$) 23.6 g/100 mL, magnesium sulfate ($MgSO_4 \cdot 7H_2O$) 24.6 g/100 mL; micronutrients: H_3BO_3 2.86 g/L, $ZnSO_4 \cdot 7H_2O$ 0.22 g/L, $MnCl_2 \cdot 4H_2O$ 1.82 g/L, $CuSO_4 \cdot 5H_2O$ 0.09 g/L, MoO_3 0.01 g/L. Then to obtain Hoagland's solution, the following amount of each stock solution should be poured in a glass cylinder: 2 mL of magnesium sulfate, 5 mL of calcium nitrate and of potassium nitrate, 1 mL each of the other stock solutions, and 50 mg of FE-DTPA and then deionized water up to 1000 mL should be added. For use in the test, Hoagland's solution is reduced to 1/4 of its strength.
3. The hydroponic apparatus (Figs. 7 and 8) consists of a basal 30 mm diam. transparent polystyrene tube closed at the bottom and opened at the top, 70 mm long with the top connected with an adhesive tape to an inverted tube, opened on the bottom and closed at the top, 100 mm long. The bottom tube has a 1 mm diameter hole at its base to allow a small drainage to keep soil moisture at its field capacity, and a 5 mm hole on one side to serve for inoculation and watering. The top inverted cover tube has two 5 mm diam holes on the wall for aeration and another on its closed top for watering.
4. The bottom tube is for one-fourth filled with a mixture of pure silica sand and of an absorbent synthetic polymer, which is moisted and planted with a pre-germinated rice seed. The assembled tube is inserted into a holding rack in a tray and

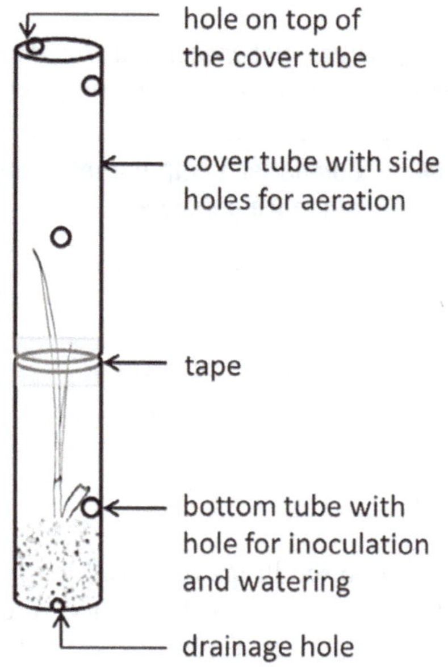

Fig. 7 Schematic representation of the hydroponic apparatus, used at the International Rice Research Institute (Manila, Philippines) for resistance screening of rice lines to *Meloidogyne graminicola*. (See text for more details. From Ref. [18], Permission by Taylor & Francis)

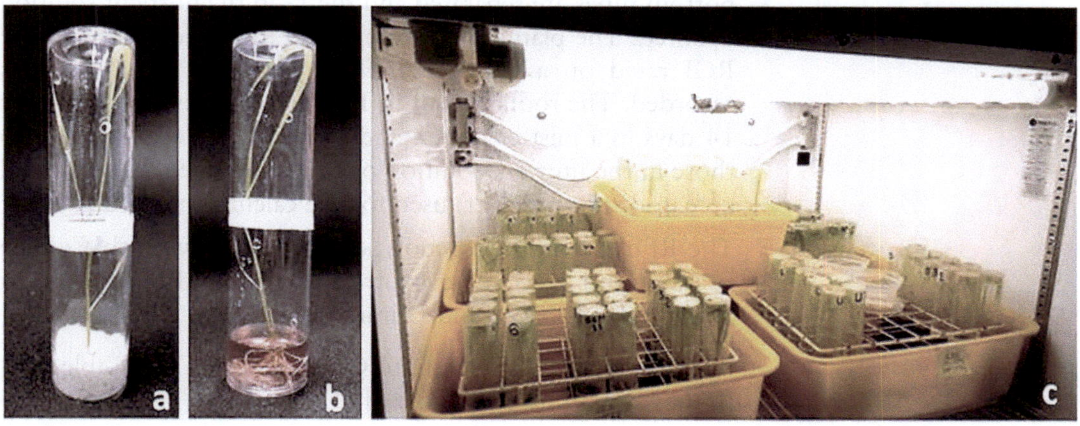

Fig. 8 Hydroponic apparatus and experimental setup of the hydroponic system to evaluate the host response of rice genotypes to infection of *Meloidogyne graminicola*, used at the International Rice Research Institute (Manila, Philippines). (**a**) Plant in sand + adsorbent polymer substrate; (**b**) Plant in nutrient solution; (**c**) Experiment setup in an indoor growth chamber at 29/26 °C (day/night temperature) with 12-h photoperiod. (From Ref. [18], Permission by Taylor & Francis)

maintained in a growth chamber at 26–29 °C (night–day) and 12 h light. Usually, there are eight replicates per entry, arranged according to a completely randomized block design.

5. The bottom tubes are inoculated with 50 J2, 5 days after planting. The nutrient solution is added twice a week. One week after inoculation, the top and bottom tubes are separated

Table 2
Assessment of the host response of *Oryza glaberrima* and *O. sativa* genotypes to *M. graminicola* based on a comparison with the response of susceptible and a resistant reference rice genotype

Comparison with the susceptible reference genotype	Comparison with the resistant reference genotype	Host response of the genotype tested
Significantly[a] different	Not significantly different	Resistant
Not significantly different	Significantly different	Susceptible
Significantly different	Significantly different	Partially resistant
Not significantly different	Not significantly different	Inconclusive

From Ref. [18], with kind permission of Taylor & Francis
[a]According to Dunnet's t test ($p < 0.05$)

and rice seedlings removed from the silica in the bottom tube. Rice roots are gently washed free of the adhering soil mix and seedlings transferred into clean bottom tubes without draining the hole and containing 15 mL of the nutrient solution alone. The two tubes are reassembled and seedlings left to grow for 16 more days.

6. Twenty-three days after inoculation, equivalent to the time required by the nematode to complete one generation, the bottom tubes are separated and the J2 in the nutrient solution counted. The plants are also removed from the tubes and the RGI rated on a 0–5 scale. Then, the fresh root weight is recorded. The roots are cut in 1 cm long pieces, incubated for 14 days in a mist chamber, and the J2 counted and added to those in the nutrient solution of the same tube to obtain total nematode per tube (Pf) used for the calculation of Rf.
7. The Rf of each tested genotype is compared with that of the resistant and susceptible controls. Data are then statistically analyzed by ANOVA and Dunnet's test.

Plants were rated according to Cabasan et al. [18] (Table 2). In the method by Nkurunziza [21]:

1. Rice seeds are germinated in Petri dishes lined with wetted tissue paper at 30 °C for 4 days.

   Then each germinated seed is transplanted in PVC tubes containing sand and absorbent polymer which are kept at 28 °C under a light regime of 12 h per day and 70–74% RH and fertilized three times per week with Hoagland's solution (10 mL per tube).
2. After 14 days, rice seedlings are inoculated with 200 eggs and/or J2 per tube. Nematode reproduction, RGI, and EMI are assessed 24 days after inoculation.

In the method by Cladius-Cole et al. [22]:

1. Three seeds are sown in each pot containing 5 dm^3 soil. After germination, seedlings are thinned to two and when they are 10 cm tall are inoculated with 5000 eggs and J2 by pipetting the nematodes in three to four holes around the stem, and then allowed to grow for 12 weeks.
2. Then, the top plant is cut when soil is still wet. Root and top plant weights and number of spikes and grain per spike are recorded.
3. Roots are then washed free of soil, scored for RGI, EMI on a 0–5 scale, chopped into 1–2 cm long pieces, and eggs extracted by the NaOCl method [15] and counted. Soil is well mixed, and a subsample collected to extract nematodes.
4. The total nematode per pot is determined and Rf calculated. Plants are rated: resistant if $Rf \leq 1$, $RGI \leq 2$, and no significant yield losses; tolerant if $Rf > 1$, $RGI \leq 2$, and no significant yield loss; hypersusceptible if $Rf \leq 1$, $RGI = 2$, and significant yield loss; and susceptible if $Rf > 1$, $RGI > 2$, and significant yield loss. Others use 2 dm^3 or smaller pots and similar evaluation [23, 24].

3.1.3 Screening of Tuber Plants

Screening tuber plants requires large pots. The evaluation of the plant reaction to RKN should be conducted no later than growth stage III (tuber initiation). RKNs parasitize tubers and reproduce inside tubers, issues making the nematode extraction for assessing Rf more difficult. Cool climate RKN species such as *M. chitwoodi*, *M. fallax*, and *M. hapla* inflict seriously economic damage to the potato industry in the cool season because of the adverse effect of their infestations on the quality of the potato tubers, which are not marketable for human consumption. The warm climate species, such as *M. arenaria*, *M. incognita*, and *M javanica*, would parasitize tuber crops mainly during warm period of the year (Fig. 9), as it occurs with sweet potato. Some of the cool climate species, such as *M. chitwoodi* and *M. fallax*, are listed as quarantine organisms by many countries, including those belonging to the European and Mediterranean Plant Protection Organization, and are banned in many international markets.

The screening of potato cultivars is facilitated by the use of tubers as planting material. In this case, tubers are induced to sprout, cut into 15 g pieces containing one sprout, and kept in a tray under light (to avoid elongation of the sprout) for at least 10–15 days, at room temperature to facilitate the hardening of the scars of their sectioned sides. Hardened propagative tuber pieces are then planted in 10–15 cm diam. pots containing around 800–1000 cm^3 sterilized sandy soil.

Screening offsprings from breeding programs is more complicated and laborious. In this case, seedlings are obtained from true

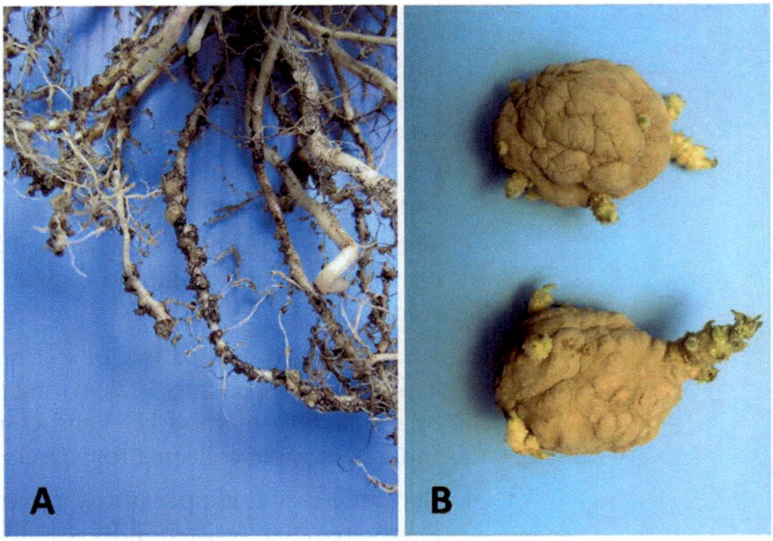

Fig. 9 Galls and egg masses induced by *Meloidogyne incognita* in potato roots. (**a**) Galled roots; (**b**) Deformed potato tubers parasitized by the nematode

seeds. Hereafter is reported a synthesis of the most used protocols in breeding centers in the United States (USA) [25] and the Netherlands [26].

1. Soak approximately 20 true seeds per entry in 0.1% gibberellic acid solution for 24 h and then place them in Petri dishes lined with wetted paper filter. Seven days later, plant germinated seeds singly in 2.5 cm diam. pots containing a sterilized mixture of 75% sand and 25% soil and fertilized with 2 g per dm^3 of a slow-release fertilizer (14-14-14) NPK/dm^3 of soil mix.
2. Four week later, transfer the plants to 10 cm diam. pots containing the same soil mixture and inoculated with 5000 eggs and J2 of *M. chitwoodi* in three holes about 2–3 cm deep. Keep plants in a greenhouse at temperature and light at 25–28 °C and at 14–16 h per day, respectively.
3. Two months after inoculation, when the nematode is supposed to have completed two generations, uproot five plants, gently wash the roots in running water, and extract the RKN eggs and J2 in 0.6% NaOCl solution using the Hussey and Barker method [15].
4. Multiply the resistant plants by rooting top cuttings (8–10 cm long) or by removing meristematic tissue to obtain in vitro cultures to confirm their resistant features.
5. The confirmation test is conducted in pots containing 2 dm^3 soil in which 40–60-day-old seedlings are transplanted. Based on the reproduction rate (Rf), plants are considered as hosts ($Rf > 1$), poor host ($1.0 > Rf > 0.1$), or nonhost ($Rf > 0.1$) or resistant.

6. In the Netherlands, seeds are sown directly in 7 cm diam. pots. Vigorously growing plants are inoculated with 600 eggs and J2 of each RKN species 3 weeks after seed germination.
7. Seven weeks after inoculation, the plants are uprooted, stained in a 0.15% phloxine B solution, and the egg masses counted.
8. Plants with less than six egg masses are considered resistant and, therefore, are used to obtain cuttings for rooting. These rooted cuttings are used for confirmation of the resistance and screened toward the remaining RKN species.
9. In both countries, besides test entries, resistant and susceptible potato or tomato are grown at the same time to verify the suitable conditions of the greenhouse for RKN infestation and reproduction.

3.1.4 Screening of Perennials

The general screening rules for perennials are the same as those of the annual plants, which require large containers and a long evaluation period.

- Olive trees (Fig. 10): We suggest the method proposed by Nico et al. [27] for screening olive trees to several *Meloidogyne* spp. The procedure is as hereafter.

1. Transplant 6-month-old olive plant stocks in pots containing 1 dm^3 of sterilized soil mixture (sand/clay loam, 2:1) and inoculate with 1000 eggs and J2 juveniles from a culture of an identified RKN species reared on tomato cv. Roma and extracted with the NaOCl method [15].

Fig. 10 Olive seedling parasitized by *Meloidogyne incognita*. (**a**) Galled roots; (**b**) Olive plantlet produced from in vitro culture (**c**). (Courtesy of N. Sasanelli)

2. Keep pots (ten replicates per plant–nematode species combination) in a growing chamber at 25 ± 1 °C, 90% RU, and 14 h light per day.
3. Water plants every day with 100 mL water and fertilize once a week with 100 mL of nutrient solution [19].
4. Four months after inoculation, cut plant at soil level, wash roots free of soil, and rate RGI according to a 0–6 scale (0 = non-gall, 1 = 1–10% galled roots; 2 = 11–20%; 3 = 21–40%; 4 = 41–70%; 5 = 71–90%; 6 = 91–100%). See also other scales described in Subheading 3.1.1.
5. Weigh root and top plant and extract nematodes from soil by flotation and centrifugation and from roots by maceration–centrifugation [16]. Count extracted RKNs and calculate the Rf.

Sasanelli et al. [28] suggest using the following:

1. Grow rooted cuttings for 10 months in pots containing clean soil and then transplanted in clay pots containing 0.5 dm^3 of steam-sterilized sandy soil infested with 12 eggs of a RKN species per cm^3 soil. This inoculum level can be obtained by adding to the soil in the pots aliquots of a sterilized sandy soil highly infested with entire egg masses of the nematodes [29, 30].
2. Arrange pots according to a randomized block design on benches in a greenhouse set at 26 ± 2 °C. Then, plants are left to grow for 90 days when they are uprooted.
3. Extract RKNs from 500 cm^3 soil with Coolen's method [13] and from all the roots by grinding them three times in a blender containing 1% NaOCl for 20 s. The ground root suspension in NaOCl solution is then sieved through a 250 μm aperture sieve nested on a 5 μm aperture sieve. Extract also eggs and J2 from soil particles retained on the latter sieve by Coolen's method [13] and then counted and added to those from soil to calculate the Rf.
4. RGI was rated according to a 1–6 scale based on the number of galls in the entire root system (1 = no gall; 2 = 1–20 galls; 3 =11–30, 4 = 31–70, 5 = 71–90, 6 = >90).
5. The reaction of different cultivars to RKN is based on the Rf values. According to this classification, olive lines are rated highly resistant (Rf 0), resistant (Rf < 1), moderately resistant (Rf 1–2), moderately susceptible (Rf 2.1–5), susceptible (Rf 5.1–10), and highly susceptible (Rf >10).

To reduce the time required by the test, Sasanelli et al. [31], in another screening, used rooted explants obtained from shoots in vitro (Fig. 10) and described the methods for their production.

- *Prunus* spp. trees

In Spain, Pinochet et al. [32] assessed the resistance of *Prunus* rootstocks to *Meloidogyne hapla*, *M. incognita*, and *M. javanica*. The authors used the following:

1. Transplanted seedlings or rooted cuttings in 1 dm^3 pots containing sandy soil and maintained in a screenhouse.
2. Pots were inoculated 30 days after transplanting with 5000 eggs and J2 of the populations of these species per pot.
3. The reaction to these RKN species of the tested rootstocks was compared with that of known resistant and susceptible rootstocks (Figs. 11 and 12). Six months later, the test was discontinued, the roots were weighed, and the RGI for each plant is determined (according to a 0–6 scale).
4. Then, eggs and J2 s from soil and roots were extracted, total number per pot recorded, and Rf calculated. Nematodes were extracted from soil by combining Cobb's sieving method with the centrifugal flotation in a sugar solution, and from the roots by macerating the roots in a blender containing a 0.2–0.3% NaOCl solution and then sieving through sieves of 150, 74, and 25 μm aperture.

Fig. 11 Reaction of roots of peach rootstocks to parasitization by *Meloidogyne arenaria*. Left, susceptible response of rootstock GF 677 which shows severely galled roots induced by the nematode; right, resistant response of rootstock GF 577, which shows healthy roots after exposure to nematode infestation. (Courtesy of D. Esmenjaud, INRAE Sophia Antipolis, France)

Fig. 12 Screening of peach rootstocks for resistance to *Meloidogyne* sp. in greenhouse. (Courtesy of D. Esmenjaud, INRAE Sophia Antipolis, France)

5. The reaction to the nematodes was rated according to Taylor and Sasser [33] as highly resistant (HR), resistant (R), moderately resistant (MR), tolerant (T), and susceptible (S).

In Florida (USA), Qiu et al. [34] assessed the reaction of peach rootstocks to *M. arenaria* and *M. floridensis*. The authors used the following:

1. Six-month-old rooted cuttings were transplanted into 25 cm diam clay pots containing a mixture of sterilized sand and perlite (3:1).
2. The plants were inoculated 1 month later with 2000, 5000, or 10,000 eggs per pot.
3. The experiment was terminated 2 months later, when plants were uprooted and RGI and EMI determined using a 0–5 scale according to Taylor and Sasser [33].
4. To obtain proof of the resistant status of plants not showing egg masses, their roots were stained in a 20% (v/v) food coloring solution for 20 min to enhance the detection of the egg masses and females that may have been missed on destained roots [35].
5. Eggs were also extracted from roots and soil. Total population per pot and g roots determined and Rf calculated.
6. The plants were designated as good host (Rf $\geq$ 1), poor host (Rf > 0.1–1), and nonhost (Rf $\leq$ 1).

In Georgia (USA), Nyczepir et al. [36] screened the reaction of peach rootstocks to *M. incognita* and *M. javanica*. These authors used the following:

1. Seeds of peach tree rootstocks were sown in 10 cm diam pots containing a mixture of sand and vermiculite.
2. After germination, 17-day-old seedlings were inoculated with 3000 eggs and J2 of a RKN species that had been reproduced on tomato cv. Rutger and extracted using the NaOCl method of Hussey and Barker [15].
3. Seedlings were arranged in a greenhouse bench, kept at a temperature ranging from 21 to 35 °C, and watered and fertilized with a soluble 14-14-14 NPK fertilizer whenever needed.
4. The test was terminated 110 days after inoculation and the following data were recorded: number of galls per root, number of egg masses per root, RGI and EMI values according to a 0–5 scale [37], and dry root weight.
5. Plants were rated as immune (RGI and EMI = 0), resistant (RGI and EMI 1–2), and susceptible (RGI and EMI ≥ 3).

In Chile, Meza et al. [37] screened the reaction of stone fruit rootstocks to several root-knot nematode species, including *M. ethiopica*. The authors used the following:

1. Three-month-old peach trees (50 cm tall) were planted in October (spring) in 4 dm^3 plastic pots containing sterilized sandy soil.
2. Four weeks later, pots were inoculated singly with10,000 eggs and J2 of the selected RKN species. Plants were arranged in a greenhouse set at 27±2 °C and 16 h light per day, fertilized with a slow-release fertilizer (15-9-12 NPK plus microelements), and watered as required.
3. Plants were harvested 5 months after inoculation when the top plant was cut, the roots washed free of soil, and number of galls estimated visually. Then, they were cut in small pieces, weighed, and ground in a blender containing a 0.2% NaOCl solution at 14,500 rpm. Ground roots in NaOCl solution were screened through nested sieves of 20, 200, and 500 meshes. RKN J2s were extracted from soil by combining the sieving and decanting method with Baerman's funnel method.
4. The plants were considered immune (no galls and no nematodes i), highly resistant (very low galling and Rf < 0.1), resistant (low galling and Rf between >0.1 and <1), and susceptible (abundant galling and Rf > 1).

Yagci et al. [38] screened peach and nectarine rootstocks to different Turkish populations of *Meloidogyne* spp. The authors used the following:

1. Rootstocks, 15–20 cm tall and with 10–20 leaves, were grown in 3 dm^3 pots containing a sterilized soil and sand mixture.
2. Rootstocks were inoculated with 3000 eggs and J2 per pot, distributed in 3 cm deep holes. Pots were maintained in a growth chamber at 25 ± 2 °C and 14 h light per day.
3. Four months after inoculation, plants were harvested and RGI assessed on a 0–5 scale.
4. Plants rated 0–2 were considered resistant and those rated >2 susceptible.

For a review on resistance of peach rootstocks and other woody plants to *Meloidogyne* spp., see Esmenjaud et al. [39], Saucet et al. [40], and Esmenjaud [41].

– Sugarcane

In Australia, Stirling et al. [42], Bhuyan et al. [43], and Bhuiyan and Garlick [44] suggest the following procedure for screening sugarcane to *M. javanica* (Fig. 13):

1. Cut defoliated stalks of sugarcane test lines in one budded set. Dip sets in water at 50 °C for 30 min to eliminate fungi and bacteria and place them in trays containing wetted vermiculite for about 2 weeks at 31 °C to facilitate rooting.

Fig. 13 Screening of sugarcane to *Meloidogyne javanica* in greenhouse. Left, potted sugarcane plants to be used for resistance screening. Right, inoculation of sugarcane plants with 5000 nematode eggs and J2 per plant, using a calibrated pipette. (Courtesy of S. Bhuiyan, Queensland Micro- and Nanotechnology Centre, School of Engineering and Built Environment, Griffith University, Nathan, Australia)

2. Transfer vegetated sets about 6 cm tall into 700 cm^3 pots containing sterilized washed white sand and keep them on benches in greenhouse.
3. When young plants have developed roots, about 4 weeks later, inoculate each pot with 5000 eggs in three to four holes.
4. Evaluate the plant reaction 12 weeks later by recording the following variable: root weight, top plant weight, RGI and EMI according to a 0–5 rating scale, number of nematode eggs and J2 in roots and soil, total nematode per pot/plant (total eggs and J2 on the roots + total eggs and J2 in the soil), and per g of roots (number of nematodes in the roots/root weight).

Numbers of RKNs per plant and per g roots were shown to be the most reliable indicators to discriminate the different test lines.

– Coffee

Villain et al. [45] suggest the following procedures for screening coffee seedlings to RKN species:

1. Germinate coffee seeds in 18 g vermiculite saturated with 45 mL water in closed boxes (Fig. 14).
2. Soon after germination, transplant them in small PVC tubes of 110 mL (26 mm internal diameter and 210 mm long), suspended in plastic boxes and filled with 100 cm^3 of sieved (2 mm

Fig. 14 Coffee seedings growing in small PVC tubes containing an organic substrate inoculated with *Meloidogyne paranaensis.* Tubes are suspended in hermetic plastic boxes placed on a bench occupying a confined space in a greenhouse. An automatic control system regulates light, temperature, humidity, water, and nutrient supply. (From Ref. [45], Copyright Springer Nature)

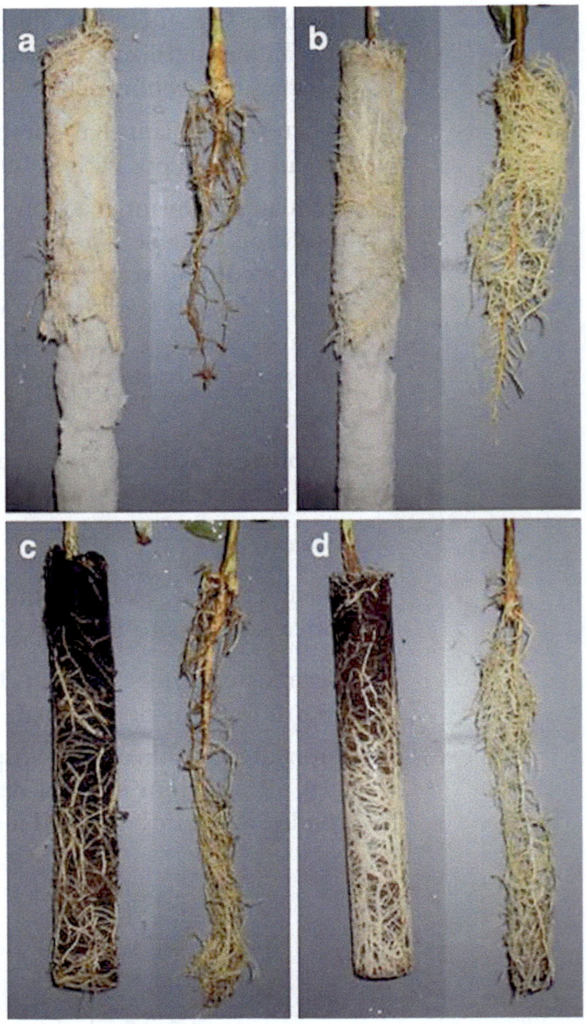

Fig. 15 Reaction of root systems of susceptible coffee cv. Caturra and resistant cv. Nemaya to *Meloidogyne paranaensis* in infested inert and organic substrates. (**a**) Galled roots of the susceptible cv. Caturra growing in inert substrate; (**b**) Healthy roots of the cv. Nemaya growing in inert substrate; (**c**) Galled roots of the susceptible cv. Caturra growing in organic substrate; (**d**) Healthy roots of the cv. Nemaya growing in organic substrate. (From Ref. [45], Copyright Springer Nature)

aperture sieve) organic substrate. The tubes have the bottom closed with a stainless steel net of 0.25 mm aperture and lined with a transparent polyethylene sheet 5 mm longer than the tubes for the easier extraction of all substrate and plant roots at the end of the test (Fig. 15).

3. Inoculate seedlings in two intervals, 750 J2 2 weeks after transplanting and 750 J2 2 days later, for a total of 1500 J2 obtained by the mistifier chamber.

Fig. 16 Reaction of a susceptible coffee cultivar to *Meloidogyne exigua*. (Note small galls induced by the nematode in the feeder roots of the seedlings)

4. Uproot plants 8 weeks later and extract RKNs using the mist chamber. Collect RKNs three times, every 5 days, and obtain the total number of nematodes from the three counts.
5. Evaluate the plant reaction to RKNs (Fig. 16) by comparing statistically the total nematode numbers extracted from plants.

The method is rather rapid and requires little space.

To screen coffee to *Meloidogyne konaensis*, Aoki et al. [46] suggest the following procedures:

1. Transplant germinated coffee seeds in 15 cm diam. clay pots containing steam-sterilized soil and silica sand.
2. Then, inoculate 30 cm tall (10 leaf stage) plants with 3000 eggs and J2 distributed in three to four holes 3 cm deep around the plant stem.
3. Eight months after inoculation, record eggs and J2 from soil and roots, root and top plant weights, and RGI and EMI and determine Rf.
4. Plants having Rf < 1 are considered resistant.

– Grapevine

In California (USA), a large breeding program was funded in 2010 to obtain grapevines cvs. and rootstocks resistant to several nematode species, including different virulent populations of *Meloidogyne* spp., *Xiphinema index*, and other ectoparasitic nematodes. Ferris et al. [47] suggest the following procedures:

1. Inoculate rooted cuttings transplanted in 15 cm diam. pots with 500 or 1000 of selected RKN populations.

2. Three months after inoculation, cut and weigh top plant. Stain roots in a solution of 0.1% of erioglaucine for easier observation of the egg masses.
3. Extract RKNs from soil by a modified Baermann's funnel technique or by a flotation method combined with a centrifugation technique. Then, express RKN populations on the tested genotypes as % of those of the control cv. Colombard using their reproduction rate (RI or Rr).
4. Tested genotypes are considered resistant if RI or Rr is <10%, moderately resistant from 10 to <30%, moderately susceptible from 30 to 50%, and susceptible >50%.

– Pistachio tree

In Iran, Madani et al. [48] suggest the following procedures for screening cvs. of *Pistacia vera* and wild *Pistacia* to populations of *Meloidogyne* spp.:

1. Soak overnight seeds of *Pistacia* spp. sterilized for 5 min in a 1% sodium hypochlorite solution and cover them with a layer of wet cotton tissues to facilitate germination at 23–28 °C. Apply fungicide every 2 days to prevent development of fungi.
2. After 1 week, transfer germinated seeds into 20 × 25 × 10 cm plastic pots containing a sterile mixture of sand, perlite, and peat moss (50:25:50 v/v) and cover them with a 1 cm thick layer of sand. Keep pots in a greenhouse at 25–27 °C.
3. After seed germination, transplant two leaf stage seedlings in 15 × 25 cm clay pots containing 500 cm^3 of river sand, vermiculite, and peat moss (50:25:25 v/v) and keep them at 22–28 ° C. Inoculate seedlings with RKNs after 1 week.
4. Seeds of wild species are to be soaked overnight in water to detach the green overcoat and then sown in containers filled with river sand and wrapped in aluminum foils. Wrapped containers should be maintained at 4 °C for 30–40 days until germination.
5. Inoculate with 10,000 eggs and J2 seedlings of cultivated pistachio (*P. vera*) when they reach the 3–4-leaf stage, and those of wild pistachio (*Pistacia* spp.) when they are 3 months old. Water plants whenever required and add 0.5% Hoagland's solution once a week
6. Discontinue the test 135 days after inoculation and extract RKNs from roots and soil.
7. Evaluate plant reaction based on RGI and Rf. Plants should be considered resistant if RGI $\leq$ 2 and Rf $<$ 1 and tolerant if RGI $\leq$ 2 and Rf $>$ 1; otherwise, they are to be rated as susceptible.

– Other tree fruits

In Brazil, Freitas et al. [49] screened 89 genotypes from 25 fruit species to assess their host status to *Meloidogyne enterolobii*. The authors used the following:

1. Plants obtained from seeds or cuttings were transplanted in 5 dm^3 pots (plastic bags) containing a mixture of sterilized clay soil, sand, and a compost (1:1:1) and arranged in a greenhouse at 25–30 °C.
2. When seedlings were 10–15 cm tall, each plant was inoculated with 10,000 eggs and J2s of the nematodes in holes around the stem.
3. At observation time (from 3 to 8 months after inoculation, according to plant species), plants were uprooted, the roots gently washed under running water to clean them from adhering soil, weighed, stained in phloxine B, and their RGI and EMI evaluated on a 0–5 rating scale.
4. Then, RKNs were extracted from the roots by blending in a 1% NaOCl, counted, and Rf calculated.
5. Plants were considered nonhost if Rf < 0.1, poor host if Rf was between 0.1 and <1, and good host if Rf was ≥ 1.

Gomes et al. [50] used a similar procedure to assess the inheritance of resistance to *M enterolobii* of guava (*Psidium* spp.) hybrids. In these tests, the root system of the resistant plants was cut in two halves, one half was removed from the plant and used to assess its reaction to the nematode and the other half was repotted and left to grow to be back-crossed with guava cvs. for the recovery of good agronomic traits.

3.2 Screening for Resistance to CNs

Heterodera spp. and *Globodera* spp. include species that damage many staple food crops and have a persistent life stage called cyst [51, 52]. The cysts are formed at the end of the crop cycle and derive from senescent females that, after reaching maturity, die and have their body transformed in a tannic persistent sac (cyst that may contain 200–400 eggs). Eggs may remain viable inside the cyst in the soil up to 20 years. However, some species lay eggs also in a mucilaginous matrix called egg masses. Biologically, the second-stage juveniles (J2) inside the eggs contained into the cysts hatch only if stimulated by root exudates or diffusates of the plant host and after a dormancy period of 2–4 months coupled with suitable soil temperatures and moisture content. Embryonated eggs in the egg masses, instead, contain J2 that hatch as soon as soil conditions are suitable. Most of the species can be stimulated to hatch also by solutions of inorganic chemicals (Table 3), which often are preferred to root diffusates to obtain J2 in scientific experiments. The tolerance limit of cultivated crops to most CN species is around 1–2 eggs g/soil and severe yield losses may occur in soil infested

Table 3
Optimal incubation temperatures (°C) and hatching agents to obtain J2 from cysts of some *Heterodera* spp. and *Globodera* spp

Nematode species	Optimal temperature	Root exudates of host plants	Chemical agent (solution)
H. avenae-group	15	Yes	just distilled water
H. carotae	20	Yes[a]	10 mM zinc cloride
H. ciceri	20	Yes, in exudates of pea	3 mM of zinc chloride, 0.6 mM sodium metavanadate
H. glycines	25	Yes	3 mM magnesium sulphate, 3 mM zinc sulphate
H. goettingiana	15	Yes/no	none
H schachtii	25	Yes	3 mM zinc chloride[a]
G. rostochiensis *G. pallida*	21 15–20	Yes Yes	0.6 mM sodium metavanadate[a], 0.3 mM picrolonic acid

[a]Good hatch response using these hatching agents

with 20–30 eggs/g. Many CN species have populations differing in their host preferences that are indicated as pathotypes in some species and races in others. The methods used for screening resistance to some selected species of economically important and widely distributed cyst-forming nematodes are listed below.

3.2.1 Screening for Resistance of Cereals to *Heterodera avenae* Group

This group of CNs includes 12 species associated with winter cereals. They are widely distributed in temperate and some warm regions and cause severe damages especially to wheat, barley, oats, and rye [53]. The most important species are *Heterodera avenae*, *H. filipjevi*, and *H. latipons* (Fig. 17). Moreover, each species may have different populations, called pathotypes, which show different degrees of virulence toward different cereal species or cultivars/lines of the same species [54, 55]. Therefore, when screening for resistance to these CNs, it is necessary to identify exactly the species of the nematode and its pathotypes (*see* Subheading 3.4). They complete a generation per growing season of the host crop, do not form egg masses, and at the end of the crop season, undergo a dormancy stage. However, it seems that dormancy is induced by the cool autumn–winter temperature after harvest of the host crop in temperate geographic regions and, instead, by hot summer temperatures in warm regions with Mediterranean climate. The dormant stage is represented by the J2s inside the eggs packed into the cysts. These dormant J2s do not initiate any root

Fig. 17 Damaging effect of the cereal cyst nematodes to wheat. (**a**) Severe yellowing of wheat in a field infested by *Heterodera avenae* in Sardinia, Italy; (**b**) Weaving and uneven growth of wheat in a field infested by *H. latipons* in Syria

infestation when inoculated in soil in screening tests for resistance without any hatching stimulus. CN populations used in these tests should contain embryonated eggs with J2s that are ready to hatch; otherwise, J2 hatching must be induced by storing cysts at 4–5 °C for about 2–4 months.

The following protocol is suggested:

1. Collect infested soil from the field or from a pot culture (not very dry). If it is required, store in a cool place, at 4–5 °C, for 2–4 months. Alternatively, store cysts at cool temperature after extraction from the soil.
2. Extract cysts by processing the infested soil using Cobb's sieving and decanting method or the Fenwick can and collect the cysts by filtering the soil in water suspension trough a 710 μm aperture sieve (25 meshes) nested on a 250 μm sieve (60 meshes).
3. Gather cysts and soil debris on a corner of the 250 μm sieve using a laboratory squash bottle or a gentle water spray and transfer them in a beaker with a stream of water.
4. Then, pour the water suspension with cysts on a funnel lined with a cotton wool milk filter or special filter that allows a fast percolation of the water.
5. Finally, count the cysts trapped by the filter and separate them from soil debris.
6. For the hatch test, place cysts (100–200 cysts per sieve according to sieve size) with a small brush in small 200 μm aperture plastic sieves using a stereomicroscope. Then, immerge the small sieves containing the cysts in a 0.5 NaOCl solution for 10 min and soon after, rinse them three to four times in distilled water.

7. Put the small sieves with cysts in small Petri dishes containing distilled water just to reach cyst level. As cysts of *H. avenae* hatch easily in water, there is no real need to use root diffusates to hatch.
8. Incubate cysts at 10–15 °C and collect emerged J2 every other day for immediate use or to be stored in a refrigerator. Most of the J2 emerge from cysts during the first 2–3 weeks of incubation.
9. Germinate seeds of each genotype separately in Petri dishes lined with a disk of wetted common filter paper at 20–25 °C for about 3–5 days.
10. Sow germinated seeds singly in tubes containing the sterile soil mix (six to ten tubes/replicates per genotype). Controls should include inoculated tubes without plants and with plants to determine J2 mortality in the soil and their penetration and development in the roots.
11. One week after emergence, inoculate each plant with 400 J2 (Pi) in three holes around the stem base.
12. Arrange all tubes on benches in a greenhouse, or better in a growing chamber, set at 20 °C and 12–14 h light per day.
13. About nine weeks after inoculation, verify the formation of females/cysts in the roots in the control tubes.
14. When females are present, reduce watering that should be interrupted 2 weeks after to induce soil desiccation.
15. When the soil is completely dry, uproot plants, remove plant top, and mix soil and roots thoroughly to detach cysts from the roots into the soil.
16. Then, dry soil containing root fragments at 40–50 °C to facilitate the floating in water of the cysts, which should be extracted using a Fenwick can.
17. Count and soon after crush extracted cysts in water to release and count the eggs (Pf).
18. Then calculate Rf = Pf/Pi, and assess RI.
19. Based on the number of CN females and cysts per plant, classify genotypes as follows: resistant (R) = number of females equal or smaller than that on resistant control; moderately resistant (MR) = number of females slightly greater than that on the resistant control; moderately susceptible (MS) = number of females significantly greater than that on the resistant control but smaller than that on the susceptible control; susceptible (S) = female number as that on susceptible control; and highly susceptible (HS) = female number greater than that on susceptible control [56–59].

However, in case of screening of a large number of genotypes, the assessment of the resistance can be limited to the observation of presence of females and cysts on roots.

1. Therefore, use transparent tubes, and about 90 days after inoculation, verify the formation of white females and cysts on the external root ball.
2. Grade root parasitization by comparing that of tested genotypes to susceptible and resistant controls using a scale, where the maximum grade (100%) corresponds to a parasitization similar to that of the susceptible control and 0 to the same parasitization level or less than that of the resistant control.
3. This less accurate screening method has the advantage to save resistant plants from screened progenies of a breeding program. In this case, transplant plants rated resistant in 15 20 cm diam. pots containing non-infested soil and keep them to be crossed according to the breeding program or become self-pollinated, while discard the susceptible plants.
4. If the plant reaction to the CN is not clear according to the root ball observation, free the roots from adhering soil under a stream of tap water, avoiding damaging the root tissues and transplant the selected resistant plants for additional screening.

Most of the researchers prefer the use of small containers (special tubes) for screenings and CN J2s for inoculation. This has the advantage of reducing the amount of inoculum and space that are needed. Using larger pots containing 800–1000 cm^3 soil is very effective, but in this case, it is suggested to inoculate the soil with cysts that can be extracted from an infested field or infested pots or by mixing not infested with infested soil.

3.2.2 Screening for Resistance of Soybean to Heterodera glycines

This species is known as the soybean cyst nematode (SCN) and is widespread wherever soybean and other host plants are cultivated [60]. Originated from the Far East countries, today, this species can be found in Japan, Korea, China, and many other countries in the same area and in the Americas. It is very prolific (more generations per crop cycle) and, therefore, very damaging (Fig. 18). Many virulent strains modify their feeding habits with time and in the same place. These populations are indicated as races [61, 62]. Also, not all nematode populations classified as belonging to the same race or HG type show the same virulence. Besides soybean, this species parasitizes also haricot bean and other crop plants in the genera *Phaseolus* and *Vigna*. Most probably, it is the nematode that is most involved in breeding programs for the selection of soybean resistant lines. The economic importance of soybean in many areas of the world has prompted research to develop rapid screening programs that can be carried out in limited space.

Fig. 18 Damaging effect of the soybean cyst nematode, *Heterodera glycines*, to soybean in a nematode-infested field in Tennessee, USA

The agreed procedure is reported hereafter.

In 2008, Niblack et al. [63], in a meeting specifically organized in the USA to discuss the "Assessment of Resistance to the Soybean Cyst Nematodes," proposed the following protocol (named SCE08) for the assessment of the reaction of soybean lines to SCN. The correct implementation of this protocol requires the use of: (*i*) three different SCN populations classified as the same HG type, (*ii*) at least three plants per replicates, and (*iii*) standard methods and environmental conditions. The protocol includes the following guidelines for screening soybean lines in a greenhouse:

1. Disinfect the seed surface (e.g., by soaking seeds in 1% NaOCl followed by two or more rinses in sterilized distilled water).
2. Germinate disinfected seeds in sterilized germination paper for 3 days.
3. Transplant seedlings of uniform size and vigor into a pasteurized sandy soil mix. Do not use soil-less mixes or soil amendments such as peat moss or vermiculite; they will cause problems with extraction and counting females and cysts.
4. Use plastic containers of 100–500 cm^3 volume.
5. Arrange plants in a randomized complete block or completely randomized experimental design.
6. One day after transplanting seedlings, extract eggs and J2 from cysts maintained for at least five generations on a susceptible cultivar, such as Lee 74, Hutcheson, Essex, or Williams 82.
7. Eggs can be sterilized with NaOCl sodium hypochlorite or differently.

8. As standard inoculum, use a water suspension of eggs and J2 to be applied at the rate of 20 eggs + J2/cm^3 soil.
9. Set greenhouse temperature at 27–28 °C and 16 h day length.
10. Allow plants to grow for 28–30 days and at harvest, uproot the plants and extract nematode females and cysts from the roots by rubbing the roots under pressurized water spray over a 25 mesh sieve (710 μm) nested on 60 mesh sieve (250 μm).
11. Count collected females and cysts from the sieve using a stereoscope at about 10× magnification. This is a crucial step because the number of females and cysts of the tested genotypes should be compared to that of the designated susceptible lines to obtain the female index factor (FI).
12. More recommendations include a list of all HG type indicator lines (PI 548402 (Peking), PI 88788, PI 90763, PI 437654, PI 209332, PI 89772, PI 548316 (Cloud) available from the US Department of Agriculture (USDA) Soybean Germplasm Collection). The seeds of the listed lines can be obtained by the USDA for research purposes in the USA and abroad by addressing any request to the curator of the line collection, Dr. Mahan Adam, ARS (e-mail Adam.Mahan@usda.gov).
13. In the screening, besides the standard susceptible control check Lee 74, also include cvs. Essex, Hutcheson, or William 82.
14. To consider the test reliable, the numbers of females per plant root system of all these susceptible lines should be 100 or more.
15. Calculate the female index (FI) that should be expressed as the average of the three nematode populations used in the screening.
16. Then, label the cultivar according to the level of resistance as defined in Table 4 and by indicating the HG type population to which is resistant.

Table 4
Female index thresholds and corresponding ratings for the evaluation of the reaction of soybean lines to the soybean cyst nematode *Heterodera glycines*

Female index	Genotype rating	Acronym on label
<10	Resistant	R
10 to <30	Moderately resistant	MR
30 to <60	Moderately susceptible	MS
60 or higher	Susceptible	S

From Ref. [61], as adapted according to Ref. [63], permission by APS

Fig. 19 Physiological disorders induced by the sugar beet cyst nematode, *Heterodera schachtii*, to sugar beet and procedures for breeding nematode resistant lines. (**a**) Sugar beet plants showing severe wilting in sunny hours in a field infested by the nematode in England; (**b**) Wild *Beta* species used to introgress resistance to the nematode in cvs. of sugar beet in California, USA; (**c**) Screening breeding materials in aluminum foil cylinder for resistance to the nematode in California, USA; (**d**) Promising offspring plants saved and transferred to larger pots to produce seeds in California, USA. (Greenhouse photos taken at the USDA-ARS, Salinas, California, USA)

This protocol is the result of the cooperative work of experts other than the authors mentioned above and is a very good guideline to characterize cultivars/lines at the final stage of the breeding program. It is basically and largely used in the USA [64–66] and in other countries [67–69] (*see* **Note 2**).

3.2.3 Screening for Resistance of Sugar Beet to Heterodera schachtii

This species is known as the sugar beet cyst nematode (SBCN) and is widespread all over the world wherever sugar beet and other host crops are cultivated. The nematode has a rather wide host range and damages sugar beet (Fig. 19), table beets, and species of the botanical families Chenopodiaceae and Cruciferae, such as cauliflowers, broccoli, radish, and other wild species. Breeding programs were undertaken [70, 71], and others are underway [72], on the resistance of sugar beet and the cover crops radish and mustard (*Brassica* spp.) to this nematode. These breeding programs for cover crops have significant agronomic relevance because these cover crops are also used as nematicidal plants in bio-fumigation. Therefore, appropriate screening techniques are needed. Besides, differences in population virulence toward resistant genes have been reported [73], and, therefore, it is suggested that more populations be used in breeding programs.

Before starting, realize that seeds of wild beets may require a treatment to germinate. The following treatments have provided satisfactory results:

1. Divide seeds in three groups, namely, small, medium, and large, and place in cuvettes.
2. Then add sulfuric acid just to cover the seeds that should be exposed to the acid for 3, 6, and 8 h, according to their dimension. Soon after, rinse seeds in water, place them on a towel on the table, and remove the burned coat by rubbing the seeds on the towel.
3. Examine seeds with the help of a lens or a stereoscope for presence of holes in the coat. Repeat the treatment for a shorter time if no holes are observed.
4. Place ready treated seeds for 20 min in a glass of water and add a little of soda powder to neutralize the acid. Then, transfer seeds first in fresh water to eliminate soda and then on a towel on a table to dry. Seeds are now ready for sowing or storing for several months. They will not lose germinability.
5. Germinate seeds of cultivated sugar beet by sowing them 1 cm deep in a vessel of about 45 cm × 25 cm × 12 cm half filled with a wetted mixture of sterilized river or quartz sand with 30% vermiculite.
6. Fertilize germinated seeds with Hoagland's solution or a similar fertilizer containing microelements.
7. When sugar beet seedlings reach the stage of two to three true leaves, transplant them in pots of 10–14 cm diam. containing a sterile sandy soil (85–90% sand) that has previously been infested with cysts containing a number of eggs corresponding to a soil density of ten eggs per cm^3 or g soil.
8. When screening a large number of plants, many authors prefer screening the seedlings in tubes rather than in pots to occupy less space in the greenhouse. Tubes can be of aluminum foil (Fig. 19) or of transparent polystyrene for easy observation at evaluation time. Most of the tubes used today are available on the market, such as Cone-Tainer SC10 (Fig. 20), 3.5 cm up diam. and 20 cm long (Stuewe and Sons, Tangent, OR, USA), containing about 150–300 g soil.
9. To obtain the required inoculum, we suggest reproducing the nematodes in pots (20–25 cm diam.) before the screening. Pots should contain sterile soil and placed in greenhouse or on a bench outdoors or in a screenhouse. This system reduces the risk of introducing, into the soil used for the screening, pests

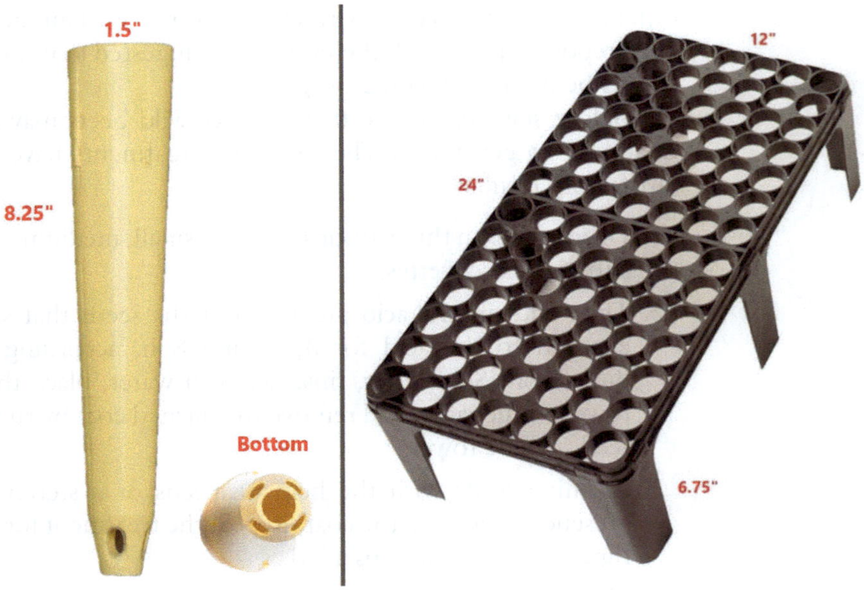

Fig. 20 Tools useful for screening different plant species for resistance to plant parasitic nematodes. (Note standard tubes marketed as Cone-Tainer SC 10 and a holding rack)

and diseases. Alternatively, collect infested soil from the rhizosphere of severely infested crops at the end of their cycle. Discard the top 5 cm of soil and collect soil from the deep layers (5–20 cm deep) of the soil profile, surrounding the plant rhizosphere. Screen collected soil through a large 4–5 cm aperture sieve of about 80 cm × 150 cm to eliminate crop residues, stones, and large clods, and store it in the shade outdoors so that cysts can be exposed to winter and summer temperatures to facilitate J2 hatching and avoiding direct sun exposure.

10. Mix the collected nematode-infested soil in a concrete mixer if possible and divide the infested soil in well-separated large piles. Then, take four to six soil samples, each of 200 g, from each pile to be used to extract cysts using a Fenwick's can. Separate soil debris from the cysts that are to be crushed for the determination of the number of their eggs that should be referred to 1 g soil. The average number of eggs of these four to six soil samples represents the average number of eggs in the soil of the pile where the samples were collected.

11. Mix sterile and infested soils in the appropriate proportion to obtain the desired nematode infestation level of the soil that will be used to fill pots for the test. The preparation of these soil mixtures for multiple pots requires the uniform mixture of the infested and sterile soils that should be mixed in a concrete mixer for about 5 min. Mixing separately the soil of each pot is more accurate but tedious and time consuming. On the other hand, when mixing the soil for many pots to obtain a low

Fig. 21 Device for extracting cysts from large quantities of soil. (**a**) Device like that constructed at Nematology Department, University of California, Riverside, CA, USA, but with a larger can, which is used at CNR, Bari, Italy, and ICARDA, Aleppo, Syria; (**b**) Device used at USDA-ARS, Salinas, California, USA

density, there is the risk that some pots may receive more inoculum and some less than planned, simply because the eggs cannot be equally distributed in the soil since they are packed into the cysts. Pots receiving just a single cyst more may have 100 times more eggs than pots missing that cyst. However, if the inoculum level is not too low and the pots/tubes are not too small, this risk is reduced. The mass screening using cysts, although accurately enough, must be considered indicative. Any observed resistant reaction in the tested plants must be confirmed in a more accurate test in which J2 are used rather than cysts. We recommend the use of a large-sized can for the extraction of the cyst from a large amount of soil. These devices (Fig. 21) of different sizes were constructed as a model in the Nematology Department, University of California, Riverside, USA, and have been used successfully for the extraction of cysts from a large amount of soil infested by *Heterodera avenae*, *H. carotae*, *H. ciceri*, *H. goettingiana*, *Globodera rostochiensis*, and *G. pallida* in nematological research units at the CNR Institute of the first author in Bari, Italy, and ICARDA (International Center for Agricultural Research in the Dry Areas) in Aleppo, Syria. The large cans in these devices require accurate cleaning process after each cyst extraction to avoid contamination of subsequent soil samples. Cysts and soil debris extracted by a large amount of soil instead of being collected on a 710 μm aperture (25 meshes/inch) sieve nested on 250 μm aperture (60 meshes/inch) sieve are collected in a large sieve consisting of a 5 L pail with the bottom replaced by a nylon net of 210 μm aperture (72 meshes/inch). At the end of every extraction from 10 dm^3 of soil, cysts and soil debris collected on the net of the pail are transferred with a brush in a tray and left to dry in the shade. Dried cysts and soil debris stored in the tray are sieved through the 710 μm sieve to remove larger soil debris from cysts, which are stored in the shade or at 4–5 °C if vernalization is necessary. This system avoids the storage of large quantity of infested soil in a cool room because the extracted cysts and debris can be stored in a refrigerator. However, the extraction of cysts requires considerable manpower that can be reduced by using a small sowing machine (Fig. 21) to load the infested soil into the large apparatus.

12. Before inoculation, dilute and thoroughly mix cysts and debris in sterile river sand (an amount of 10–30 kg of sand may be required for large volume of cysts and debris).
13. Then, take six to ten subsamples of 10 g each and extract cysts just by pouring the subsample into the 710 μm aperture sieve nested on the 250 μm aperture sieve and spraying with pressurized water the subsample to separate the large soil debris from the cysts, which are collected on the 250 μm aperture sieve. Then, collect, count, and crush the cysts to estimate their egg content. Refer number of eggs to g of the cyst–sand mixture and quantify the cyst–sand mixture to be mixed with the sterile soil in the pots to obtain the desired inoculum density per pot.
14. The same procedures can be used also to screen for resistance or host status of *Brassica* spp. to *H. schachtii* [74].
15. Leave plants to grow for about 60–70 days. Then, harvest plants and count numbers of females or cysts, estimate their egg content, and calculate Rf and RI (=Rr) or FI. A shorter growth cycle of 40 days can be adopted if the evaluation will be based on the observation of white females on the roots, rated on an arbitrary scale, such as that used for *H. ciceri* on chickpea in the following section. The simple separation of the nematode-infested plants (susceptible) from those non-infested (immune or resistant) is a useful and sufficient assessment in case of screening tests with large number (several thousands) of entries, because it does not destroy resistant plants that can be transplanted subsequently in larger pots. The observation of the females on the roots can be done after gently washing the roots, or on the external part of the root ball. In both cases, plants can be considered resistant if

Fig. 22 Damaging effect of the chickpea cyst nematode, *Heterodera ciceri*, to chickpea. (**a**) Nematode-infested field showing a large patch of stunted and chlorotic plants; (**b**) White females and brown cysts of the nematode attached to chickpea roots

nematodes (expressed as females and cysts or eggs per plant) on the roots are absent, less than 10% of those on the susceptible control, or less than those on the resistant control.

Good results were also obtained in Chile, Italy, and Venezuela by mixing infested soil with sterile soil [75–77].

3.2.4 Screening for Resistance of Chickpea to Heterodera ciceri

This species known as the chickpea cyst nematode, *Heterodera ciceri*, was first found in North Syria [78] and thereafter in Turkey, Lebanon, and Jordan where it causes severe damage to chickpea (Fig. 22), lentil, pea, grass pea, and, in some cases, alfalfa. So far, no cultivar of the mentioned host plants is resistant to *H. ciceri*, but resistance has been found in the wild species *Cicer reticulatum* and *C. bijugum* and *C. pinnatifidum* of which *C. reticulatum* is crossable with the cultivated *C. arietinum* and is used to transfer the resistance genes into cultigens at ICARDA. The nematode completes only a generation per year on chickpea with a growth cycle spanning from late autumn to late spring. The tolerance limit of chickpea to the nematode is of 1 egg/g soil [79]. A screening method for resistance was developed at ICARDA in Aleppo, Syria, by Di Vito et al. [80] and is reported hereafter.

1. Reproduce the nematode in pots on a susceptible chickpea cultivar in a plastic-house 3–4 months before the screening test (in the previous autumn if the test is conducted during end of winter–spring months) and store cysts in the soil and roots in a shaded area. At ICARDA, usually 400–600 buckets containing each 4 dm^3 of sterilized soil are sown with six to eight seeds of the susceptible chickpea cultivar or lines before each screening. In general, two screenings per year are planned:

Fig. 23 Different steps to screen chickpea offsprings for resistance to the chickpea cyst nematode, *Heterodera ciceri*. (**a**) Sieving the nearly dried soil remnant with cysts obtained after cyst extraction to eliminate large debris and contaminants; (**b**) Mixing the sieved soil material containing cysts with sandy soil; (**c**) Weighing the amount of inoculum to be mixed with potting soil; (**d**) Emptying the concrete mixer after mixing potting soil with inoculum and fertilizer; (**e**) Filling Jiffy pots and buckets with infested soil before sowing germinated chickpea seeds; (**f**) Benches in a plastic-house with chickpea genotypes grown in soil infested with *H. ciceri* and ready to be rated for their reaction to the nematode; (**g**) Feeder roots of chickpea, grown outside the Jiffy pots, on which the rating of the nematode infestation will be done after gently washing; (**h**) Genotypes of chickpea rated resistant and transplanted singly in buckets and further grown to produce seeds. (Note the tutors holding the plants)

the first from mid-November to end of January–mid-February and the second from end of February–early March to mid-May. In each run, seedlings from 2000 to 3000 seeds are screened.

2. In October, extract cysts from the infested soil using a can similar but larger than that used at the Department of Nematology, University of California, in Riverside (Fig. 21), as described at the end of the methods for *H. schachtii*.
3. Mix all the inoculum extracted (nematode cysts, soil debris), in general 2–3 kg, with sterilized river sand, about 30–40 kg (Fig. 23).
4. Estimate the density of the nematode in this infested sand by collecting six to eight subsamples, each of 10 g, which are processed for cyst extraction on a 25 mesh sieve (710 μm aperture) nested on a 60 mesh sieve (250 μm aperture) by spraying the subsample with pressurized water.
5. Transfer cysts retained on the 60 mesh sieve to a milk filter paper and separate them from the remaining debris. Then, crush cysts in water in a grinding tube to release eggs and J2, and dilute them soon in 100 or 500 mL of water. Mix eggs and J2 in water in a beaker, place 1 mL of the egg suspension in water in a counting dish, and determine, by using a

stereomicroscope (at 25 × magnification), the nematode number, which should be referred to as 1 g of the infested sand.

6. Calculate and weigh the appropriate amount of nematode-infested sand to be mixed with the amount of potting soil in each pot to obtain an initial density of 15–20 eggs + J2/g.
7. Prepare doses of fertilizer for each mixing.
8. Mix a large amount of potting soil with the appropriate quantity of inoculum (nematode-infested sand) and fertilizer in a cement mixer for 5 min to fill all the pots required for the screening test. It may be required several loads, each of 50–60 dm^3 soil to be mixed.
9. Use the infested and fertilized soil to fill buckets or pots with 3–4 dm^3 soil (up to 3 cm from the edge) and the special small pots of peat moss (Jiffy pots).
10. Bury a group of three Jiffy pots seeded singly with one seed of the chickpea test line in each large pot or bucket.
11. Bury additional groups of three Jiffy pots seeded with the seed of the chickpea susceptible control in two to three buckets to follow the development of the nematode.
12. Water pots as required. Usually, no supplemental applications of fertilizer and agrochemicals to control pests and diseases are required.
13. About 2 months after seed germination, uproot some susceptible control plants and verify root growth outside the wall of the Jiffy pots and presence of nematode females protruding from the root surface.
14. Soon after their detection, evaluate the plant reaction to *H. ciceri* as described below.
15. Uproot plants grown in the Jiffy pots from the buckets, avoiding damaging the roots grown outside their walls; dip roots protruding from the Jiffy pots in a pail containing water, trying to remove as much as possible the adhering soil, and avoiding detaching females; observe by the naked eye the number of cysts and females on the entire root system and assess infestation intensity and rate plant reaction according to a scale of 1–5, based on the number of cysts and females in the roots.
16. In this scale, plants having 0–2 females in the root system (scoring 1) are rated as highly resistant, those with 3–5 (scoring 2) resistant, those with 6–20 (scoring 3) moderately resistant or moderately susceptible, those with 21–50 (scoring4) susceptible, and those with >50 (scoring 5) very susceptible.
17. Transplant plants rated highly resistant and resistant singly in buckets and keep them for seed production. Plants can be tutored with a special trellis designed at ICARDA to protect their canopy (Fig. 23).

Fig. 24 Damaging effect of potato cyst nematodes to potato. (**a**) Field infested with the pale potato cyst nematode, *Globodera pallida*, showing a large patch of stunted and chlorotic plants in Siracusa, Italy; (**b**) Screening potato genotypes in pots; (**c**) Yellow females of the golden potato cyst nematode, *Globodera rostochiensis*, attached to potato roots; (**d**) White females of *G. pallida* attached to potato roots

18. Discard the other susceptible plants and use part of the soil to fill buckets for reproduction of *H. ciceri* if necessary (*see* **Note 3**).

3.2.5 Screening for Resistance to Globodera rostochiensis *and* G. pallida

These two species, commonly known as potato cyst nematodes (PCN), are specifically indicated as the golden and pale potato cyst nematodes, respectively. They originated from South America and were imported into Europe from where they spread all over the world with parasitized potato seed tubers. Today, they occur in all major potato-growing areas where they cause impressive yield losses, sometimes complete crop failure (Fig. 24). Fortunately, in 1952, Ellenbay [81] found resistance to *G. rostochiensis* in *Solanum tuberosum* ssp. *andigena*. This finding promoted numerous investigations to search for more sources of resistance to the nematodes and breeding programs to transfer the resistance from *S. tuberosum* ssp. *andigena* to cultigens of *S. tuberosum* ssp. *tuberosum*. This resulted in the release in England (UK), in 1967, of the first potato cultivar, Mary Piper, resistant to *G. rostochiensis*. Although in the UK this cultivar still remains resistant to *G. rostochiensis*, because only the pathotype Ro1 occurs, outside the UK, this cultivar failed in preventing the parasitization of other populations of this nematode species. It was soon realized that by natural selection, some populations of *G. rostochiensis* having different virulence have evolved in this species. These variants were classified as pathotypes. Subsequent studies have shown that within *G. rostochiensis* populations, some females were white creamy (pale) in color rather than yellow (golden). The results of morphological and phylogenetic analyses of the populations with pale females indicated that they were representatives of a new species that was described as *G. pallida*. Detailed information on the taxonomy of potato cyst nematodes has been reported by Subbotin et al. [51, 52]. Further field studies provided evidence that *G. pallida*, like *G. rostochiensis*, also comprises different pathotypes. So far, we know eight pathotypes of PCNs in Europe (five of *G. rostochiensis* and three of *G. pallida*) and three additional different pathotypes, all belonging to *G. pallida* in South America. Lately, a new species of *Globodera* infesting potato, *G. ellingtonae*, has been described from Oregon, USA [82], and later also found in Idaho (USA), Argentina, and Chile, but its importance is not well elucidated because it has been confused with *G. rostochiensis*. This great diversity of pathotypes in the species of potato cyst nematodes is still stimulating searching for more sources of resistance and breeding programs to obtain cultivars that besides being resistant to one or more pathotypes are also suitable to different agroecological areas and diversified demand by the food market [83–90]. Therefore, there is a great need for screening protocols that are reliable and easy to implement. Hereafter, we describe the most important and suitable protocols obtained from the abovementioned articles and authors' own experience [77, 91–93].

The appropriate screening procedures according to the planting material available are listed below:

– Starting from seeds

1. Place seeds in small trays containing sandy soil softened by adding vermiculite and 20% by volume of peat and cover them with 0.5–1 cm soil. Keep trays on benches in a greenhouse at 18–20 °C and 16 h light per day and water and fertilize the soil at least once a week.
2. When seedlings reach the stage of two leaves, transplant them carefully in the small pots containing similar soil mixture but with less peat and vermiculite.
3. When seedlings become 5–7 cm tall, transfer them in larger pots (10–14 cm diam) containing soil infested with 5 or 20 eggs and J2/g or cm^3 soil. The lowest initial nematode density should be used when the reaction of the genotypes is based on the determination of the Rf of PCN, and the largest when this reaction is based on the determination of the number of PCN females on the root system.

- Starting from in vitro cultures
  1. When plantlets are 3–4 cm tall, torn the culture tube gently with a knife to separate the very fragile plantlets and free their roots from the medium using a soft stream of tap water.
  2. Then transplant the plantlets in the 5 × 5 cm pots, containing the same soil as that used for the seedlings from seeds mentioned above. Tutor plantlets with a small stick, fertilize, and keep the pots in a shaded area at 18–22 °C for about one week and then, in a sunny area for hardening.
  3. When plants are 7–8 cm tall, transplant them in the 10–14 cm diam. pots containing infested soil with the same nematode density mentioned above for the seedlings from seeds. Tutor plants with a stick, fertilize every 10–15 days, and water as required.
  4. If the planting material from in vitro cultures is already at the seedling stage, seedlings should be hardened in a greenhouse and then transplanted in the 10–14 cm diam. pots as explained before. For information on how to obtain in vitro culture, *see* Vinterhalter et al. [94].
- Starting from tubers
  1. If not already sprouting, stimulate tubers to sprout by maintaining them at 20–30 °C and high humidity (80–90%) or by applying specific hormones as directed. When sprouts are about 1 cm long, cut tubers in pieces of 13–15 g, each having a sprout in the middle (not on the border) and store them for about a week at room temperature and under light to facilitate healing scar formation on the sectioned tuber and avoid elongation of the sprouts.
  2. Plant sprouted tuber pieces at a depth of 3–5 cm in the infested soil at the same nematode density mentioned above and contained in large pots. Fertilize and water as required.
- Starting from rooted cuttings

  It may happen that the planting material available for screening is not sufficient to provide the planned number of replicated plants or that there is the need to save and reproduce soon the resistant genotypes. Planting materials can be obtained preferably from cuttings. In this case, we suggest propagating, at first, the plants/tubers available in pots of 20–25 cm diam. The following procedures are recommended for the production of potato cuttings and their screening for resistance to PCN when they are fully grown in developed plants.

Fig. 25 Determination of *Globodera* spp. pathotypes. (**a**) Mother potato plants used to obtain propagative cuttings; (**b**) Apical stem cuttings; (**c**) Axillary stem cuttings; (**d**) Apical stem cuttings dipped in powder containing indole butyric acid to stimulate rooting; (**e**) Tray with planted cuttings in potting soil and another ready to be planted; (**f**) Trays with rooting cuttings. Note the tray in the background with a glass cover to reduce evaporation and excessive loss of water during the rooting process; (**g**) Rooted apical stem cuttings; (**h**) Rooted axillary stem cuttings

1. When the plants have already several shoots 6–8 cm long, remove and use them as propagative cuttings. Soon after, apply fertilizer to stimulate the plant to sprout again and produce at least ten additional cuttings. As a precautionary measure, treat the plant canopy with fungicides and pesticides to prevent diseases and infestations of noxious arthropods and to allow the productions of healthy cuttings.
2. Before rooting, basal leaves are removed, leaving each cutting with the apical two leaves (*see* Fig. 25 for details). Then, cuttings are placed in a beaker containing water for half an hour and then dust the base of the stems with a powder containing indole butyric acid (IBA), which is commercially sold in garden shops or obtainable as active ingredient from chemical companies.
3. Plant cuttings in a tray (size depending on the amount of work) filled 2/3 with a wetted sterile potting soil consisting of 1/3 of peat and 2/3 of river sand.
4. Insert cuttings into holes made with a planting stick in rows 5–10 cm apart from each other, and protect them from drying by appressing soil around their base and by spraying water on their leaves.

5. Cover the tray with a transparent glass or plastic lid 1 mm thick (preferred). To favor some air exchange, two wood wedges 0.5 mm thick are inserted between the glass lid and tray walls in both sides of the tray.
6. Place the tray with the cuttings on a bench well illuminated and away from direct sunlight in a greenhouse set at 18–20 °C.
7. Every 4–5 days, spray the cuttings in the tray with water to avoid excessive evaporation.
8. Two weeks later, when cuttings start sprouting, indicating root elongation from their shoot into the soil, apply a diluted fertilizing solution (not exceeding 2‰) in the tray. At this time, cuttings should be well rooted and ready to be transplanted into large (10–14 cm diam) pots containing infested soil with nematode density mentioned above.
9. Keep pots in greenhouse, fertilize every 10–15 days, and water as required.
10. Sixty–seventy days after transplanting in the infested soil, examine plant root system for presence of females protruding from the root surface. Once the suitable conditions of the greenhouse for the nematode development are verified, carry out the assessment of the reaction of the screened genotypes and then separate the susceptible genotypes that are heavily infested by the nematode females (Fig. 24) from those that are resistant and non-infested or with the roots colonized by only a few females. This assessment can be done by visual inspection of the root system.
11. At any case, before using the resistant plants in further breeding programs, we suggest confirming the resistant reaction using both phenotype and genotype characterization. Although we believe that the evaluation of the plant reaction to the nematode based on the visual assessment of the number of females on the root ball is reliable, more data on the influence of the tested genotypes on the reproduction of the nematode provide a more complete characterization of the reaction of the genotype to the nematode colonization.
12. Therefore, when several females have already turned to brown cyst, allow the plants to grow for 2 more weeks, remove their top, and allow the desiccation of the soil by discontinuing watering.
13. Then, uproot plants and place their root ball and surrounding soil in a tray and mix energetically roots and soil to detach cysts from the roots. Remove a subsample of the soil containing root fragments for processing and store the remaining portion for further eventual analyses.

14. Extract and count cysts and eggs from subsamples with the procedures described above for *H. ciceri*.
15. Calculate Rf (reproduction factor) and Rr (reproduction rate) referred both to susceptible and resistant control.
16. If it is necessary, at any developmental stage of the plants, all nematode stages can be extracted from the roots and counted, using an appropriate method such as Coolen's method (16).
17. The phytosanitary authorities of some countries have prepared specific protocols that should be followed at the completion of a breeding program, to better and univocally characterize the reaction of a genotype that will be released as new cultivar. In EU, according to the directive 2007/33/CE of June 11, 2007, among other guidelines, the test to ascertain the susceptibility to both potato cyst nematode species has to be conducted in pots containing at least 1 kg soil infested with five eggs + J2 (cm^3), using besides local populations, also Ro1 (population Ecosse), Ro5 (population Harmerz), Pa1 (population Scottish), and Pa3 (population Chavornay), all maintained for 4 months at 4 °C before use. The test should be run for 3 months, cysts and eggs from the soil extracted and counted, and the relative susceptibility of the test cultivar expressed as % of the control susceptible cv. Desirée, and according to a 1–9 scale where 1 (very susceptible) corresponds to a relative susceptibility of >100 and 9 (very resistant) to a relative susceptibility of <1 (*see* **Note 4**).

3.3 Assessment of Plant Tolerance

Since tolerance is very much related to nematode reproduction and yield performance, a reliable assessment can be obtained when screening tests are conducted in conditions as close as possible to those of the field. Hereafter, indications are given to conduct tests of tolerance to root-knot and cyst-forming nematode species in both greenhouse and field conditions for plants that produce fruits, seeds, and herbal parts.

3.3.1 In Greenhouse

The procedures for this test are listed below:

1. Determine nematode population density of the infested soil as described above for other cyst nematodes.
2. Mix appropriate aliquots of infested soil and fertilizer with the sterilized soil to obtain the desired population density in the soil of the pots to be used for the test. We suggest comparing the reaction of the tested plant with three different population densities of the nematode listed in Table 1.
3. At the same time, plant lines to be tested along with the susceptible, resistant, and tolerant cultivars to be used as

controls for comparison. Moreover, each cultivar must be sown also in non-infested soil to compare the yield decrease of the cultivars at each nematode density.

4. Arrange pots on a greenhouse bench according to an experimental design with eight to ten replicated pots of each tested line and nematode density combination.
5. Allow plants to grow for an entire crop cycle (3–5 months) to reach maturity.
6. According to the plant tested, harvest and weigh the yield consisting of fruit periodically, as they reach size and maturity grade, and in the case of seeds at the end of crop cycle.
7. Then cut top plant at soil level and weigh.
8. Uproot the remaining belowground portion of the plant and place it in a tray where the roots are gently cleaned from adhering soil, using a mild water spray, if necessary, and avoiding the detachment of nematodes protruding from the root surface.
9. Remove the excess of water by wrapping gently the roots with absorbent paper and record their weight. Then, examine moist roots and assess root galls and egg masses for root-knot nematodes and female index for cyst nematodes. In case of root-knot nematode infestation, it is convenient to extract eggs and J2 using the NaOCl method of Hussey and Barker [15] at 1% concentration.
10. Extract nematodes from the soil by processing a subsample of 100–500 cm^3 and count life stages as described above for other root-knot and cyst nematode species.
11. Determine total number of nematodes per pot, calculate Rf and Rr, and record total yield and plant weight.
12. Finally, conduct statistical analysis of the different variables recorded. If nematode variables (Pf, Rf, Rr) of the tested cultivar are similar to those of the susceptible cultivars, while agronomic variables (weight of root, top plant, and marketable yield) are close to the resistant cultivar, then the tested cultivar can be considered as tolerant. Further confirmation under field conditions of the reaction of the tested cultivar could give a measure of the agronomic value of the observed tolerance.

3.3.2 Under Field Conditions

The protocol and assessment are basically the same as those described for the screening conducted in greenhouse.

1. Prepare two large amounts of soil, one infested and the other steam-sterilized to fill the microplots, since these tests require at least three inoculum levels and eight to ten replicates per treatment.

2. To obtain the necessary amount of inoculum, reproduce the nematode during the preceding growing season in similar microplots that should be covered with an appropriate lid to protect the nematodes from high temperature in summer and flooding from excessive rain.
3. Before planting, mix soil from the infested microplots using a concrete mixer. Then take six to ten subsamples to extract nematodes and determine the nematode population level per cm^3 soil. Based on this value, mix proportions of infested and sterilized soil to obtain the required inoculum density.

These microplot experiments are very laborious and require manpower and organizational skill. To obtain the required nematode inoculum of a cyst nematode, we suggest using infested soil or nematode cysts that should be extracted in advance as explained for the screening of chickpea to *H. ciceri*. For root-knot nematodes, alternatively, chopped roots infested by the already determined number of eggs, and J2 in the attached large egg masses should be mixed with sterilized sandy soil and can be used as source of inoculum, as it is frequently done at first author's nematology laboratory [29, 30] (*see* **Note 5**).

3.4 Assessment of Nematode Virulent Populations on Plants Carrying Resistance Genes

All living organisms are characterized by a very large genetic variability and diversity. The huge genetic diversity in plant species has been for more than a century a source of invaluable germplasm that is used to improve the yield and growth of many agricultural crops. Crops are adversely affected by many pests and diseases that cause severe yield losses. Today, exploring plant biodiversity, especially among wild relatives of crops, has resulted in the identification of sources of resistance to diseases and pests, including PPNs that are being transferred by crossing pollination for introgression into many cultigens. Unfortunately, the use of sources of resistance has exerted genetic pressure on PPNs that has resulted in the selection of virulent populations able to parasitize and damage plants that were considered resistant cultivars. Moreover, within the same nematode species, populations having different reproduction and feeding habits have been observed, which have been indicated as different races or pathotypes. However, there has been some disagreement in the classification of nematode populations in races or pathotypes. In the past, the populations of *Meloidogyne* spp. were considered races, whereas those of most cyst-forming nematodes and especially populations of potato cyst nematodes *Globodera* spp. were indicated as pathotypes. Today, it is preferred to differentiate nematode populations in races and pathotypes when it is demonstrated a gene-for-gene relationship [8]. However, in case of classification in races, as this is based on the use of different host plants, the term host race should be used. A more complex case is that of the *Heterodera avenae* group, in which the classification is based on

Table 5
Host differential test to discriminate the races of the four most common root-knot nematode species (*Meloidogyne* spp.)

Meloidogyne species and race	Standard host plant species and "cultivar"					
	Tobacco "NC 95"	Cotton "Deltapine"	Pepper "Early California wonder"	Water melon "Charleston Gray"	Peanut "Florunner"	Tomato "Rutgers"
M. incognita						
Race 1	–[a]	–	+	+	–	+
Race 2	+	–	+	+	–	+
Race 3	–	+	+	+	–	+
Race 4	+	+	+	+	–	+
M. arenaria						
Race 1	+	–	+	+	+	+
Race 2	+	–	–	+	–	+
M. javanica	+	+	±	+	–	+
M. hapla	+	–	+	–	+	+

Proposed by Taylor and Sasser [33]
[a]+ and – indicate reproduction or not of the nematode population on the host plant, respectively

the capability of a population to reproduce or not on a host plant containing one of the three major resistance genes and on the genotypes of three different host plant species (barley, oat, and wheat) [95].

Identification of these races and pathotypes is instrumental for controlling nematodes with nonhost or resistant crop cultivars, which have been developed to manage the four major and economically important *Meloidogyne* species (Table 5), *M. arenaria*, *M. hapla*, *M. incognita*, and *M. javanica*, and other selected economically important and widely distributed cyst-forming nematodes, such as the cereal cyst nematode *H. avenae* group (Table 6), the soybean cyst nematode *H. glycines* (Table 7), and the potato cyst nematodes *G. rostochiensis* and *G. pallida* (Table 8, Figs. 25 and 26). For these nematode species, standard differential host plant species and cultivars or plant introduction have been designed to identify races and pathotypes and are reported in the mentioned tables.

However, the occurrence of races or pathotypes has been reported in many other root-knot and cyst-forming nematode species. Mojtahedi et al. [96] were able to distinguish two races of *Meloidogyne chitwoodi*, of which race 1 reproduces on wheat, carrot, and tomato and not on pepper and alfalfa, while race 2 reproduces

Table 6
Virulence groups and pathotypes of populations of the cereal cyst nematode *Heterodera avenae* group according to whether they reproduce or not on the cultivars listed in the International Test Assortment (ITA) for Defining Cereal Cyst Nematode Pathotypes[a]

		H. latipons									*H. hordecalis*					
***Heterodera* sp:**		*H. avenae*											*H. australis*	*H. filipjevi*		*H. avenae*
Virulence Group:		Group 1									Group 2		Group 3			
Patotype:		Ha11	Ha21	Ha31	Ha41	Ha51	Ha61	Ha71	Ha81	Ha91	Ha12	Ha22	Ha13	Ha23	Ha33	Ha43
ITA cultivar and species																
Cultivar (other names)	**Species**															
Barley (*Hordeum* spp.)																
Emir (KVL785)	*H. vulgare*	S[b]	S	NA	S	NA	R	S	S	NA	S	S	S	S	S	S
Varde	*H. vulgare*	S	S	NA	S	NA	S	S	S	S	S	S	S	S	S	S
Ortolan	*H. vulgare*	R	R	R	R	R	R	R	R	R	S	S	S	S	S	S
KVL191	*H. pallidum*	R	R	R	NA	S	S	S	NA	S	R	NA	NA	S	NA	S
Siri	*H. vulgare*	R	R	R	S	S	S	R	R	R	R	R	S	S	S	S
Morocco (CI3902)	*H. vulgare*	R	R	R	R	R	R	(S)	R	R	R	R	R	(R)	R	R
Morocaine C. I.8341		R	R	NA	NA	R	NA	NA	NA	S	R	NA	R	(R)	NA	(S)
Bajo Aragon 1–1	*H. vulgare*	R	NA	NA	R	NA	R	(S)	R	R	R	R	S	S	R	S
Herta	*H. vulgare*	S	S	R	NA	R	NA	R	NA	(R)	S	NA	S	NA	NA	R
Martin 403–1	*H. vulgare*	R	NA	NA	R	NA	R	R	S	R	R	S	R	S	S	R
Dalmatische	*H. vulgare*	(R)	NA	NA	S	NA	NA	S	S	S	S	S	S	(R)	NA	S
La Estanzuela	*H. vulgare*	NA	NA	NA	NA	NA	NA	S	NA	S	NA	NA	NA	R	S	R
Harlan 43	*H. vulgare*	R	NA	NA	NA	NA	NA	R	NA	(S)	R	NA	NA	(R)	S	R

(continued)

Table 6 (continued)

Heterodera sp:		*H. latipons*									*H. hordecalis*					
		H. avenae											*H. australis*	*H. filipjevi*		*H. avenae*
Virulence Group:		Group 1									Group 2		Group 3			
Patotype:		Ha11	Ha21	Ha31	Ha41	Ha51	Ha61	Ha71	Ha81	Ha91	Ha12	Ha22	Ha13	Ha23	Ha33	Ha43
ITA cultivar and species																
Cultivar (other names)	Species															
Oat (*Avena* spp.)																
Nidar II	*A. sativa*	S	NA	NA	(R)	NA	S	R	R	R	S	R	S	S	S	NA
Sun II	*A. sativa*	S	R	R	R	R	S	R	R	R	S	R	S	S	S	S
Pusa Hybrid BS1	*A. sativa*	R	R	NA	R	R	R	R	R	R	R	R	S	S	S	S
CI3444; 640318-40-2-1																
Silva (KVL1414)	*A. sativa*	(R)	NA	NA	R	NA	(R)	R	R	R	(R)	R	(R)	S	S	R
1376 (CC4658)	*A. sterilis*	R	R	NA	R	R	R	R	R	R	R	R	R	R	R	R
IGV.H.72-646	*A. sterilis*	R	NA	NA	R	NA	R	R	R	S	R	R	S	S	S	S
Ty.72-42-1	*A. sterilis*	R	NA	NA	R	NA	NA	R	NA	NA	R	NA	NA	S	S	NA
Wheat (*Triticum* spp.)																
Capa (KVL8067)	*T. aestivum*	S	S	NA	S	NA	S	S	S	S	S	S	S	S	S	S
Loros × Koga (63/1.7.15.12)	*T. aestivum*	R	R	NA	R	NA	(R)	R	R	S	R	R	(R)	S	S	R
Iskamish-K-2-light	*T. aestivum*	S	NA	NA	R	NA	(R)	(S)	R	(R)	S	S	S	S	S	R
AUS 10894	*T. aestivum*	R	NA	NA	R	NA	R	R	R	S	R	S	(R)	S	S	R
Psathias	*T. durum*	S	NA	NA	S	NA	NA	R	R	(S)	S	R	S	R	R	S

[a]This table is based on data reported in Table 2 by Smiley et al. (2011) [59] and in Table 1 by Cui et al. (2015) [58] who reported the identification of *H. avenae* pathotype Ha 91. In Group 1, the pathotypes Ha 41 and Ha 51 are representatives of the species *Heterodera latipons* according to Databat et al. (2020) [55], who screened four populations of *H. latipons* from Turkey. In Group 3, pathotype Ha13 represents *Heterodera australis*; Ha 23 and Ha 33 are representatives of *Heterodera filipjevi* and Ha 43 represents a population of *H. avenae* from China [58]. Genotypes Ha 13 and Ha 23 represent *Heterodera hordecalis*

[b]Reaction to pathotypes: *R* resistant, *(R)* moderately resistant, *(S)* moderately susceptible, *S* susceptible, *NA* no observation

Table 7
Example of HG Type designation of different populations of the soybean cyst nematode, *Heterodera glycines*, according to Niblack et al. [62], based on female index (FI) values on the indicator lines (1–7) expressed as % of females developed on the susceptible cv. Lee 74

Nematode population	Soybean indicator lines (Plant introductions = PI)							HG Type
	PI548402 (1)[a]	PI88788 (2)	PI90763 (3)	PI437654 (4)	PI209332 (5)	PI89772 (6)	PI548316 (7)	
	Female index (FI)							
Soybean 9	0	7	0	1	8	7	9	0
Soybean 10	4	5	0	0	6	3	nt[a]	0–
Soybean 11	70	65	53	81	45	35	80	1.2.3.4.5.6.7
Soybean 12	3	0	15	26	48	51	33	3.4.5.6.7
Soybean 13	40	21	44	nt[b]	0	0	nt[a]	1.2.3–
Soybean 14	23	5	nt[a]	33	61	29	4	1.4.5.6–
Cow pea 15	42	48	nt[a]	nt[a]	nt[a]	nt[a]	nt[a]	1.2–
Cow Pea 16	18	nt[a]	0	6	8	33	21	1.6.7–
French Bean	0	9	8	41	53	nt[a]	nt[a]	4.5–
Vigna	67	42	0	52	0	93	73	1.2.4.6.7

Names of the nematode populations and FI values are artificial and not from experiments. The designation of HG Type of 0 given to population Soybean 9 is due to the fact that all indicator lines were resistant to this population and had FI values <10%; the HG Type 1234567 designation given to population Soybean 11 indicates that all indicator lines were susceptible (FI >10%) to this nematode population; the HG Type 123- designation given to population Soybean 13 indicates that this population reproduced on the indicator lines 1, 2, and 3. The dash symbol (–) following the numbers indicates that the reaction of one or more of the other designated soybean lines to this population was not assessed

[a]In brackets the standard number of each of the seven indicator lines

[b]nt = not tested

on wheat, alfalfa, and tomato and not on pepper and carrot. Although these two races parasitize cultivated potato (*Solanum tuberosum*), they do not reproduce on the wild potato *Solanum bulbocastanum*, but more recently a new race from Prosser (WA, USA) has shown capability to reproduce on this *Solanum* species and has been designated as race 1 (Roza) [97].

Pokharel et al. [98] tested ten nematode populations of *M. graminicola* from Southeast Asia and one from Florida (USA), and noted that all populations had the same host range, but the isolate from Florida did not cause any damage to some rice cultivars, indicating that *M. graminicola* consists of at least two races.

Lance et al. [71] found clear differences in the virulence of three populations of the sugar beet cyst nematode *H. schachtii* toward plants of *Beta* spp., and Mueller [73] clearly demonstrated

Table 8
Host plant differentials to identify virulence groups and pathotypes of the potato cyst nematode, *Globodera rostochiensis* and *G. pallida*, populations according to whether they reproduce or not on the selected *Solanum* species and accessions

	Nematode species:	*G. rostochiensis*					*G. pallida*						
	Virulence group:	Ro1		Ro3		Ro5	Pa1		–	–	–	Pa2/3	
	European pathotypes:	Ro1	Ro4	Ro2	Ro3	Ro5	Pa1		–	–	Pa2	Pa3	–
Host plant differentials:	South American pathotypes:	R1A	R1B	R2A	R3A	–	P1A	P1B	P2A	P3A	P4A	P5A	P6A
Solanum species and accessions	Ploidy, resistance gene												
S. tuberosum ssp. *tuberosum*	4×, (minor)	±[1]	+	+	+	+	+	+	+	+	+	+	+
S. tuberosum ssp. *andigena* CPC 1673	4×, H1,on chromosome5	–	–	+	+	+	+	"	"	"	+	+	"
S. kurtzianum KTT 60.21.19	2×, K1 K2 A and B	–	(+)	–	(+)	(+)	+	+	+	+	+	+	+
S. vernei GLKS 58.1642.4	2×, Quantitative	–	+	–	–	+	+	+	+	–	+	+	+
S. vernei Vt 62.33.3	2×, Quantitative		–	–	–	+	–	+	–	–	–	+	+
Ex. *S. multidissectum* hybrid P55/7	2×, 1 + polygenes H2	+	+	+	+	+	–	±	–	+	+	+	+
S. tuberosum ssp. *antigena*	H3 + polygenes	+	"	"	"	"	(–)	"	"	"	(–)	(–)	"
CIP 280090.10	Quantitative	"	"	–[2]	–	"	"	"	"	"	–	–[2]	+
S. vernei hybrid 69.1377/94	2×, Polygenes	–	–	–	–	–	–	"	"	"	–	–	"
S. vernei hybrid 63.346/19	2×, Polygenes	–	–	–	–	–	+	"	"	"	+	+	"
S. spegazzini	2×, Fa = H1	–	–	+	–	–	"	"	"	"	"	"	"
S. spegazzini	2×, b + 2 minor	+	–	+	–	–	"	"	"	"	"	"	"
	Glo1 on chromosome7	(–)	+	+	+	+	"	"	"	"	"	"	"

Adapted from Subbotin et al. (2010a) [51]
[1]Reaction to pathotypes: + = compatible interaction: nematode reproduction, potato susceptible; – = incompatible interaction: no nematode reproduction, potato resistant; (–) = partial or uncertain interaction; " = no information
[2]The first author has observed good reproduction (+) of Italian populations of both *Globodera* species on this clone

Fig. 26 Potato clones used for determination of *Globodera* spp. pathotypes. (**a**) Tubers of potato clones of different sizes, shapes, and colors; (**b**) Bench in a greenhouse with selected potato clones growing in pots containing soil infested with populations of *Globodera* spp. for the identification of the nematode pathotypes

the occurrence of two races of *H. schachtii* named Schac 2 and Schac 1, 2.

Di Vito et al. [99] analyzed eight populations of *Heterodera ciceri*, two from Syria, one each from Lebanon and Jordan, and four from Turkey. It was found that one population from Syria and one from Turkey showed different host preference and reproduced also on alfalfa, whereas the other populations failed to parasitize this leguminous forage.

Moreover, there is clear evidence that virulent populations of root-knot nematodes emerge under the genetic pressure caused by the persistent use of a resistant cultivar [100]. In greenhouse experiments, Verdejo-Lucas et al. [101] tested 29 populations of *M. javanica*, *M. incognita*, and *M. arenaria* collected from cultivated fields and found that 48% of them were virulent toward the *Mi* gene. The occurrence of *Meloidogyne* spp. populations virulent to the *Mi* gene is now widespread throughout the Mediterranean area and elsewhere making the use of resistant tomato cultivar problematic. The same phenomena have been reported for *M. incognita* toward the *N* gene in pepper [102] and are even more exacerbated in fields infested by other damaging and polyphagous root-knot nematode species, such as *M. enterolobii*, *M. ethiopica*, and *M. floridensis*, because these species are able to overcome the resistance introgressed in both annual and perennial crop plants [34, 36, 50]. Similar problems occur in populations of potato cyst nematodes, *G. rostochiensis* and *G. pallida*, *H. glycines*, and *H. schachtii*, and provide evidence of the importance of a correct identification of these species for a successful nematode management in agriculture.

3.4.1 Main Schemes to Determine Different Virulence Groups

Tables 5, 6, 7 and 8 report plants used in international tests to identify races and pathotypes of different nematode groups and species.

The suggested procedures are hereafter.

1. Basically, the methods are the same as those described for the assessment of the feeding preferences of different nematode species. The identification of virulent groups is based on the reproduction of the nematodes on differential plants, such as reproduction rate or relative infestation rate (% of nematode numbers in the differential plant compared to those in a susceptible plant). If the reproduction of the nematode on the differential plant is <1 or the relative infestation rate is <5% or also >3%, the plant is rated resistant (nonhost); if these variables are larger, the plant is considered susceptible (good host). Based on the host or nonhost status of the differential plants to the nematodes and on the schemes presented in the table, the nematode population can be assigned to a specific race/ pathotype.
2. The identification of races of *Meloidogyne* spp. is based on the reproduction of the nematode, and therefore, the assessment of the host status of each differential plant (Table 5) must be done when most of the females on the roots have produced egg masses full of eggs that can be extracted from the entire root with the NaCOl method [15]. This assessment is carried out 45–60 days after inoculation when plants are kept at the optimal temperature. An inoculum level of one to two eggs and J2 per cm^3 soil is suggested.
3. In case of potato cyst nematodes, *Globodera* spp., the identification of the nematode pathotypes is also based on the reproduction of the nematode population. However, the eggs of these species are enclosed in brown cysts, which are formed in the root system and may fall into the soil usually 80–90 days after planting. Therefore, cysts must be extracted from soil containing root fragments using the Fenwick can. Extracted cysts are crushed to release the eggs that are counted to calculate the reproduction rate of the nematode as described above for *H. ciceri*. Plants allowing a reproduction rate of the nematode <1 are considered resistant and those with reproduction rate >1 susceptible. An inoculum level of five eggs and J2 per g or cm^3 soil is suggested.
4. For all other cyst nematodes, generally, the host status of the tested differential plants is based on the female index (% female number on the tested differential plants compared to susceptible control plant). Therefore, we suggest using an inoculum density of five to ten egg/g soil and extract females from the roots by maceration–centrifugation methods or just detaching

females from the roots placed on a 710 μm sieve nested on a 250 μm sieve, using pressurized water spray, and counting females and cysts remained on the 250 μm aperture sieve. Usually, the test may take 40–50 days.

Greenhouse space, equipment, and technical skills are required for these tests because it may be necessary to propagate the differential plants received from gene banks or research centers and assess their reaction to the infestation of the used nematode populations. For potatoes, see instruction given for screening to *Globodera* spp.

4 Notes

1. At times, specialized equipment for the conduction of the experiments to assess plant reaction to PPN might not be readily available. In that case, materials produced for other uses may be modified and adapted. Many items must be built on site or with the help of a skilled craftsman, and the procedures suggested hereafter must be adapted to the space and equipment available. Moreover, before starting large-scale screenings, it is necessary to become familiar with the method chosen and to anticipate difficulties when a method is applied for the first time. Very often, breeding programs require careful planning and cooperation between breeders and nematologists.
2. However, during a breeding program, thousands of plants must be tested, and those resistant to the nematode must be saved and grown for further crosses and production of seeds, making necessary a nondestructive handling of the plants under evaluation. We suggest also considering the possibility of using only one population of the same HG type and entire cysts as inoculum at initial and intermediate stages of the breeding program, to speed up the screening work. Since the evaluation must be done without destroying the plant, we suggest using transparent containers that allow the assessment of the number of females on the root ball to avoid the destruction of the plants that can be saved. On the other hand, since seed companies sell their seeds in the international market, we suggested that at the final stage of the screening and before marketing new cultivars, they should be tested for their resistance toward all HG type populations.
3. The genetic traits of wild species and local varieties should be checked before conducting any screening test. Rarely, these entries are pure because of the great genetic variability of chickpea seeds. Seeds of the same germplasm code do not have the same resistance traits and may contain only one with

resistance traits. Therefore, the plants rated as resistant must be saved and transplanted to produce seeds to be used for further assessment of their resistance traits, before being used as new resistant lines in a breeding program.

4. We would like to point out that cuttings used in these tests must be obtained from plants in full vigor; otherwise, they will fail to produce root. Using soil infested with 15–20 eggs and J2 of the nematode per cm^3 or g soil, as an initial density, will certainly result in a very good number of nematode females on the root ball, which facilitates the discrimination of resistant from susceptible genotypes. However, we suggest having extra plants of susceptible and resistant genotypes to check the development of females on the root balls and the most appropriate time for plant reaction evaluation.
5. Different parameters are used to assess the tolerance of perennials to nematodes. The assessment of such tolerance requires several years of observations, until the plant start yielding. In this case, the inoculation of a rather low number of nematodes is preferred because after a few generations, the nematode will increase its population levels toward which a plant may show tolerance (satisfactory yield performance) or intolerance (poor yield). Therefore, the reaction that perennials show is related to the ceiling reached by the population of the nematodes rather than the initial population inoculated.

Acknowledgments

We wish to thank all colleagues who have provided suggestions and photographs, journals who have granted us permissions to reuse their photos, and Alberto Troccoli for assistance with the figures.

References

1. Rich JR, Dunn RA, Noling JW (2004) Nematicides: past and present uses. In: Chen ZX, Chen SY, Dikson DW (eds) Nematology: advances and perspectives: nematode management and utilization, vol 2. CABI, Wallingford, pp 1179–1200
2. Ahmar S, Gill RA, Jung K-H, Faheem A, Qasim MU, Mubeen Mand Zhou W (2020) Conventional and future outlook. Int J Mol Sci 21:2590. https://doi.org/10.3390/ijms21072590
3. Pratap A, Das A, Kumar S, Gupta S (2021) Perspectives on introgression breeding in food legumes. Front Plant Sci 11. https://doi.org/10.3389/fpls.2020.589189
4. Owen K (2022) A triumph of tolerance: managing the threat to wheat production by the root lesion nematode *Pratylenchus thornei* in the subtropical grain region of eastern Australia. In: Sikora RA, Desaeger J, Molendijk LPG (eds) Integrated nematode management: state-of-the-art and vision for the future. CABI, Wallingford, pp 13–19
5. Sikora RA, Molendijk LPG, Desaeger J (2022) Integrated nematode management and crop health: future challenges and opportunities. In: Sikora RA, Desaeger J, Molendijk LPG (eds) Integrated nematode management: state-of-the-art and vision for the future. CABI, Wallingford, pp 3–10

6. Roberts PA (1990) Resistance to nematodes: definitions, concepts, and consequences. In: Starr JL (ed) Methods for evaluating plant species for resistance to plant-parasitic nematodes. Society of Nematology, Hyattsville, pp 1–15
7. Roberts PA (2002) Concept and consequence of resistance. In: Starr JL, Bridge J, Cook R (eds) Plant resistance to parasitic nematodes. CABI, Wallingford, pp 23–41
8. Trudgill DL (1991) Resistance to and tolerance of plant parasitic nematodes in plants. Annu Rev Phytopathol 29:167–192
9. Andrews M, Andrews ME (2017) Specificity in legume-rhizobia symbioses. Int Int J Mol Sci 18:705. https://doi.org/10.3390/ijms18040705
10. Warschefsky EJ, Klein LL, Frank MH, Chitwood DH, Londo JP, von Wettberg EGB, Miller AJ (2016) Rootstocks: diversity, domestication, and impacts on shoot phenotypes. Trends Plant Sci 21:418–437. https://doi.org/10.1016/j.tplants.2015.11.008
11. Sharma SB, Jain KC, Lingaraju S (2000) Tolerance to reniform nematode (*Rotylenchulus reniformis*) race A in pigeonpea (*Cajanus cajan*) genotypes. Ann Appl Biol 136:247–252
12. Reuther M, Lang C, Grundler FMW (2017) Nematode-tolerant sugar beet varieties resistant or susceptible to the beet cyst nematode *Heterodera schachtii*. Sugar Ind 142:277–284
13. Perry RN, Moens M, Starr JL (eds) (2009) Root-knot nematodes. CABI, Wallingford, p 488
14. Smith PG (1944) Embryo culture of a tomato species hybrid. Proc Am Soc Hortic Sci 44:413–416
15. Hussey RS, Barker KR (1973) A comparison of methods of collecting inocula of *Meloidogyne* spp., including a new technique. Plant Dis Reptr 57:1025–1028
16. Coolen WA (1979) Methods for the extraction of *Meloidogyne* spp. and other nematodes from roots and soil. In: Lamberti F, Taylor CE (eds) Root-knot nematodes (*Meloidogyne* spp.) – systematics, biology and control. Academic, London, pp 317–329
17. Jenkins WP (1964) A rapid centrifugal-flotation technique for separating nematodes from soil. Plant Dis Reptr 48:692
18. Cabasan MTN, Fernandez L, De Waele D (2016) Host response of *Oryza glaberrima* and *O. sativa* rice genotypes to the rice root-knot nematode *Meloidogyne graminicola* in a hydroponic system under growth chamber. Arch Phytopath Plant Prot 48:740–750. https://doi.org/10.1080/03235408.2016.1140608
19. Hoagland Dr. Arnon DI (1950) The water-culture method for growing plants without soil (Revised by Arnon DI). California Agricultural Experiment Station, Berkeley CA, USA. Circular 347:32
20. Hothen SD, Marley KA, Larson RA (2003) Photochemistry in Hoagland's nutrient solution. J Plant Nutr 26:845–854. https://doi.org/10.1081/PLN-120018569
21. Nkurunziza R (2018) Analysis of resistance in rice (*Oryza sativa* L.) genotypes LD24 and Khao Pahk Maw to root-knot nematodes (*Meloidogyne* spp.) and root-lesion nematode (*Pratylenchus zeae*). Master thesis 2017–2018, University of Ghent, Belgium
22. Cladius-Cole AO, Salam AO, Mande S (2018) Host status of interspecific upland rice progenies (*Oryza sativa* × *Oryza barthii*) to *Meloidogyne incognita*. J Rice Res 6:4. https://doi.org/10.4172/2375-4338-1000198
23. Pokharel RR, Abawi GS, Duxbury JM (2011) Greenhouse evaluation of rice and wheat germplasm for resistance to *Meloidogyne graminicola* with comments on evaluation indices and proposal of a new one. Nematol Medit 39:157–168
24. Mattos VS, Leite RR, Cares JE, Gomes ACMM, Moita AW, Lobo VLS, Carneiro RMDG (2019) *Oryza glumaepatula*, a new source of resistance to *Meloidogyne graminicola* and histological characterization of its defense mechanism. Phytopathology 109:1941–1948. https://doi.org/10.1094/PHITO-02-19-0044-R
25. Graebner RC, Brown CR, Ingham RE, Hagerty CH, Mojtahedi H, Quick RA, Hamlin LL, Wade N, Bamberg JB, Sathuwalli V (2018) Resistance to *Meloidogyne chitwoodi* identified in wild potato species. Am J Potato Res 95:679–686. https://doi.org/10.1007/s12230-018-9674-9
26. Janssen GJV, van Norel A, Verkerk-Bakker JR (1996) Resistance to *Meloidogyne chitwoodi, M. fallax* and *M. hapla* in wild tuber-bearing *Solanum* spp. Euphytica 92:287–294
27. Nico AI, Jimenez-Diaz RM, Castillo P (2003) Host suitability of the olive cultivars Arbequina and Pictual for plant–parasitic nematodes. J Nematol 35:29–34
28. Sasanelli N, Fontanazza G, Lamberti F, D'Addabbo T, Patumi M, Vergani G (1997) Reaction of olive cultivars to *Meloidogyne* species. Nematol Medit 25:183–190

29. Di Vito M, Greco N, Carella A (1986) The effect of *Meloidogyne incognita* and importance of the inoculum on eggplant. J Nematol 18:487–490
30. Greco N, Di Vito M (2009) Population dynamics and damage levels. In: Perry RN, Moens M, Starr JL (eds) Root-Knot Nematodes. CABI, Wallingford, pp 246–274
31. Sasanelli N, D'Addabbo T, Dell'Orco P, Mencuccini M (2000) The in vitro use of olive explants in screening trials for resistance to the root-knot nematode, *Meloidogyne incognita*. Nematropica 30:101–106
32. Pinochet J, Verdejo-Lucas S, Marull J (1989) Evaluacion de siete patrones de prunus a tres especies de *Meloidogyne* en España. Nematropica 19:125–134
33. Taylor AL, Sasser JN (1978) Biology, identification and control of root-knot nematodes (*Meloidogyne* spp.). North Carolina State University Graphics, Raleigh, p 111
34. Qiu S, Maquillan MND, Brito JA, Beckman TG, Dickson DW (2021) Susceptibility of Flordaguard peach rootstock to a resistant-breaking population of *Meloidogyne floridensis* and two populations of *Meloidogyne arenaria*. J Nematol 53. https://doi.org/10.21307/jonem-2021-111
35. Thies JA, Merril SB, Corley EL (2002) Red food coloring stain: new safer procedures for staining nematodes in roots and egg masses on root surface. J Nematol 34:179–181
36. Nyczepir AP, Bekman T, Gand Reighard GL (1999) Reproduction and development of *Meloidogyne incognita* and *M. javanica* on Guardian peach rootstock. J Nematol 31: 334–340
37. Mezza P, Soto B, Rojas L, Esmenjaud D (2016) Identification of *Meloidogyne* species from the central valley of Chile and interaction with stone fruit rootstocks. Plant Dis 100:1358–1363
38. Yagci KG, Devran Z (2019) Reaction of peach and nectarine root-stocks to different populations of *Meloidogyne incognita* (kofoid & White 1919) and *Meloidogyne javanica* (Treub, 1885). Turk Entomol Derg 43:171–178
39. Esmenjaud D, Voisin R, van Ghelder C, Bosselut N, Lafargue B, Di Vito M, Dirlewanger E, Poëssel JL, Kleinhentz M (2008) Genetic dissection of resistance to root-knot nematodes *Meloidogyne* spp. in plum, peach, almond, and apricot from various segregating interspecific prunus progenies. Tree Genet Genomes 5:279–289. https://doi.org/10.1007/s11295-008-0173-x
40. Saucet SB, Ghelder CV, Abad P, Duval H, Esmenjaud D (2016) Resistance to root-knot nematodes *Meloidogyne* spp. in woody plants. New Phytol 211:41–56. https://doi.org/10.1111/nph.13933
41. Esmenjaud D (2021) Deciphering resistance to root-knot nematodes in *Prunus* for rootstock breeding: sources, genetics and characterization of the *Ma* locus. Horticulturae 7: 564. https://doi.org/10.3390// horticulturae7120564
42. Stirling GR, Cox MC, Ogden-Brown J (2011) Resistance to plant parasitic nematodes (*Pratylenchus zeae* and *Meloidogyne javanica*) in *Erianthus* and crosses between *Erianthus* and sugarcane. Int Sugar J 114: 230–237
43. Bhuiyan SA, Croft BJ, Stirling GR, Wong E, Jackson P, Cox M (2016) Assessment of resistance to root-lesion and root-knot nematodes in Australian hybrid clones of sugarcane and its wild relatives. Australas Plant Pathol 45: 166–173
44. Bhuiyan SA, Garlick K (2021) Evaluation of root-knot nematode resistance assay for sugarcane accessions lines in Australia. J Nematol 53. https://doi.org/10.21307/jofnem-2021-006
45. Villain L, Aribi J, Rèversat G, Anthon F (2010) A high-throughput method for early screening of coffee (*Coffea* spp.) genotypes for resistance to root-knot nematodes (*Meloidogyne* spp.). J Plant Pathol 128:451–458. https://doi.org/10.1007/s10658.010-9671-4ì+
46. Aoki S, Sipe B, Astorga C, Nagai C (2012) Resistance of semi wild *Coffea Arabica* L. from Ethiopia to a root-knot nematode, *Meloidogyne konaensis*. Nematropica 42:131–136
47. Ferris H, Zheng L, Walker MA (2012) Resistance of grape rootstock to plant-parasitic nematodes. J Nematol 44:377–386
48. Madani M, Akhiani A, Mamazdadeh M, Kheiri A (2012) Resistance evaluation of the pistachio rootstocks to *Meloidogyne* spp. in Iran. J Appl Hortic 14:134–138
49. Freitas VM, Silva JGP, Gomes CB, Castro JMC, Correa VR, Carneiro RMDG (2017) Host status of selected fruit crops to *Meloidogyne enterolobii*. Eur J Plant Path 148:307–319. https://doi.org/10.1007/s10658-16-1090-8

50. Gomes VM, Ribeiro RM, Viana AP, de Souza RM, Santos EA, Rodrigues DL, de Almeida OF (2016) Inheritance of resistance to *Meloidogyne enterolobii* and individual selection in segregating populations of *Psidium* spp. Eur J Plant Path 148:699–708. https://doi.org/10.1007/s10658-016-1128-y
51. Subbotin SA, Mundo-Ocampo M, Baldwin JG (2010) Systematics of cyst nematodes (Nematode: Heteroderina). In: Hunt JD, Perry RN (eds) Nematology monographs and perspectives 8A. BRILL, Leiden/Boston, p 351
52. Subbotin SA, Mundo-Ocampo M, Baldwin JG (2010b) Systematics of cyst nematodes (Nematode: Heteroderinae). In: Hunt JD, Perry RN (eds) Nematology Monographs and Perspectives 8B. BRILL, Leiden/Boston, 512
53. Smiley RW, Maafi ZT, Dadabat AA, Peng D, Iqbal S, Sa S, Jones MGK, Waeyenberge L (2017) Cereal cyst nematodes: a complex and destructive group of *Heterodera* species. Plant Dis 101:1692–1720. https://doi.org/10.1094/PDIS-03-17-0355-FE
54. Cui J, Jiao Y, Zhou B, Ren H, Li H, Liu S, Jiang S, Meng H, Li M, Dadabat AA, Peng D (2020) Pathotype, resistance classification, and seed-coating control of *Heterodera avenae* and *H. filipjevi* in the north China plain. Plant Dis 104:3230–3238. https://doi.org/10.1094/PDIS-02-20-0258-RE
55. Dadabat AA, Imren M, Özer G, Mokrini F, Duman N, Paulitz T (2020) Genetic and pathogenic variation in *Heterodera latipons* populations from Turkey. Nematology 22:1–10. https://doi.org/10.1163/15685411-bja-10029
56. Dadabat AA, Ferney G-BH, ErginbasOrakci G, Dreisigacker IM, Toktay H, Elekchoglu HI, Mekete T, Nicol JM, Ansari O, Ogbonnaya F (2016) Association analysis of resistance to cereal cyst nematodes (*Heterodera avenae*) and root lesion nematodes (*Pratylenchus neglectus* and *P. thornei*) in CIMMIT advanced spring wheat lines for semi-arid conditions. Breed Sci 66:692–702. https://doi.org/10.1270/jsbbs.15158
57. Pariyar SR, Dadabat AA, Siddique S, Erginbas-Orakci G, Elashry A, Morgounov A, Grundler FMW (2016) Identification and characterization of resistance to cereal cyst nematode *Heterodera filipjevi* in winter wheat. Nematology 18:377–402
58. Cui J-k, Huang W-k, Peng H, Liu S-m, Wang G-f, Kong L-a, Peng D-l (2015) A new pathotype characterization of Daxing and Huangyuan population of cereal cyst nematode (*Heterodera avenae*) in China. J Integr Agr 14:724–731. https://doi.org/10.1016/s-2095-3119(14)60982-5
59. Smiley RW, Yan G, Pinkerton JN (2011) Resistance of wheat, barley and oat to *Heterodera avenae* in the Pacific Northwest, USA. Nematology 13:539–552. https://doi.org/10.1163/138855410X531862
60. Schmitt DP, Wrather JA, Riggs RD (eds) (2004) Biology and management of soybean cyst nematode. Schmitt & Associates of Marceline, Marceline, p 262
61. Schmitt DP, Shannon G (1992) Differentiating soybean cyst nematode race and resistance response of soybean. Crop Sci 32:275–277
62. Niblack TL, Arelli PL, Noel GR, Opperman CH, Orf JH, Schmitt DP, Shannon JC, Tylca GL (2002) A revised classification scheme for genetically diverse populations of *Heterodera glycines*. J Nematol 34:279–288
63. Niblack TL, Tylca GL, Arelli P et al (2009) A standard greenhouse method for assessing soybean cyst nematode resistance in soybean: SCEo8 (standardized Cyst Evaluation 2008). Online. Plant Health Prog 10:1. https://doi.org/10.1094/PHP-2009-0513-01-RV
64. Acharya K, Tande C, Byamukama E (2017) Assessment of commercial soybean cultivars for resistance against prevalent *Heterodera glycines* populations of South Dakota. Plant Health Prog 18:156–161. https://doi.org/10.1094/PHP-03-17-0017-RS
65. Wen L, Yuan C, Herman TK (2017) Accessions of perennial *Glycine* species with resistance to multiple types of soybean cyst nematode (*Heterodera glycines*). Plant Dis 101:1201–1206. https://doi.org/10.1094/PDIS-10-16-1472-RE
66. Howland A, Monning N, Mathesius J, Nathan M, Mitchum MG (2018) Survey of *Heterodera glycines* population densities and virulence phenotypes during 2015-2016 in Missouri. Plant Dis 102:2407–2410. https://doi.org/10.1094/PDIS-04-18-0650-
67. Kim M-S, Sung M-K, Kim M-W, Seo H-J, Kim D-G, Chung J-I (2013) Investigation of soybean cyst nematode *Heterodera glycines* type and evaluation of resistance on soybean varieties and germplasm in Korea. Korean J Crop Sci 58:161–168. https://doi.org/10.7740/kjcs.2013.58.2.161
68. Hua C, Li C, Hu Y, Mao Y, You J, Wang M, Chen J, Tian Z, Wang C (2018) Identification of HG types of soybean cyst nematode *Heterodera glycines* and resistance screening on

soybean genotypes in northeast China. J Nematol 50. https://doi.org/10.21307/jofnema-2018-007

69. St-Amour VTB, Mimee B, Torkamaneh D, Jean M, Belzile F, O'Donoughue LS (2020) Characterizing resistance to soybean cyst nematode in PI 494182, an early maturing soybean accession. Crop Sci 60:2053–2069. https://doi.org/10.1002/csc2.20162
70. Savitsky H, Price C (1965) Resistance to sugarbeet cyst nematode (*Heterodera schachtii*) in F_1 tetraploid hybrids between *Beta vulgaris* and *Beta patellaris*. JASSBT 15:370–373
71. Lance W, Müller J, De Bock TSM (1993) Virulence in the beet cyst nematode (*Heterodera schachtii*) versus some alien genes for resistance in beet. Fund App Nematol 16: 447–454
72. Wright AJD, Bussell JS, Stevens M, Back MA, Sparkes DL (2018) Physiological differences between sugar beet varieties susceptible, tolerant or resistant to the beet cyst nematode, *Heterodera schachtii* (Schmidt) under uninfested conditions. Eur JAgr 98:37–45. https://doi.org/10.1016/j.eja.2018.05.005
73. Müller J (1998) New pathotypes of the beet cyst nematode (*Heterodera schachtii*) differentiated on alien genes for resistance in beet (*Beta vulgaris*). Fund App Nematol 21:519–526
74. Nielsen EL, Baltensperger DD, Kerr ED, Rife CL (2003) Host suitability of rape seed for *Heterodera schachtii*. J Nematol 35:35–38
75. Greco N, Moreno IL (1992) Influence of *Globodera rostochiensis* on yield of summer, winter and spring sown potato in Chile. Nematropica 22:165–173
76. Jimenez N, Crozzoli R, Greco N (2000) Effect of *Globodera rostochiensis* on the yield of potato in Venezuela. Nematol Medit 28: 295–299
77. Greco N, Carputo D, Frusciante L, Russo G, Brandonisio A (2005) Screening of clones of *Solanum* spp. for resistance to potato cyst nematodes, *Globodera rostochiensis* and *G. pallida*. Nematol Medit 33:163–170
78. Vovlas N, Greco N, Di Vito M (1985) *Heterodera ciceri* sp. n. (Nematoda: Heteroderidae) on *Cicer arietinum* from northern Syria. Nematol Medit 13:239–252
79. Greco N, Di Vito M, Saxena MC, Reddy MV (1988) Effect of *Heterodera ciceri* on yield of chickpea and lentil and development of this nematode on chickpea in Syria. Nematologica 34:98–114
80. Di Vito M, Greco N, Singh KB, Saxena MC (1988) Response of chickpea germplasm line to *Heterodera ciceri*. Nematol Medit 16:17–18
81. Ellenbay C (1952) Resistance to potato root eelworm *Heterodera rostochiensis* Wollenweber. Nature 170:1016
82. Handoo ZA, Carta LK, Skantar AM, Chitwood DJ (2012) Description of *Globodera ellingtonae* n. sp. (Nematoda: Heteroderidae) from Oregon. J Nematol 44:40–57
83. Castelli L, Bryan G, Block VC, Ramsay G, Fillips MS (2005) Investigation of resistance specificity amongst fifteen wild *Solanum* species to a range of *Globodera pallida* and *G. rostochiensis* populations. Nematology 7: 689–699
84. Faggian R, Powsel A, Slater AT (2012) Screening for resistance to potato cyst nematode in Australian potato cultivars and alternative solanaceous hosts. Australas Plant Pathol 41:453–461
85. Limantseva L, Mironenko N, Shuvalov O, Antonova O, Khiutti A, Novikova L, Afanasenko O, Spooner D, Gavrilenko T (2014) Characterization of resistance to *Globodera rostochiensis* pathotype Ro1 in cultivated and wild potato species accessions from the Vavilov Institute of Plant Industry. Plant Breed 133:660–665
86. Withworth JL, Novy RG, Zasada IA, Wang X, Dandurand LM, Kuhl JC (2018) Resistance of potato breeding clones and cultivars to three species of potato cyst nematode. Plant Dis 102:2120–2128. https://doi.org/10.1094/PDIS-12-17-1978-RE
87. Mwangi JM, Niere B, Daub M, Finchk MR, Kiewnick S (2019) Reproduction of *Globodera pallida* on tissue culture-derived potato plants and their potentials use in resistance screening process. Nematology 21:613–623
88. Milczarek D, Tatarowska B, Plich J, Podlewska-Przetachiewicz A, Flis B (2020) *Solanum gourlayi* – a source of cyst nematode resistance in potato breeding. Potato Res 63: 589–595. https://doi.org/10.1007/s11540-020-09459-9
89. Koroleva AK, Sivolapova AB, Voinova MS, Ivanova AS, Meleshin AA, Manankov VV (2021) Assessment of resistance of potato hybrids from the Russian Potato Research Center's collection to *Globodera rostochiensis*. IOP Conf Ser Earth Environ Sci 941:012028. https://doi.org/10.1088/1755-1315/941/1/012028

90. Silvestre R, Dandurand LM, Zasada IA, Franco J, Kuhl JC (2021) An assessment of potato cyst nematodes, (*Globodera* spp.) research from the Andean region of South America. Part 2: search for resistance in potato. Nematropica 51:106–130
91. Greco N, Brandonisio A, De Cosmis P (2007) Pathotypes and heterogeneity of Italian populations of *Globodera rostochiensis* and *G. pallida*. Nematol Medit 35:137–142
92. Greco N, Di Vito M, Parisi B, Ranalli P, Brandonisio A, Catalano F (2007) Resistance of new Italian potato breeding clones to cyst and root-knot nematodes. Nematol Medit 35:227–235
93. Molinari S, Greco N, Zouhar M (2010) Superoxide dismutase isoelectric focusing patterns as a tool to differentiate pathotypes of *Globodera* spp. Nematology 12:751–758
94. Vinterhalter D, Dragićević I, Vinterhalter B (2008) Potato *in vitro* culture techniques and biotechnology. Fruit Veg Cereal Sci Biotech 2(Special issue 1):16–45
95. Yuan H-x, Sun J-w, Yanh W-x, Xing X-p, Wang Z-y, Riley IT, Li H-I (2010) New pathotype of *Heterodera avenae* (cereal cyst nematode) from winter wheat in Zhenzhou, Henan, China. Australasian Plant Pathol 39: 107–111
96. Mojtahedi H, Santo GS, Wilson GH (1988) Host tests to differentiate *Meloidogyne chitwoodi* races 1 and 2 and *M. hapla*. J Nematol 20:468–473
97. Mojtahedi H, Brown CR, Riga E, Zhang LH (2007) A new pathotype of *Meloidogyne chitwoodi* race 1 from Washington State. Plant Dis 91:1051
98. Pokharel RR, Abawi GS, Duxubury JM, Smat CD, Wang X, Brito JA (2010) Variability and recognition of two races in *Meloidogyne graminicola*. Australas Plant Pathol 39:326–333
99. Di Vito M, Greco N, Malhotra RS, Singh KB, Saxena MC, Catalano F (2001) Reproduction of eight populations of *Heterodera ciceri* on selected plant species. Nematol Medit 29:79–90
100. Verdejo-Lucas S, Cortada L, Sorribas FJ, Ornat C (2009) Selection of virulent populations of *Meloidogyne javanica* by repeated cultivation of *Mi* resistance gene tomato rootstocks under field conditions. Plant Pathol 58: 990–998
101. Verdejo-Lucas S, Talavera M, Andrés MF (2012) Virulence response to the *Mi* gene of *Meloidogyne* populations from tomato in greenhouse. Crop Prot 39:97–105
102. Thies JA (2011) Virulence of *Meloidogyne incognita* to expression of N gene in pepper. J Nematol 43:90–94

Chapter 4

Detection of Resistance, Susceptibility, Tolerance, and Virulence in Plant–Nematode Interactions: Part II: Migratory and Semi-endoparasitic Nematodes

Nicola Greco

Abstract

This chapter is a continuation of Chap. 3. Initially, protocols for the screening of several host plants to their major migratory and semi-endoparasitic nematodes are presented. Then the problems related to assessment of tolerance to these nematodes are described, followed by the determination of nematode races. The main plant–nematode interactions considered are annuals and perennials to *Pratylenchus* spp.; banana to *Radopholus similis*; potato to *Nacobbus aberrans*; several crop plants, including onion, alfalfa, clovers, and potato, to *Ditylenchus dipsaci*; broad bean to *D. giga*; potato and sweet potato to *D. destructor*; peanut to *D. africanus*; rice to *D. angustus* and *Aphelenchoides besseyi*; wheat to *Anguina tritici*; different plants to *Rotylenchulus reniformis*; and citrus to *Tylenchulus semipenetrans*. Schemes to identify races or biotypes are only presented for *D. dipsaci* and *T. semipenetrans*. The occurrence of pathotypes in other nematode species is also discussed. Finally, comments are made on ectoparasitic nematodes.

Key words Resistance, Tolerance, Virulence, Root-lesion nematodes, *Pratylenchus* spp., *Radopholus similis*, *Ditylenchus* spp., *Aphelenchoides* spp., *Rotylenchulus reniformis*, *Tylenchulus semipenetrans*, *Nacobbus* spp.

1 Introduction

Migratory-endoparasitic nematodes enter plant tissues and move through, causing a range of symptoms including parenchymal tissues that feel spongy on touching, distortions and rotting of the damaged plant parts (*Ditylenchus dipsaci*), decoloration (*D. angustus*, *Aphelenchoides besseyi*) of the aerial plant parts, and root browning (*Pratylenchus* spp. and *Radopholus similis*). These nematodes complete their life cycle inside the host plant but never establish a permanent feeding site.

Author Nicola Greco is now retired.

Sergio Molinari (ed.), *Plant-Nematode Interactions*, Methods in Molecular Biology, vol. 2756,
https://doi.org/10.1007/978-1-0716-3638-1_4,

Semi-endoparasitic nematodes, instead, have an intermediate behavior between sedentary-endoparasitic nematodes (*Meloidogyne* spp., *Globodera* spp., *Heterodera* spp.) and migratory-endoparasitic nematodes. They never enter entirely into the host plant tissue but introduce only the anterior part of their body, with the rest remaining outside. As soon as they find a suitable site, they establish a permanent feeding site, become sedentary, and start to feed. The posterior part of their body then becomes saccate and the life cycle is then completed.

The genus *Nacobbus* is covered here because it belongs to the family Pratylenchidae but is in a different subfamily: Nacobbinae for *Nacobbus* and Pratylenchinae for *Pratylenchus*. Its ethology is quite different from *Pratylenchus* as all young stages of *Nacobbus* spp. (male and immature females) are migratory and infective, while mature females become saccate, sedentary, and completely embedded in the root. Their presence is obvious because of the galls they form, so they are known as false root-knot nematodes.

As migratory nematodes do not establish a permanent feeding site in the host plant tissue, the development of resistance in plants to these nematodes is not common [1]. Notwithstanding, thanks to continuous research efforts, resistance has been found in many plants. As a result, the control of *D. dipsaci* in alfalfa and clovers, of *D. angustus* and *Aphelenchoides besseyi* in rice, and of *Pratylenchus* spp. in wheat is mostly based on the use of resistant and tolerant cultivars. Consequently, the protocols and suggestions provide information on how to identify resistance and tolerance to these nematodes and virulence in nematode populations.

Definitions given in Chap. 3 and by Roberts [2, 3] and Trudgill [4] also apply here.

2 Materials

2.1 Tests to Assess the Reaction to Nematodes of Plants from Gene Banks or Breeding Programs

As this may involve the evaluation of several thousands of planting materials, adequate equipment, large nematode populations of different virulence groups, and well-trained personnel are necessary.

2.2 Root-Lesion Nematodes: Screenings for Resistance of Annual Plants to Pratylenchus spp.

In case of legumes, such as chickpea, soybean, and French bean, plants must be inoculated with the specific root nodulation bacterium [5]. Populations of the *Pratylenchus* species toward which to screen, reared in advance on susceptible plants such as chickpea or soybean or, better, on carrot disks.

Seeds of the genotypes to screen, including well-known resistant and susceptible varieties or lines to be used as susceptible and resistant controls and also, in case of the susceptible control, to be sure that the nematodes reproduced well. Particular plant containers should be

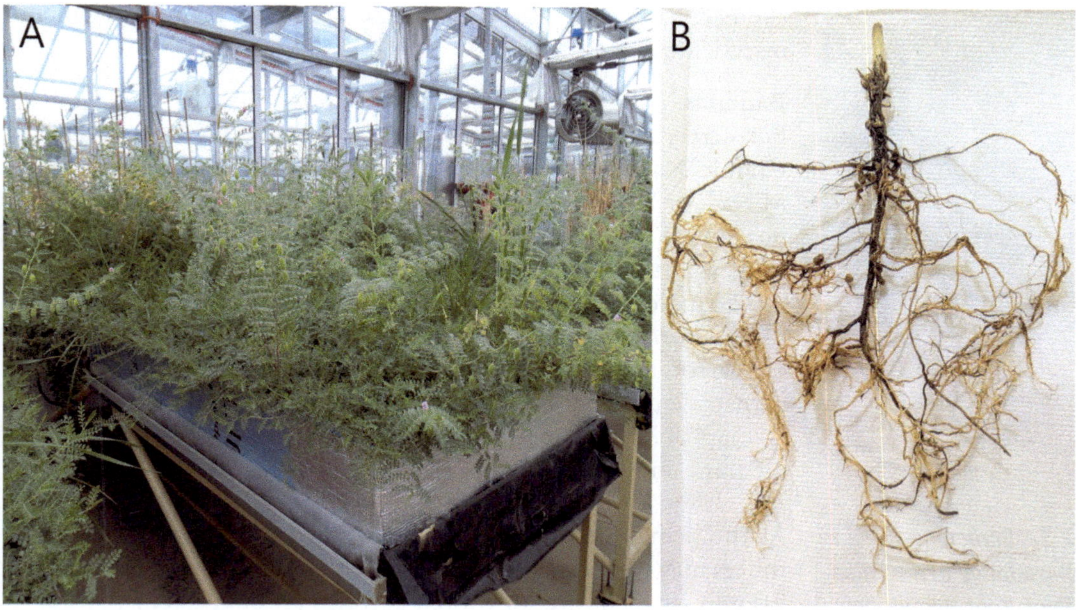

Fig. 1 (**a**) Greenhouse bench with a system bottom watering system of chickpea pots. (**b**) a chickpea roots showing brown spot caused by *Pratylenchus* spp. parasitization. (Figures kindly provided by the Crop Nematology Group, Centre for Crop Health, University of Southern Queensland, Australia, with kind permission of John Thompson PSM)

used (containing about 100–350 g soil), or otherwise common pots of 10–12 cm diam. containing 800–1000 cm^3 soil.

Sterilized soil mix (containing at least 80–90% sand), trays, and potting soil for seed germination.

For optimal equipment and methodology, it is suggested to read articles by Reen et al. [6], Thomson et al. [7], and Rostad et al. [8] for the work conducted at the Crop Nematology Group, Centre for Crop Health, University of Southern Queensland, Australia, where scientists have been investigating for decades the resistance in chickpea and wheat to *Pratylenchus thornei* and *P. neglectus* (Fig. 1). Their greenhouses are equipped for constant temperature and continuous irrigation to provide optimal conditions for both plant growth and nematode development, to avoid any constraint and be sure that results, especially symptoms, are solely due to plant response to nematodes.

In general, the tendency is to use a small container to save space and to inoculate plants with one to two nematode specimens/g or cm^3 soil, while others prefer about eight to ten nematodes per g soil.

2.2.1 Screening Perennials to *Pratylenchus* spp.

***Prunus* spp.**

Seeds, rooted cuttings, in vitro culture plants of the genotype to screen, at least a genotype known as resistant and one susceptible to serve as control for comparison. One or more populations of the nematode to be used for screening, reared on carrot disk. Pots containing 1–2 dm^3 soil; sterilized sandy soil.

Fertilizer, preferably a soluble 20-20-20 NPK + microelements; insecticide and acaricide.

Bananas

For screening, a susceptible control cultivar is needed, such as Grande Naine, a resistant cultivar, as are Lisan Jari Buaya, Gros Michel, or Yangambi Km^5, and plants of the other cultivars or cross progenies. As banana plants are rather large, to save space, it is suggested to use in vitro-produced plantlets.

Nematode populations to be used for screening. Both nematodes (*Pratylenchus goodeyi* and *Radopholus similis*) reproduce well on carrot disk, and this method is suggested to increase their nematode populations.

Jars of 300 mL, containing Murashige and Skoog (MS) rooting medium, in which plantlets are transferred.

2.3 Screening for Resistance to the False Root-Knot Nematode, Nacobbus aberrans

As the nematode infects both roots and tubers, the use of potato tubers may not be suggested in the screening. Instead, seeds, plants from in vitro culture, or rooted cuttings are suggested.

Therefore, have available seedlings of potato cvs. or other plants to screen and some cvs. known to be susceptible and resistant to the nematode. For potato, see also Chap. 3 on screening for resistance to *Globodera* spp.

Nematode populations to be used for screening. Pots of the desired size.

2.4 Screening for Resistance to Ditylenchus spp.

2.4.1 D. dipsaci

All these species prefer rather low temperature (15–17 °C) and high relative humidity caused by fog, dew, rain, and sprinkler irrigation to infect and reproduce. *Ditylenchus angustus*, instead, is a warm climate nematode (28–30 °C) and damages rice in eastern Asian countries. All the mentioned species of *Ditylenchus* can be reared on carrot disks and on callus of the host plants or obtained directly from fresh or dried plant materials.

The required material for screening resistance to *D. dipsaci* consists of the nematode populations to be used for screening. Realize that populations infecting white clover or red clover may not damage alfalfa and vice versa.

Seeds of the plant species to screen and, whenever available, also seeds of a very susceptible cultivar, a very resistant cultivar, and a partially resistant one. Of strawberry, use cool-stored plants or, better, rooted stolons from mother plants.

Preparations of rhizobium nitrogen-fixing bacteria, specific of the legumes to test [5].

Petri dishes or trays for germinating seeds.

Containers for plants, such as tubes, clay or plastic pots, or Styrofoam polypots, according to the plant to screen.

Carboxymethyl cellulose and a very soluble fertilizer.

Sterilized sandy soil, small plastic sheets, or plastic cages for covering plants, useful in greenhouse to limit evaporation and maintain high air humidity soon after nematode inoculation.

2.4.2 Screening for Resistance of Broad Bean to D. gigas

Required material includes seeds of broad bean to assess seeds of a very susceptible and a very resistant cultivar/line to serve as resistant and susceptible controls. The "INRA 29H" line is resistant, except for some Moroccan populations of the nematode.

Trays to germinate seeds.

Populations of the nematode from your geographical area of interest.

Carboxymethyl cellulose.

Laboratory tubes 25 cm long and 2.5 cm in diameter; or small pots/polypot 4 × 4 cm wide. Other materials needed are perlite or vermiculite, soil mix, and growth cabinet.

2.4.3 Screening Potatoes and Sweet Potatoes for Resistance to D. destructor

The material you need for screening is tubers of the cultivars/lines/seeds to screen. Include also tubers of a very susceptible cultivar, such as Désirée, and of a resistant one to serve as controls. If the planting material available is not sufficient to obtain six to ten replicate pots per entry, multiply it.

2.4.4 Screening of Peanuts for Resistance to D. africanus

Seeds of genotypes to be evaluated, microplots, root nodulation rhizobium, fungicide (seed treatment), in vitro *D. africanus* cultures.

2.4.5 Screening of Rice for Resistance to Ditylenchus angustus

The material you need includes seeds of rice genotypes to evaluate.

Seeds of cvs. or line known to be resistant, tolerant, and susceptible to the nematode, to serve as controls.

Sterilized soil typical of rice crop; pots; fertilizer to use in large pots or Hoagland solution for small tubes.

2.5 Screening of Rice for Resistance to Aphelenchoides besseyi

Before screening for resistance to *Aphelenchoides* spp., be sure that the species of your nematode population has been identified correctly. Then have available seeds of the test varieties, of a known susceptible, resistant, and tolerant variety.

Equipment to sterilize seeds in hot water.

Sterilized potting soil containing at least 70% sand.

Population(s) of the nematode to be used for screening that can be obtained from the field or reared on fungi (*Botrytis cinerea* or *Alternaria* spp.) or on carrot disks.

Clay pots of 20–25 cm diam.

Fertilizer, such as a soluble 20–20-20 NPK plus microelements or a similar one.

2.6 Screening of Wheat for Resistance to *Anguina tritici*

The required material is seeds of the genotypes to test nematode populations.

The best way to store and obtain nematodes for the test is from infected galled brown seed, crushed, and put in water in Petri dishes.

After a few hours, thousands of second-stage nematodes will be actively swimming and can be used for inoculation purposes.

2.7 Screening for Resistance to *Rotylenchulus reniformis*

Seeds of the test genotypes.

Seeds of cvs. known to be susceptible and resistant to be included in the test for comparison.

A commercial preparation of the root nodulation *Rhizobium* in case of legume plants, such as soybean and different beans.

Sterilized sandy soil.

2.8 Screening of Citrus for Resistance to *Tylenchulus semipenetrans*

Have available seeds or seedlings of the genotype to evaluate, at least a susceptible and a resistant control, free from nematodes.

Populations of the nematode of which is known the biotype.

Sterilized sandy soil for easier washing of the roots before extracting the nematodes at the end of the screening test (clay soil will stick to the roots, especially to the egg mass matrix of the nematode and thus will make the nematode suspension dirty and more difficult to count under the stereoscope).

Adequate pot size, about 1–3 dm^3 volume.

2.9 Assessment of Plant Tolerance to Nematodes

High inoculum levels (at least three), plus non-inoculated plants, and large pots (in greenhouse) or microplots are suggested to assess the degree of tolerance at different nematode pressures.

3 Methods

3.1 Screening for Resistance to Root-Lesion Nematodes (*Pratylenchus* spp., *Radopholus similis*)

Root-lesion nematodes, *Pratylenchus* spp. and *Radopholus similis*, are migratory endoparasites that invade and damage cortical parenchymal tissues in which they cause cavities that externally appear as brownish necrotic areas on roots. The necrotic areas can be small in light infections and cover most of the root system in severe infection (Fig. 2c, d). The genus *Pratylenchus* [9] includes about 70 species, each attacking a large number of cultivated and wild plant species. Both annuals and perennials are hosts (Fig. 2a, b) and several are staple crops (cereals, soybean and other legumes, and potatoes). The most economically important species are *P. vulnus*, which attacks mostly perennials; *P. neglectus* and *P. thornei* damaging mainly cereals and legumes; *P. penetrans* infecting potato and many other crop plants mostly in temperate regions and date palm in the Algerian oasis; *P. brachyurus*, widespread throughout the tropics on a large number of different crop plants, including legumes; *P. coffeae* infecting coffee in all major cropping areas,

Fig. 2 Effect of root-lesion nematodes, *Pratylenchus* spp., on different crop plants. (**a**) poor growth of wheat in a field infested by *P. neglectus*. (**b**) a patch of poorly growing potato in a field, in Norway, infested by *P. penetrans*. (From Ref. [89]. (**c**, **d**) brownishments caused by *Pratylenchus vulnus* on roots of olive and peach, respectively)

banana and plantain, and in the USA on citrus; *P. goodeyi* mainly on banana and plantain in Eastern African and Asian countries; *P. loosi*, a severe parasite of tea; *P. scribneri*, a parasite of many host plants especially in the USA; and *P. zeae*, noxious to corn, sorghum, and sugarcane. Sometimes difference in virulence between populations of the same species has been reported [10, 11].

Within the genus *Radopholus*, the important species is *R. similis*, noxious to citrus and banana (citrus race) or banana but not citrus (banana race).

3.2 Screenings for Resistance of Annual Plants to Pratylenchus spp.

1. If wild species are used, such as *Cicer* spp., as they may take longer than *C. arietinum* to germinate, cut seeds at the end of endosperm to favor seed imbibition and germination.
2. In case of spring wheat, for vernalization, before germination, store seeds at 4 °C for 3–4 months.
3. For germination, put seeds 2–3 cm deep in small trays containing nursery soils or prepare a soft mixture of sand (50%),

vermiculite (30%), and peat (20%), wet soil, and keep trays at 20–25 °C; use higher temperature for seeds of summer crop plants (soybean, French bean) and lower temperature for cool season crop plant seeds (chickpea, broad bean).

4. Three to four days later, transfer germinated seeds in the final container half filled with potting soil; add the specific root nodulation bacterium, in case of legumes; and add other soil to fill plant container up to 2 cm of the ridge. Have extra plots inoculated and sown with the susceptible control and others inoculated and not sown but treated as all planted pots.
5. Five to eight days later, when plants are 5–8 cm tall and you feel that roots are exploring the soil (check extra plants), inoculate pots by pouring in three to four holes, 2–3 cm deep, the water–nematode suspension to have the desired inoculum per g or cm^3 soil. With time, the experience will suggest you which is the best amount of inoculum to use, to have at evaluation time 40–60% of the roots necrotized in the control plants.
6. Arrange the plant container according to an experimental design, such as a randomized block design, in a greenhouse at 18–22 °C for cool season nematodes and 24–27 °C for warm tropical season nematode species.
7. Arrange containers singly if rather large (10–12 cm diam. pots) or if small (containing 300–400 cm^3 soil) better sunk in the sand on a large bench, equipped with automatic heating and irrigation systems [8].
8. Irrigate as needed (if no automatic system is available) and fertilize every 10–15 days with a solution at 2‰ of a 20-20-20 NPK + microelement fertilizer. Fertilizer can be applied trough the automatic irrigation system or manually plant by plant pouring the same amount of solution per pot.
9. Soon after inoculation, extract nematodes from nonplanted pots to determine the proportion of the nematode recovered by the extraction method you have chosen. Thereafter, every other week and for 2–3 months, extract nematodes from other inoculated nonplanted pots, to have a measure of the nematode decline in the absence of host plants.
10. Three to four months after inoculation of cool season crop plant–nematode combination and 2–3 months for warm season crop plants and nematodes, check a few control plants for the extent of root symptoms; if they are extended enough (to 40–60% of the root system) (Fig. 3), do the final evaluation.
11. Therefore, stop irrigation and when soil is not much wetted, uproot plants in a tray; cut plants at soil level; free the roots from adhering soil, first by shaking and then by gently washing under a tap water stream; and extract and count nematodes.

Fig. 3 Root of faba bean showing severe necrosis by *Pratylenchus* spp. Genotypes showing so extensive root necrosis are ready to be evaluated

12. Estimate the severity of the symptoms expressed as % of root necrotized. Root necrosis is a typical symptom of *Pratylenchus* infection if no other pests and diseases occur as it is supposed when using sterile soil under good greenhouse conditions and if the screening has been conducted correctly. This author, in a large experiment at ICARDA main station (Aleppo, Syria), including about 300 plots, at mid-crop cycle of chickpea (early May) was able to identify plants from plots treated with the nematicide–insecticide aldicarb, applied at the rate of 5 kg a.i./ha at planting, just looking at the root; plants from treated plots had roots rather clear, indicating little or no *Pratylenchus thornei* damage, while those from nontreated plots were largely necrotized.
13. Weigh top plants, wash and dry roots between two sheets of laboratory paper, cut them in 0.5–1 cm long pieces, extract nematodes using Young's incubation method, Baermann's funnels, the mistifier chamber (all requires several days), or Coolen's centrifugation method (will take about 1/2 h for a set of four to six samples). If roots are too much, just take a subsample of 5 or 10 g. Then count nematode specimens in the water suspension.
14. Moreover, mix the soil thoroughly, take a subsample of 100–200 g or cm^3, and extract nematodes using one of the methods available to you. Count nematodes, refer to total nematodes in the soil of each pot, and add this number to nematodes per root system to obtain total nematodes per genotype.

15. Now calculate Rf dividing final nematodes per genotype (Pf) by the number of nematodes inoculated per pot (Pi). Also, determine the reproduction rate (Rr), also called reproduction index (RI), as a % of the Pf of each genotype compared to Pf of the susceptible control.

$$\text{Rr (RI)} = \text{Pi genotype}/\text{Pi susceptible control} \times 100$$

16. For the evaluation of the reaction of genotype to the nematode, independent investigators use different rating scales. In the USA, Thompson et al. [12] rated root of wheat at harvest on a 1–5 scale as follows: 1 = little to no root browning of all the root system; 2 = approximately one fourth of the root system with browning; 3 = approximately half of the root system browning; 4 = approximately three fourths of the root system with browning; and 5 = the entire root system with browning. This author believes that, especially for legume, as a preliminary screening of a large numbers of plants, rating on the base of the extent of root browning (necrotic) is reliable and speedy. However, at the final stage of a breeding program, additional information on Pf, Rf, Rr, nematodes per g of roots, plant weight (roots and top), and comparisons with well-known susceptible and resistant plants, if there are, are useful for a better characterization of a genotypes, as in the case of those to be released as new cultivars. For this, in Australia, Thomson et al. [7] assess resistance/susceptibility according to final number of nematodes/kg soil (including roots). Then they perform statistical analysis combining the results of a large number of experiments conducted in previous years by a factor analytic linear mixed model. The results are expressed as an empirical best linear unbiased predictor (eBLUP) of combined Pf in units of $\log_e$ (*Pratyle*nchus/kg soil), based on the scores for the significant principal components. This eBLUP is a quantitative variable that is used to compare the relative resistance/susceptibility/kg soil. The eBLUP range of combined $\log_e$ (Pf) is then converted into nine subranges to classify each crop cultivar into the ordinal alpha scale required by the Grains Research and Development Corporation (GRDC), to compile a guide to farmers for extension purposes, to choose the most useful cultivar. This scale is R = resistant, R-MR = resistant to moderately resistant, MR = moderately resistant, MR-MS = moderately resistant to moderately susceptible, MS = moderately susceptible, MS-S = moderately susceptible to susceptible, S = susceptible, S-VS = susceptible to very susceptible, and VS = very susceptible [13].

17. Finally, if the resistant plants are to be saved for further use, it is suggested one of the following two options: (*i*) evaluate the plants based on the extent of necrosis on the roots externally of the root ball, transplant plant rated resistant in larger container with non-infested soil, and leave them to produce seeds and (*ii*) for sowing seeds, use peat container (Jiffy pots), as explained for screening chickpea to *Heterodera ciceri* (*see* Chap. 3), and do the evaluation of the roots that have crossed the Jiffy pots to grow in the infested soil in the larger pot. Transplant resistant plants in larger pots filled with non-infested soil.
18. Potatoes are also attacked by *Pratylenchus* spp. *P. penetrans* (Fig. 2b) is reported to cause severe damages to potato in Europe, North America, and Australia. So far, no potato cultivar resistant to *Pratylenchus* spp. has been identified and, therefore, no specific protocol has been designed to screen potato for resistance to *Pratylenchus* spp. Should it be necessary, the protocols presented for other nematodes can be adapted to *Pratylenchus* spp. However, Figueiredo et al. [14] demonstrated that differences do exist in the reaction of potatoes to *P. penetrans*, and the protocol used by these authors can be adapted to specific needs (*see* **Note 1**).

3.2.1 Screening for Resistance of Perennials to Pratylenchus *spp.*

Root-lesion nematodes attack also trees, vines, and bushy plants. Resistance has been identified in wild relatives that are now being used as rootstocks, but more researches are being conducted to increase resistance, especially in stone fruit rootstocks, grapes, sugarcanes, and bananas. In general, the procedure is similar to that described for screening against root-knot nematodes, *Meloidogyne* species, except that for the inoculation and extraction of the nematodes and the symptoms to evaluate. However, hereafter are suggestions to screen *Prunus* spp. rootstocks toward *P. vulnus* and bananas for resistance to *Pratylenchus coffeae*, *P. goodeyi*, and *Radopholus similis*, all nematodes of warm climates.

3.2.2 Screening for Resistance of Prunus *spp. Rootstocks to* Pratylenchus vulnus

1. Treat seeds of the genotypes in a 5% oxychloride copper solution for 24 h and then rinse in running tap water to eliminate the fungicide, dry seeds between paper towels, and sow in a tray containing perlite.
2. Keep tray at 4 °C for 45 days and then transfer to room temperature for germination and irrigate whenever needed.
3. If you have a supply of plants derived from in vitro cultures, transplant them in the small pot and leave them to harden avoiding direct sunlight.
4. If starting from cuttings, prepare cuttings of about 20 cm long, deep them in a hydroalcoholic solution containing 2000 ppm of indole butyric acid (IBA) for 6–10 s and then move in small

pots (200 cm^3) containing a pasteurized rooting soil mix of sand and peat (3:1 v/v), and maintain these pots at 18–22 °C in a well-illuminated (avoid direct sunlight) and moistened area (pots can be covered with a transparent cage of plastic, glass, or plexiglass, and check for humidity). Instead of the hydroalcoholic solution, the bottom end of the cuttings can be wetted and dipped in a powder containing IBA that can be found in a gardener shop.

5. When 20–30 cm tall, transplant plants in the final pots, containing 1–2 kg sterile soil (about 80% sand and 20% field soil) and have six to ten replicated plants per genotype and some more extra control plants to follow the development of plants and nematode symptoms.
6. Arrange the pots on benches in a greenhouse at 24–26 °C according to a completely randomized design or a randomized block design.
7. Four week later, extract nematodes from the carrot disk cultures, dilute in water to have 500–1000 specimens (all developmental stages) per ml suspension, and inoculate each pot with about two nematodes/cm^3 soil, by pouring the suspension in four holes 3–4 cm deep, close the hole with soil, and add some water on the hole to favor adhesion of the soil to nematodes.
8. Thereafter, irrigate as required (avoid water leaking from the pots) and fertilize every 15–20 days with a 2 g fertilizer/L solution.
9. Four to six months after inoculation, check root and symptom development (Fig. 2d) for the right time to evaluate plant reaction (at least 50% of the roots necrotized in the susceptible control), uproot plants in a tray, cut the top plant part, free roots of the adhering soil first by hand in the tray and then under running tap water, and eliminate excess water in between two sheets of laboratory paper.
10. Then rate roots for presence and extension of necrosis/brownishments using an arbitrary scale, such as a 0–10 scale according to % of roots necrotized.
11. Weigh roots; cut them in 0.5–1 cm long pieces; mix; take a subsample of 5–10 g and extract nematodes using one of the methods available to you, preferably the maceration–centrifugation method of Coolen [15]; count nematodes (all stages); and refer to the total amount of roots per pot.
12. Mix thoroughly the soil in the tray; take a subsample of 200–500 cm^3 soil, according to the extraction method you are familiar with; extract and count nematodes; and refer to the total amount of soil per pot. Then add nematodes in the soil to those in the roots per pot to obtain the total final nematode density per pot (Pf).

13. Calculate Rf and Rr and analyze statistically these variables and % necrotic roots.
14. Generally, genotypes are classified as host or nonhost as follows: NH (nonhost) = Pf < Pi; PH (poor host) = Pi < Pf <1.5 times Pi; and H (host) = Pf > 1.5 times Pi per plant. Based on the degree of resistance, plants can be classified as highly resistant (HR) when no nematodes are found in both soil and roots; resistant (R) = Rf > 1 and a few nematodes in the roots; moderately resistant (MR) = Rf slightly more than 1 and several nematodes in the soil and roots; moderately susceptible MS = Rf, nematodes in soil and roots, significantly more than resistant control and less than in susceptible control; and susceptible (S) = Rf, and nematodes in soil and roots nonsignificantly different from susceptible control.
15. More details can be found in Marul and Pinochet [16], Stalin et al. [17], and Nyczepir and Pinochet [18].

    To screen sugarcane to *P. zeae*, Bhuyan et al. [19] followed the same procedure for screening against *Meloidogyne javanica* (Chap. 3). An interesting procedure was followed by Dos Santos et al. [20] to screen *Vitis* spp. hybrids for resistance to *P. brachyurus*.

3.2.3 Screening for Resistance of Bananas to Pratylenchus spp. and Radopholus similis

The following protocol is suggested by Elsen et al. [21]:

1. When shoots are 1 cm long, transfer two in each jar for rooting and have eight replicates per genotype arranged according to an experimental design.
2. Transfer jars for 4 weeks at 27 °C.
3. Extract nematodes from the cultures and inoculate 50 females per jar, collected with a sterile micropipette, near the stems of the banana plantlets. Root symptoms (red–brownish roots) appear 3 weeks after inoculation in 80% of the plants and in 100% of them after 7 weeks. Evaluating the plants 8 week after inoculation is estimated as the best time. Longer incubation period is not suggested as most of the nematodes leave the roots and move into the medium.
4. Extract nematodes separately from both medium and roots by the maceration and sieving technique.
5. However, I'm of the opinion that that maceration sieving followed by centrifugation will result in a clearer nematode suspension.
6. Count nematodes per root and medium and calculate the total number (if possible, enumerating them per developmental stage to have additional information) per jar.
7. If the plant reaction to the nematode will be based on the nematode reproduction, inoculation of 25 nematodes per jar

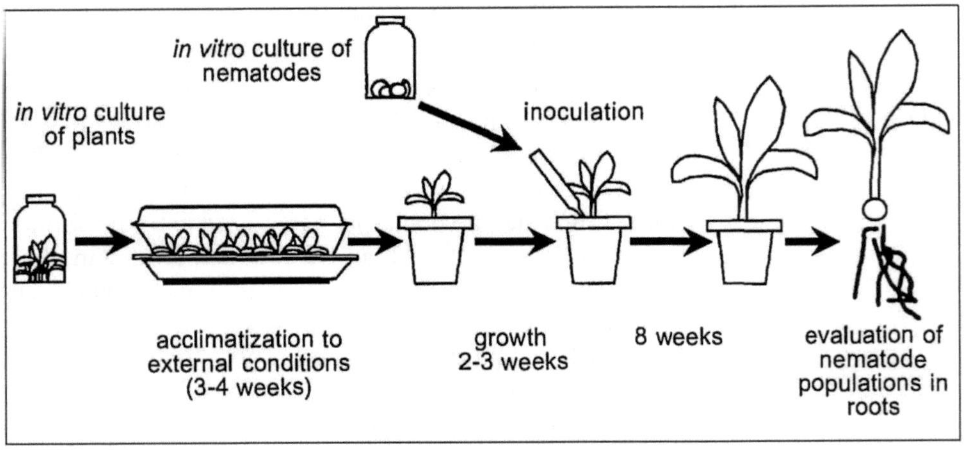

Fig. 4 Scheme for screening for resistance to nematodes of bananas starting from plantlets obtained by in vitro culture. (From Ref. [23])

is the best density. However, the author does not state if and how symptoms were estimated. If the evaluation will be based on extension of root necrosis, 50 nematodes per jar appear as the optimal inoculum. To do this, it is suggested to evaluate root brownish, as a % of root necrotic/brownish on total root size. Also, the authors do not state how they rated cultivars as resistant or susceptible. For it, this author suggests to calculate the reproduction factor (Rf) and the reproduction rate (Rr) as a % of the nematodes on the tested cultivars compared to those on the susceptible cultivar. Then, perform the statistical analysis of the data and based on it consider a cultivar susceptible if these variables are not statistically different from those of the susceptible control, resistant if not different from the resistant control, and moderately resistant/susceptible if different from both the susceptible and resistant controls.

This protocol is a very good one to screen a large number of genotypes as a preliminary screening. However, some discrepancy may occur in the evaluation of the plant reaction using in vitro tests.

At the International Institute of Tropical Agriculture (IITA), in Kampala, Uganda, Dochez et al. [22], to assess genotypes at a later stage of the breeding program, revised the previous protocol and used plants reproduced in vitro, transplanted in pots of 10 dm^3. Then hardened plants in greenhouse at 28–30 °C or screenhouse and later inoculated with nematodes reared on carrot disk. *See* also scheme in Fig. 4 by Sarah [23]. A few researchers prefer to use sword sucker-derived plants, pared of the infected roots, treat in water at 50–55 °C for 20 min to free from other pests, and then transplant in rather large pots (10 dm^3 volume) before being inoculated. Others prefer to transplant in vitro-derived plantlets in 1 dm^3 volume pot [24]. In summary, the procedure is as follows.

As populations of *R. similis* may show differences in reproduction and virulence, it is suggested to use in the screening more populations.

1. Banana plants are transplanted in 10 dm^3 (30 cm diam) pots containing sterile soil and inoculated with 1000 nematode specimens in four holes around the plants. Pots are arranged on benches according to a completely randomized design. Plants are irrigated as needed and fertilized with a compound fertilizer.
2. The test with *R. similis* can be concluded 8 weeks after inoculation of in vitro-cultured plantlets or 12 weeks after inoculation of sword sucker-derived plants. Longer time may be required in case of screening to *Pratylenchus* spp.
3. Uproot plants, cut roots, wash free them of the adhering soil, dry between two sheets of laboratory paper, and record weight.
4. To estimate percentage of root necrosis, select randomly five functional roots per plant, from each root select a segment of 10 cm, cut each lengthwise, and estimate the necrotic visible cortical tissue of the five halves. Each half may have a maximum of 20% of root necrosis so that as a maximum, the five halves will result in a maximum of 100% necrosis.
5. To determine nematode population, cut the five root segments in 0.5 cm long pieces, mix thoroughly, take a subsample of 5 g, and from these, extract nematodes (use Young's incubation method, Baermann's funnels, or maceration–centrifugation method), count nematodes in the water suspension, and refer to 100 g fresh roots. Also, extract nematodes from a soil subsample, using a method easily available to you, count, and, finally, determine total population density in the pot (both soil and roots). You can also calculate Rf and Rr (RI).
6. Do the statistical analysis of the data of nematode populations, fresh root weights, and % root necrosis. Based on the significance of the observed differences, consider a cultivar susceptible if not different from the susceptible control and resistant if not different from resistant control. The remaining genotypes must be classified as having intermediate reaction to the nematodes tested (*see* **Note 2**).

3.3 Screening for Resistance to the False Root-Knot Nematode, Nacobbus aberrans

Known as the false root-knot nematode, to differentiate it from the species of the genus *Meloidogyne*, *Nacobbus* spp. are rather common in the Americas and especially in the Andean countries, where they attack many crop plants, especially potato (Fig. 5). Populations of the nematodes seem to belong to different species and races/virulence groups. For instance, Peruvian and Bolivian populations attack potato and are believed to belong to *N. bolivianus*, while of the Argentinian populations, some attack potatoes and should be of

Fig. 5 A potato root severely galled by the false root-knot nematode, *Nacobbus aberrans*, in Bolivia. (Courtesy of J. Franco, PROIMPA, Cochabamba, Bolivia)

N. bolivianus and others not. The nematode develops from 0 to 4000 masl. Different population may show difference in optimal temperature for development [25]. Recently, Lax et al. [26], based on an integrative taxonomic approach of more than 50 nematode populations from all over America, found that within the *Nacobbus aberrans* species complex are present at least three new species, including the new species *N. celatus* (common in lowland Argentina), *N. aberrans* s.s., and *N. bolivianus*. Therefore, the observed differences within the *N. aberrans* species complex may be due to different species and then the need for precise identification to species level of the nematode population to use in the screening.

Plant seedlings of the host crop plant species in pots containing a mixture of rather sandy soil with vermiculite, about 100 cm^3 potting soil, or a rather sandy soil in 0.6 dm^3 pots and arrange them on greenhouse benches or in a growing chamber at about 25 °C and 14–16 h daylight. Have six to ten replicates per genotype and two to three times more extra inoculated pots. Of the extra pots, about half should be planted and the other be left to fallow.

1. Plant seedlings of the host crop plant species in pots containing a mixture of rather sandy soil with vermiculite, about 100 cm^3 potting soil, or a rather sandy soil in 0.6 dm^3 pots and arrange them on greenhouse benches or in a growing chamber at about 25 °C and 14–16 h daylight. Have six to ten replicates per genotype and two to three times more extra inoculated pots. Of the extra pots, about half should be planted and the other be left to fallow.
2. When roots are developed enough, about 10–20 days after planting, depending on planting materials, inoculate plants with two to four nematode specimens per cm^3 soil.

3. To obtain inoculum, extract nematode eggs from infected roots of a host plant macerated in a blender containing a 0.5% sodium hypochlorite solution [27] and then washing away the excess hypochlorite (higher concentration may reduce egg viability). Alternatively, J2 can be extracted from the infested soil using Baermann's funnel method.
4. Thereafter, irrigate as needed and fertilize every 7–10 days, so rather frequently when using soil with poor fertility, and keep pests under control.
5. Every 2 weeks, extract nematodes from a replicate fallow plot to monitor the decline of the nematode during the course of the experiment and to assess how much of the inoculated nematodes survived until harvest.
6. Also, when approaching harvest time, uproot a few control plants and check how the development of the nematodes proceed, for a better timing of the final evaluation. Of the same pot, check also the development of the root apparatus to avoid it from becoming pot bound.
7. About 90 days after inoculation, or differently if it is the case, cut top plant at soil level, uproot the plants, and wash roots gently under running tap water to become free of adhering soil. Weigh plant roots and, if you wish, also top plants (not really necessary). Of the roots, evaluate RGI (see root-knot nematodes for scale), then cut roots in 1–2 cm long pieces, mix them gently in water, dry them between sheaths of laboratory cleaning paper, and take a sample of at least 5 g if possible or the entire root if less.
8. If pots cannot be processed soon, store in a refrigerator or in a cool room for a few days until use. Extract eggs from the roots as explained before for inoculation, but now use the hypochlorite concentration at 1% to better dissolve the egg masses of the nematode. Realize that the use of the hypochlorite method alone is not sufficient because most of the nematode egg masses are embedded into the roots. Count eggs and refer them to total roots per pot.
9. Later, mix the soil and take a sample or all per pot, extract nematodes with Baermann's funnels or other methods, count them, and refer to all soil per pot. Then calculate final nematodes per pot (Pf) by summing total nematode per root to total nematode in the soil and deducting nematodes that remained in the fallow pots.
10. Finally, determine the reproduction factor according to the formula Rf = Pf/Pi and the reproduction index (RI) as the % of total nematodes in the pots of each test genotype compared to the susceptible control cultivar.

11. Based on the available variables, rate the genotype as resistant if RF <1, RI ≥10%, and EMS and RGI ≥2. Statistical analysis of the variable can be performed and genotypes considered different from resistant control if they differ significantly from them [26, 28, 29].
12. If screening potato and tomato, to save and reproduce resistant plants, of the top plant part, prepare cuttings and root them as explained for potato in the screening to *Globodera* spp. in the preceding chapter (Chap. 3). For rooting of tomato cuttings, it is enough to leave them in beakers with tap water, under room conditions for 10–15 days, and then transplant rooted cuttings in 10 cm diam. pots to acclimatize and then in a larger pot or under field conditions for further uses (*see* **Note 3**).

3.4 Screening for Resistance to Ditylenchus spp.

A few species are destructive to many crop plants. They are endoparasitic migratory nematodes of aerial parts, tubers, bulbs, and taproots. The most important species is *D. dipsaci* [30], which is very damaging to alfalfa, clovers, onion, garlic, sugar beet, carrot, strawberry, and many other crop plants (Fig. 6); it reproduces also on weeds. *D. gigas* is a severe nematode of broad bean, *D. destructor* of potato and sweet potato in Eastern Europe and Asian countries, and *D. africanus* of peanut in South Africa.

Fig. 6 Damages caused by parasitization of *Ditylenchus dipsaci*. (**a**) on alfalfa crop (From A, Steele with kind permission of SIN); (**b**) on carrots; (**c**) on onion crop; (**d**) a healthy onion plant (left) in comparison with damaged plants (right)

3.4.1 Screening for Resistance to *Ditylenchus dipsaci*

Different populations of this nematode may attack different host plants and are believed to belong to about 30 different races, each characterized by having different host rages, but several sharing more plant species as hosts. However, rather than identifying the nematode race, it is necessary to be sure that the nematode population toward which to screen attacks the test plant species. This can be done, first of all, by observing the symptoms caused by the nematodes, such as swellings, distortions, necrosis of leaves, and plant crown rotting, especially of carrot and sugar beet, and confirming the infection of the nematode by extracting nematode specimens from the damaged plant tissues. Important crop plants damaged are clovers, alfalfa, garlic, onion and other Alliaceae, bulbous flowers, broad bean, peas, spinach, oats, strawberry, and corn. However, there is evidence that in some species or cultivars of the same species, there are differences in plant susceptibility to the nematode in the same field and that, therefore, selection of the most resistant plant can be useful to control the nematode. This already occurs for alfalfa and clovers. Screenings and breeding programs are being made, the progenies of which must be evaluated for their reaction to the nematode.

Hereafter are reported the most used screening procedures, especially whenever screening a large number of plants. The reported procedures, with some modifications, can be adapted also to other plant species, including those producing taproots, such as sugar beet and carrots. However, to observe rotting at the crown level in sugar beet and carrot, longer observation time (2–-3 months) is necessary.

1. Germinate seeds of the test plant species in Petri dishes containing a disk of wetted filter paper, in the dark, in a growth cabinet at 20–23 °C.
2. After germination (3–5 days post incubation), plant germinated seeds in 1–2 cm deep holes in a container filled with sterilized rather than sandy soil containing no more than 15–20% of peat or similar material.
3. Fertilize the pots with a 2‰ solution of a composite fertilizer containing also microelements and add to the suspension the specific root nodulation bacterium if screening legume plants.
4. Set growth cabinet or greenhouse temperature at 12 °C and light at 14–16 h per day.
5. When the first two small true leaves emerge, prepare the nematode suspension and concentrate it to contain about 30–40 nematodes in a drop of the suspension and check under a stereoscope that there is approximately this number.
6. Add to the nematode suspension 1–2% of carboxymethyl cellulose, mix well, and with a hypodermic syringe or a pipette, pour a drop exactly on the meristem between the two true leaves and

cover the plants with plastic or glass covering. In case of strawberry, inoculate plants by depositing on the central sprout one to two drops of the carboxymethyl cellulose water suspension containing 50–100 nematodes, when plants start to growth after transplanting. Be sure that plants are free from nematodes.

7. Remove the covering after 3 days and increase temperature to 15–18 °C 1 week later.
8. Two weeks after inoculation, symptoms of the nematode infection, such as swellings at the plant crown level, will start to appear.
9. Two weeks later, score the plants for presence/absence of symptoms (Fig. 7). Check again for two to four consecutive weeks and save plants still symptomless for further studies if necessary.
10. To confirm plant susceptibility; cut at soil level all plants showing symptoms; weigh them; cut finely all the above plant part, including the crown; and extract and count nematodes.

For more details, consult also Williams [32], Whithehead et al. [33], Cook and Evans [34], and Storelli et al. [35] and *see* Fig. 8 (sugar beet seedling) and Fig. 9 (carrot seedlings).

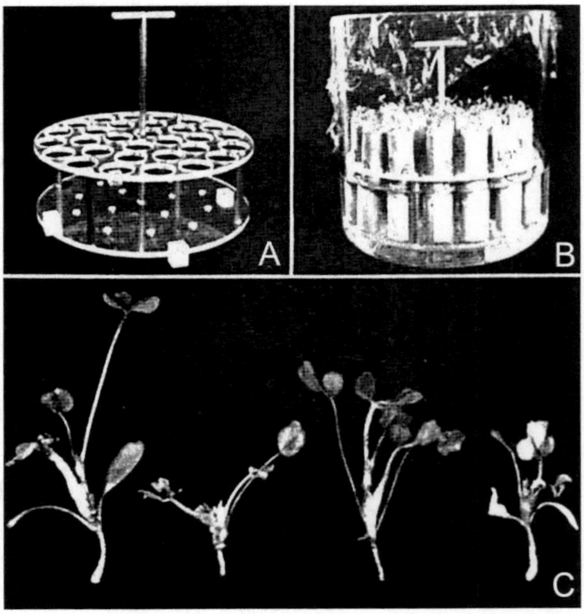

Fig. 7 Materials and equipment used to screen alfalfa for resistance to the stem and bulb nematode, *Ditylenchus dipsaci*. (**a**) Acrylic plastic rack designed to hold rolls of seedlings in inoculation tests; (**b**) Different rolls for inoculation tests, similar to those of Dijkstra [31]; (**c**) seedlings of alfalfa showing severe symptoms caused by the nematode parasitization. (From Ref. [93], copyright Springer Nature)

Fig. 8 *Ditylenchus dipsaci* infected sugar beet seedling at 14 days post-inoculation showing swollen leaf axil, a typical symptom of the nematode. (From Ref. [35] copyright Springer Nature)

Fig. 9 Seedlings of carrots showing swellings at the crown level in soil infested by *Ditylenchus dipsaci*

In the Netherlands, a mini test to screen thousands of seeds of clover and alfalfa was designed [31]. The following is the procedure:

1. Seeds of clovers are allocated in a tray for germination. The germinated seeds are laid out on a 45 cm long and 11 cm wide strip of moist filter paper. The filter is a special one that does not tear after moistening, such as a chromatographic filter.

2. On the special filter, in such a way to cover the seeds, is laid a strip of moistened normal filter paper and all rolled up. To favor in making the rolls, roll the paper around a rod about 1 cm in diameter.
3. The rolls are allocated together in trays without water and covered with a polythene plastic or similar material (Fig. 7b) and put in growing cabinet at 15–18 °C and the light is turned on for about 14–16 h per day.
4. A few days later, when the cotyledons have emerged from the seed coat, loosen the upper margin of the filter, paying attention that the special filter paper sheet remains intact on top.
5. Then rotate the rolls a little in an acute angle so that the seedlings lay on the soft filter paper.
6. Now, using a hypodermic syringe or a calibrated pipette, deposit on the cotyledons a drop of nematode suspension containing 20–30 actively moving specimens. Deposit three to four drops separately under a stereoscope at about 25× magnification to count average nematodes per drop.
7. Remove the cover 2 days after inoculation and place the rolls in a nutrient solution. A couple of days later, the first plant will start to show symptoms of the nematode infection as swellings (Fig. 10).
8. Two weeks later, the plants can be assessed for nematode reaction. Nonsymptomatic seedlings can be transplanted in larger pots (15–20 cm in diameter) and grown to produce seeds. With this technique, in a tray of 25 cm in diameter, more than 3000 seeds can be allocated.

To screen onion, a similar method was suggested with some modifications or adaptations as follow (Seinhorst, personal communications, Wageningen, the Netherlands, 1981).

1. Germinate seeds of onion in a Petri dish containing a wetted disk of a special filter paper so that it does not tear easily after wetting (Fig. 11a). Ordinary filter is not suited because it may tear easily.
2. Prepare slabs of a porous material (oasis) and strips of filter paper of the size that can accommodate in a tray. After germination, wet a slab of a porous material, and put on top of it a sheet of wetted filter paper and on this the seeds on the margin of the top smaller side. Then put on the seed another sheet of wetted filter paper and a slab of wetted oasis at a distance from the top just to cover the roots. Then continue with a sheet of filter, laying seeds on it, another slab of oasis, and so on (Fig. 11b). When a stack of sheets of filter paper and slab of oasis has been obtained, this is placed in a suitable tray and the tray is placed in a growth cabinet at 10–12 °C and the light is turned on for about 14 h per day.

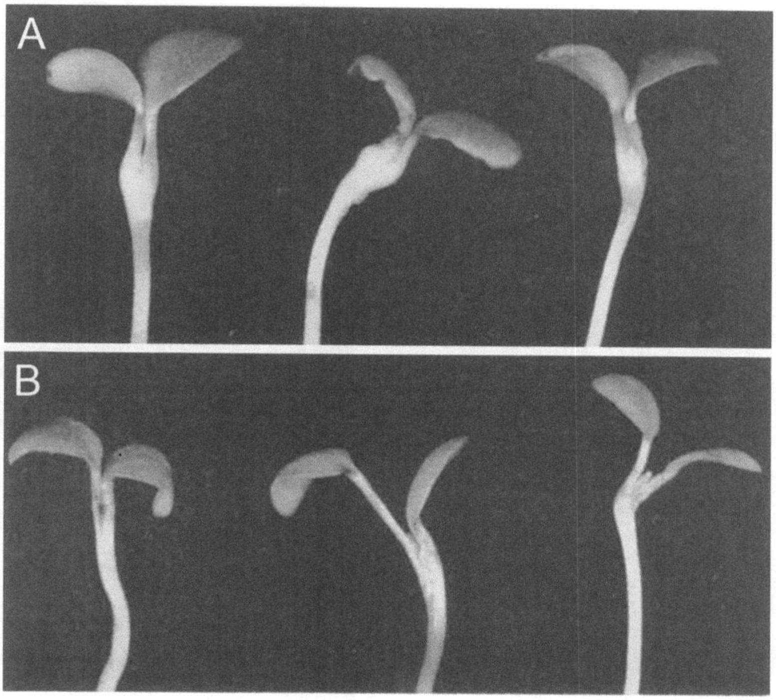

Fig. 10 Seedlings of red clover, 10 days old inoculated with the stem and bulb nematodes, *Ditylenchus dipsaci*. (**a**) susceptible seedlings showing abnormal swellings and distortions at the growing point, caused by the nematode parasitization; (**b**) resistant seedlings showing no symptoms. (From Ref. [31] copyright Springer Nature)

3. Then fine quartz sand is placed in between every two sheets of filter paper just on the seeds.
4. A suspension of nematodes is then dripped on the sand to have 20 nematode specimens per seedling.
5. From now on, when filter paper and slabs of oasis become dry, a fertilizing solution can be added in the tray.
6. At this temperature, spraying of fungicide once a week will prevent the development of fungi.
7. Three weeks after inoculation, symptoms of nematode infection (basal swellings of the seedling) will appear in about 95–98% of the seedlings (Fig. 11c). However, not 100% of the seedlings will show symptoms.
8. As no resistance gene to *D. dipsaci* is known in onion, if during the screening a few seedlings are symptomless, even after a further inoculation after transplanting, this does not ensure that resistance has been identified (*see* **Note 4**).

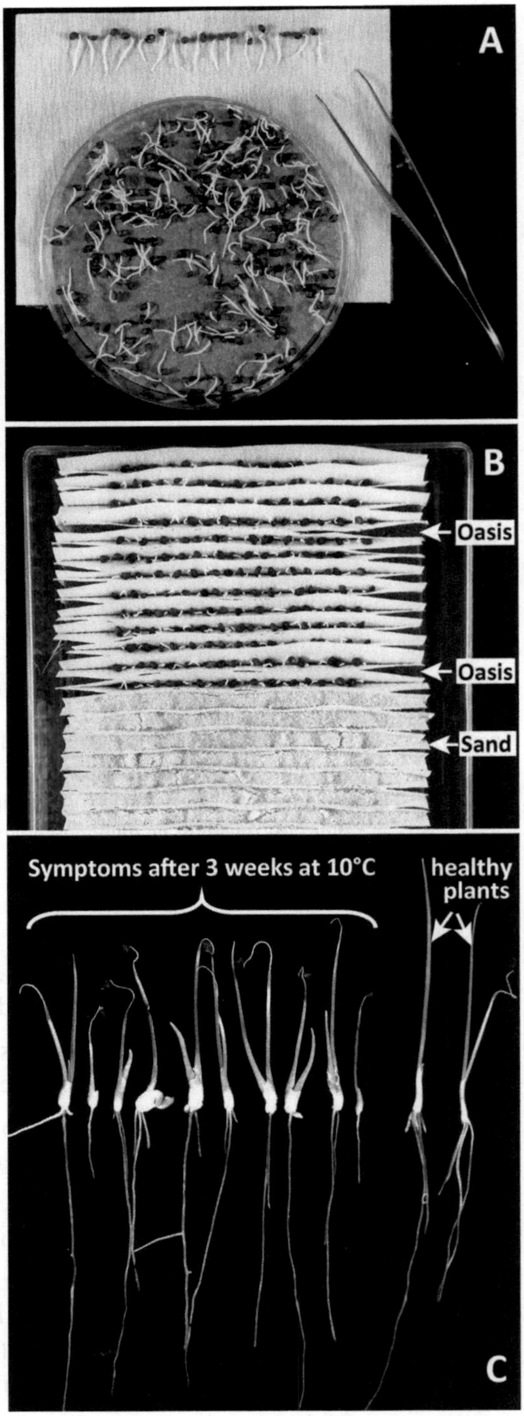

Fig. 11 Method used by JW Seinhorst to screen onions for resistance to *Ditylenchus dipsaci*. (**a**) seeds of onions germinated on a disk of filter paper put in a Petri dish; (**b**) different layers of "oasis" and filter paper in between germinated seeds of onions; (**c**) seedlings of onions, 3 weeks after inoculation, showing symptoms of the nematode parasitization (on the right healthy seedlings)

3.4.2 Screening for Resistance of Broad Bean to Ditylenchus gigas

In the past, within the *D. dipsaci* species complex, a population designated as giant race was very damaging to broad bean in the Mediterranean Basin and Europe. In 2011, this race was described as a new species and named *D. gigas* [36]. This species causes swellings of stems and leaves and stem necrosis.

Within broad bean cvs./lines, differences exist in their reaction to *D. gigas*. Hereafter is described a procedure in greenhouse to evaluate the reaction of broad bean to the nematode abstracted from Caubel and Leclercq [37] and Plowright et al. [38]. For a protocol of field screening, we suggest to consult Abbad Andaloussi [39].

1. For germination, arrange broad been seeds in a tray containing perlite or vermiculite, wet, and incubate at 23 °C.
2. Three days later, put a mixture of sand, soil, and some organic matter (at least 85–95% sand) in laboratory test tubes 140 mm long and 25 mm in diameter, or similar material, add a fertilizing solutions, and transplant in each of them a germinated seed. Identify each tube with a label or writing on it with a waterproof pen.
3. One week after planting, inoculate seedlings by depositing at the leaf axil, near the stem, a droplet (about 15 μL) of aqueous suspension of nematodes containing about 100 nematodes, obtained from artificial cultures or naturally infected plants. Best if in the nematode suspension, 1–2 g of carboxymethyl cellulose/liter has been dissolved to reduce evaporation.
4. Put the tubes in a growth cabinet set at 15 °C the first month and 18 °C the second month and set light at 14 h per day.
5. Maintain a high humidity for 2–3 days.
6. One month after inoculation, the symptoms of nematode infections will become obvious and can be rated. They consist of deformations of axillary sprout, swellings of the leaf axil and stem near the inoculation points (Fig. 12), and presence of necrosis on the stem, near the inoculation point, and to one or both sides of the stem, whose length varies but usually is limited to one merithal. The intensity of such symptoms is greatly affected by the seed genotype and nematode population.
7. Two months after inoculation, evaluate the severity of the symptoms and extract nematodes from the plant tissue. Therefore, collect the aerial plant part, cut finely, and extract and count nematodes. Nematodes can also be extracted from soil but they are not expected to be many.
8. Then compare severity of symptoms and nematodes per pot of the tested cultivars/lines.

Fig. 12 Effects of artificial inoculation of faba bean with *Ditylenchus gigas*. (**a**) Swellings at inoculation point (leaf axil) and distortions on the susceptible variety Diana; (**b**) localized necrosis, typical of resistance, on leaf stipule and stem (arrowed) of the resistant line INRA 29H. (From Ref. [37] with kind permission)

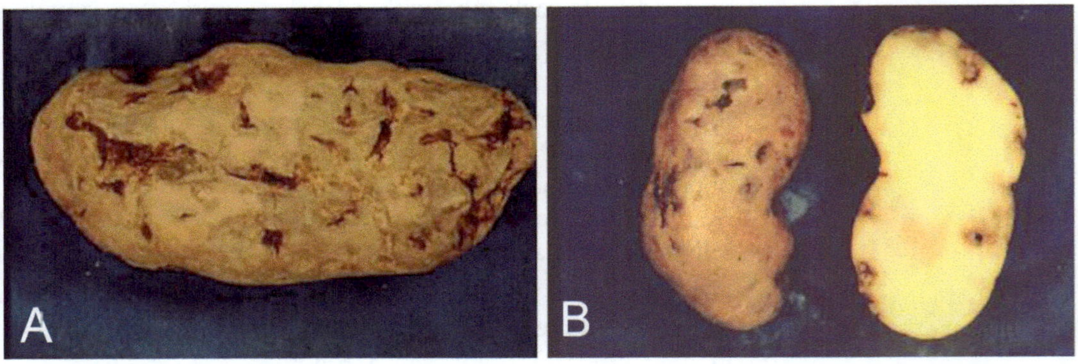

Fig. 13 Attack of *Ditylenchus destructor* on potato tuber tissues causing white mealy spots under the skin, and later an expanding dry rot, which makes the tissues dry out and the skin to shrink and crack. (From Ref. [90], with kind permission of EFSA)

3.4.3 Screening Potatoes and Sweet Potatoes for Resistance to Ditylenchus destructor

Both *D. dipsaci* and *D. destructor* can damage potatoes especially in eastern European countries, causing similar damages (Fig. 13).

The following protocol is good to screen toward both nematodes and is abstracted from Mwaura et al. [40] and Smaliev and Markova [41]:

1. The tubers of potatoes are stimulated to sprout in the dark at 20–23 °C and when the sprouts become 1 cm long are moved in the light to favor hardening.
2. Then cut tuber in pieces so that each is of about 15–16 g bearing only one sprout.
3. Prepare pot soil by mixing steam-sterilized field soil with peat mix at the rate of about 3:1 (potting soil must be rather soft, sand >70%).

4. While mixing, incorporate also a fertilizer containing NPK each at about 15–20%, 2% magnesium plus microelement, at the rate of 1–2 g per kg soil.
5. Have ready the required nematode populations reproduced on carrot disk, extract them, and adjust nematode suspension to 500 specimens per mL.
6. Fill each pot (14–16 cm in diameter) with about 1 dm^3 soil.
7. Plant in each pot a piece of sprouted tuber at 3–4 cm depth, set temperature at 18–22 °C, and light at 13 h per day and from now on irrigate as needed.
8. Keep separated groups of pots to be inoculated with different nematode species and even populations of the same species if any.
9. Two weeks after planting, when plants are just emerging from the soil, infest pots with either nematode or population by making four holes 3–4 cm deep around the plant stem pouring in each 1 mL of the nematode suspension containing 500 nematodes of mixed life stage, for a total of 2000 nematode specimens per pot. Close the holes and add a few drops or mL of water to favor adhesion of soil to nematodes.
10. Fourteen weeks after inoculation, discontinue the test and turn each group (replicates) of pots upside down in a plastic or iron tray. Depending on the organization of the work, several trays may be required.
11. Collect all tubers per pots, weigh and rate them according to numbers showing symptoms and intensity of the symptoms as seen externally and internally (*see* Fig. 13). Then peel all tubers with a knife or a device that allows having peel of 2–3 mm thick.
12. Cut all peels in small pieces, mix, and take a sample of 10 g. Be careful! Do not mix peel in a bowl containing water, as suggested to mix roots; nematodes will start to swim in water while mixing and get lost.
13. Extract nematodes using Baermann's funnel method or one of its modifications or a different method. While on the funnel for about 24 h, set the temperature at about 20 °C. Then collect and count nematodes and refer data to the total peel per pot. If necessary, store nematode suspensions at about 5 °C for a few days until record of counts has been made.
14. Mix the soil of each pot, take a subsample of the size required by your extraction method, and store it in a plastic bag kept in a cool room at 5–6 °C until processing it for nematode extraction.

15. Count nematodes, refer numbers to total soil quantity per pot, and then determine total nematode per pot (nematode in tuber and soil per pot).
16. Assess resistance based on Rf and Rs (relative susceptibility).
17. To calculate the reproduction factor, divide total nematodes extracted per pot by the number of nematodes inoculated per pot. Then classify a variety resistant if Rf <1 and susceptible if Rf >1.

    Relative susceptibility is Rs = nematode population in the pots of the test cultivar divided by the nematode population in the pots of the susceptible control cultivar × 100. Based on Rs, the following score scale can be adopted: 1 = Rs > 100%; 2 = Rs 50.1–100; 3 = 25.1–50%; 4 = 15.1–25%; 5 = 10.1–15%; 6 = 5.1–10%; 7 = 3.1–5%; 8 = 1.1–3; and 9 = <1%. In this scale, 1 represents the less resistant score and 9 the highest resistant score (*see* **Note 5**).

3.4.4 Screening for Resistance of Peanut to *Ditylenchus africanus*

This nematode is apparently endemic to South Africa, where it is seen as one of the most limiting factors in peanut production. The main effect of *D. africanus* on peanut is qualitative [42] (Fig. 14), causing peanut consignments to be downgraded. It is important to securely contain this nematode, should peanut genotypes be evaluated for resistance in countries other than South Africa.

The following protocol has been adapted from Steenkamp et al. [43]:

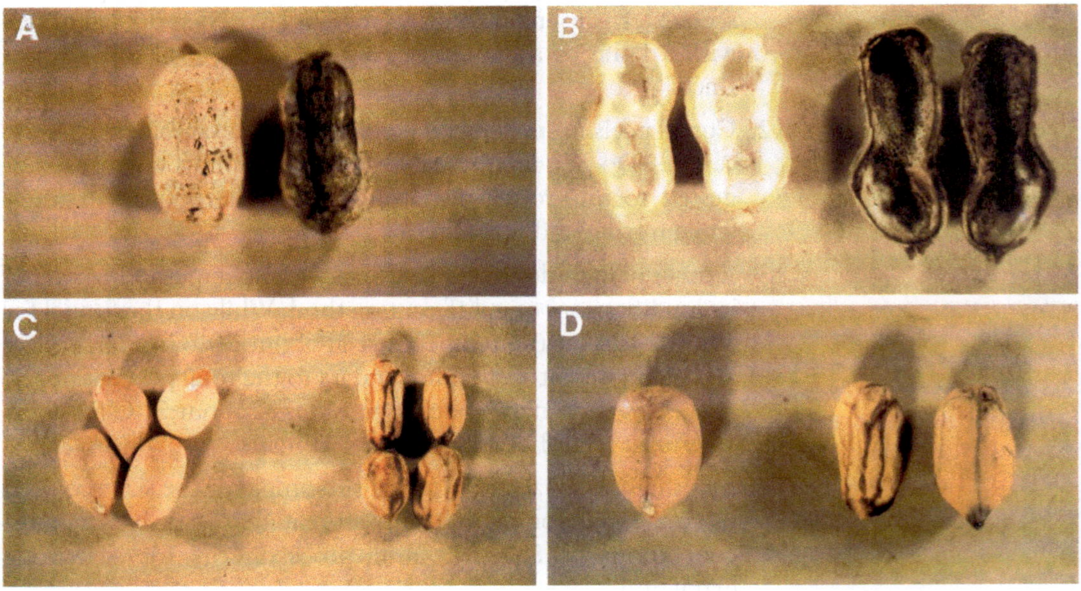

Fig. 14 Damages caused by *Ditylenchus africanus* in hull and seeds of peanuts. (From Ref. [42] with kind permission of SON)

1. Before sowing, collect a subsample of 5 g of each tested genotype and soak it in water for 24 h at about 20 °C to ascertain that they are free from *D. africanus*.
2. Before planting, fumigate microplots with a good fumigant available to you and incorporate into the soil the required fertilizer.
3. Treat seeds with a registered fungicide and *Bradyrhizobium arachis* as recommended on the label before planting.
4. Sow seeds according to inter- or intra-row recommendations for the country in which the screening is conducted. Arrange trials according to a randomized complete block design with six replicates per treatment (genotypes).
5. At sowing, inoculate seeds with 3000 *D. africanus* per plant (poured over the seed) obtained from a population reared on in vitro callus of peanut leaves.
6. At harvest, collect soil samples from each treatment. Also, collect four plants per row of each microplot, combine, and subsample to obtain 5 g hulls and 5 g kernels. Take also a sample of roots. Roots, hulls, and kernels should be cut into small pieces and soaked in tap water separately for 24 h. A sample of peanut pod/seed can be used to determine yield quality according to the grading suggested by national officials.
7. Extract nematodes from soil combining the decanting and sieving method with centrifugation method. Then count nematodes and determine Rf and also Rr compared to the susceptible control cultivar.
8. So far, no scale has been established to define a genotype resistant or not. However, as suggested in the introductive paragraphs, if numbers of nematodes and damaged pods and seeds are less than 10% of those in the susceptible cultivar, and Rf <1, the tested genotypes can be rated as resistant, otherwise can be considered as having different degrees of resistance/ susceptibility.
9. Also, compare nematodes in roots, kernels, hulls, and pegs of the test genotype with those of the susceptible cultivar, each separately (*see* **Note 6**).

3.4.5 Screening for Resistance of Rice to Ditylenchus angustus

The nematode causes a disease of rice locally called "ufra disease" in Asian countries. Its control is difficult and the only reliable management of it is the use of resistant varieties and screenings of cultivars and landraces of rice are underway.

Hereafter are reported the suggested screening procedures:

1. Fill two thirds of plastic or other material pots, measuring about 30 cm in diameter and 40 cm in height, with a typical soil from rice fields; arrange pots according to a statistical design such as in a randomized block design.

2. Germinate seeds of rice in Petri dishes containing a wet disk of filter paper at 30 °C.
3. At seed germination, 4–5 days later, sow about 30 germinated seeds per pot of about 0.5 cm deep and irrigate and fertilize pots as needed.
4. Maintain temperature at about 28–30 °C; collect a sample of stems from a nematode culture maintained on diseased rice plants. Cut each stem of the rice plants first longitudinally and then in pieces of about 0.5 cm long.
5. Then put samples on Baermann's funnels, or on a different apparatus, and leave them overnight to let the nematode move into the water. The day after, collect nematode suspension, count nematode specimens in them, and determine the average number of nematodes per stem.
6. Two weeks after sowing, raise water in the pot to the level of the uppermost rice plant node.
7. Inoculate the pots by pouring evenly on the water of each pot an amount of cut stem pieces to average 100 nematode specimens per plant.
8. Twenty-eight days after inoculation, score the plants on the base of the intensity of chlorotic discoloration at the base of the leaves and other symptoms, such as crinkling, twisting, or corrugation of leaves and panicles [44] (Fig. 15) and group them on a 0–16 scale (Table 1) to identify plant reaction.

Fig. 15 Symptoms of susceptibility of rice leaves to *Ditylenchus angustus* with severity rated on a 0 to 16 scale. (From Ref. [45], no permission required)

Table 1
Classification of rice plant reaction to ufra disease caused by *Ditylenchus angustus*, according to disease index, % tiller infection, and symptom visibility 55 days after nematode inoculation

Disease index	% tiller infection	Symptom visibility	Reaction class
0	0	No symptoms	Highly resistant
1	1–20	May or may not be visible	Resistant
3	21–40	Visible	Moderately resistant
5	41–60	Visible	Moderately susceptible
7	61–80	Visible	Susceptible
9	81–100	Visible	Highly susceptible

From Ref. [45], with kind permission of IK Press Editorial Office

9. Perform another score about 55 days after inoculation based on % of infected tiller and group entries on a 0–16 scale (Table 1) to assign each entry to a category.

   This test is suitable when testing a limited number of entries but is very reliable. It has been demonstrated that there is a very high correlation between the intensity of the symptoms and number of nematodes per plant [45], clearly indicating that evaluation of the plant reaction to *D. angustus* based on symptom expression is reliable.

However, if the numbers of the nematode infecting the plants are desired, perform another experiment as explained hereafter.

1. Instead of using large pots, use glass or plastic tubes 20 cm long and 2.5–3 cm in diameter and follow exactly the same procedure described earlier, except for filling two thirds of the tube with a suitable sandy soil; sowing only one pre-germinated seed per tube; adding a fertilizer solution, such as Hougland, or a similar one; filling the tube with water 2 weeks later; and inoculating 100 nematode specimens per tube.
2. Twenty-eight days after inoculation, discontinue the test, score the plants on the base of severity of symptoms and then cut the stems, record their weight, cut them, and extract and count nematodes as explained before. Then score plant reaction on the base of both severity of the symptoms and nematode reproduction (*see* **Note 7**).

3.5 Screening for Resistance of Rice to Aphelenchoides besseyi

The genus *Aphelenchoides* includes many species of endo-ectoparasitic migratory nematodes. Although damage is caused to many crop plant species, probably the crop most affected is rice by attacks of *Aphelenchoides besseyi* [46, 47]. The nematode parasitizes many plant species, including strawberry, and survives as quiescent anhydrobiotic stage in the rice kernel for several years. It is a seed-borne parasite and has also been reported to damage soybean and haricot beans in America. However, Subbotin et al. [48], based on morphological and molecular studies, described *A. pseudobesseyi* from populations earlier identified as *A. besseyi*.

Here only protocols to screen rice for resistance to *A. besseyi* are described; however, the same protocols can be adapted also to other plant species. For symptoms caused by the nematode to rice, *see* Figs. 16 and 17.

1. Sterilize seeds in a water bath at 55–57 °C for 15 min to eliminate nematodes that can be present in the kernel.
2. Germinate seeds in a Petri dish containing a disk of wetted filter paper at 28 °C.
3. At germination (3–4 days later), transplant a couple of germinated seed in the center of each pot containing 2–5 dm^3 sterilized soil and thin plants to one per pot after emergence.
4. Have six to ten replicated pots per genotype, arranged according to a randomized block design, on a bench in a greenhouse maintained at 25–28 °C with a 14–16 h light per day and 70–80% RH.
5. Irrigate as required and fertilize using a 2‰ fertilizer solution, every 10–15 days.
6. Some authors conduct the test in plots under field or greenhouse conditions.
7. Inoculate plants. There are several inoculation methods, each having advantages and disadvantages, and the most common are (*i*) pour nematodes (about 200) in the pots around the plant stem, (*ii*) inject nematodes (50–100) in 0.1–0.2 mL water suspension with a syringe in the leaf sheath (effective but time consuming and not comfortable) [49], (*iii*) inject nematode (200–500) suspension (0.5–1 mL) in the panicle (very reliable but rather time consuming and long-lasting test), and (*iv*) spray nematode (200–500) suspension (0.5–1 mL) on the panicle (effective but rather long-lasting test) [50].
8. Observations on nematode infection can start 2 weeks after inoculation. However, realize that nematode in plants increases continuously until flowering and dough seed stage, about 60–70 days after inoculation. This is the best time to extract nematodes, especially when designation of the plant reaction is based on Rf and % infected plants. Thereafter, nematodes decrease but remain at rather high levels. At late growth stage (since tillering, about 30 days after inoculation), it is also possible to take note of the plant showing yellow tip symptom. To simulate paddy conditions, Godoy et al. [51], after inoculation of 500 nematodes, submerge pots until plants become all yellow (about 135 days later).

Fig. 16 Symptoms on rice plants infected by *Aphelenchoides besseyi*. (**a**–**d**) white tip on Huajing 5, Nanjing 47, Ningjing 1, and Wuyujing 8, respectively. (**e**) small grains and erect panicles on Wuyujing 3 (left) and control (right). (**f**–**h**) symptomless on Wuyujing 22 (left) and control (right). (**g**) black and cracked grains of Wuyujing 22. H, healthy grains of Wuyujing 22. (From Ref. [46], no permission required)

Fig. 17 Evolution of the white tip caused by *Aphelenchoides besseyi* on rice leaves, after planting infected seeds. (**a**) early first symptom of 1 cm long leaf tip chlorosis; (**b**) 2–3 cm long leaf chlorosis after 1 week; (**c**) after 2 weeks, chlorosis on the first leaf extends to half the first leaf, while tip chlorosis starts to appear on the second leaf of the same plant; (**d**) on the third week, the entire first leaf is dried out. (From Ref. [47] with kind permission of the journal)

9. To extract nematodes, the entire plant or a sample of it is cut in small pieces and nematodes are extracted by Baermann's funnel technique or with the mistifier chamber at 25 °C for 12–24 h.
10. To evaluate the effect of the nematode on the grain, at plant maturity, spikes are collected and grains per spike, unfilled, filled, and partially filled grains, are counted. Thereafter, nematodes from them are also extracted and counted as suggested.
11. The resistance of the rice genotypes can be assessed on the base of the nematodes extracted from each plant and the development of symptoms they have caused, using the following rating scale suggested by Jamali et al. [52]:

    0 = both white tip and nematodes absent

    1 = white tip symptom absent and nematodes per plant 1–10

3 = white tip symptoms absent and nematode per plant >10

5 = white tip symptoms present and many nematodes present.

The average infection index of each genotype is calculated according to the following formula:

$$P = \sum (B \times n)/N$$

where $\Sigma(B \times n)$ is the sum of the number (n) of plants times the corresponding infection index (B) and N is the total number of inoculated plants. Based on this index, all genotypes can then be classified as 0 = immune; 0.1–1 = highly resistant; 1.1–3 moderately resistant; 3.1–4 moderately susceptible; and 4.1–5 highly susceptible (*see* **Note 8**).

3.6 Screening for Resistance of Wheat to Anguina tritici

This is the first plant parasitic nematode to have been observed. It was in 1744 when Needham while dissecting in water wheat kernel observed that something that appeared as in-animated started to move under his eyes while checking under a microscope. The nematode is today known as the wheat seed gall nematode because of the damage it causes mainly to wheat, barley, and rye (Fig. 18). Typical symptoms of the nematode parasitization are basal swelling of the plants, twisting of leaves, and distorted awns; it is also known as wheat ear cockling in which thousands of second-stage juveniles survive in anhydrobiotic state for several years. Because infected barley plants are taller, in the near east countries, the nematode is known as the disease of tall plants. It is a typical nematode of rather old-style agriculture where the use of seeds certified for being free of nematodes and diseases is limited, especially in Asian countries. It has been ascertained that wheat cultivars/populations resistant to the nematode do exist and, therefore, screenings and breeding programs are being conducted.

1. Germinate seeds of the genotype to test in a Petri dish containing a disk of wetted filter paper at 20–23 °C for 3–4 days. Seeds of winter-type wheat may require a 3–4-month period at 2–3 °C to satisfy their vernalization need.
2. Sow germinated seeds in 15–20 cm diam. Pots (one per pot) containing rather sandy soil already infested with crushed infected seed of which the nematode content has previously been determined. Usually, two to three crushed infected seeds are more than enough. Alternatively, when plants emerge from the soil, inoculate them with a water suspension containing about 5000 nematodes per pot poured around the stem.
3. Pots can be arranged in a greenhouse at about 15–20 °C, by spraying water daily during the first 7–10 days after inoculation, or better in a screenhouse or outdoors from autumn to spring.

Fig. 18 Infection of *Anguina tritici*. (**a**) plantlets of wheat showing typically early symptoms on the leaves (arrowed); (**b**) distorted spike showing black kernels; (**c**) a healthy spike of barley in comparison with spikes completely aborted (left) and partially aborted (right); (**d**) healthy (left) and deformed and brownish kernels (right); (**e**) vision under a stereoscope of thousands of second-stage juveniles (quiescent and survival stage of the nematode) coming out of infected wheat kernel, after crushing and left in water overnight. (A and B from Ref. [54] with kind permission of IK Press Editorial Office)

4. Irrigate plants whenever needed and fertilize every other week with a solution of a composite fertilizer.
5. Periodically (every 2 weeks) take note of the leaf symptoms and of their severity (Fig. 18a, b).
6. At plant maturity, record plant growth variables (height, weight, spike length, weight of grain per spike, numbers of grain per spike, number and % of galled and apparently healthy seeds, weight of 1000 seeds) [53–55].

Also, extract nematodes from seeds by crushing and putting them on Baermann's funnels, or one of its modifications, overnight and count nematodes in the water suspension after proper dilutions for an easy count under a stereoscope. With large screenings, as the observation of severe leaf symptoms is a clear indication of the susceptibility of the genotype to the nematode, symptomatic plants can be eliminated at this stage. However, a further and certain indication of plant susceptibility is the presence at harvest of galled seeds, on the spike of wheat, and of spikes with aborted flowers and arrested growth of kernel in barley (Fig. 18c).

Plant not showing symptoms of the parasitization throughout the crop cycle can be considered resistant. Then collect seeds, furtherly check for absence of galls and nematodes, and, if so, use them for future studies.

3.7 Screening for Resistance to Rotylenchulus reniformis

This is a semi-endoparasitic sedentary nematode having a very large host range that includes soybean and many other legume plants, cotton, tomatoes, and others. It is spread mainly in tropical and subtropical countries. Because of its economic importance and of difficulties for its management, many screenings and breeding programs are underway to search for new sources of resistance in the available germplasm collections and progenies.

The following suggested protocol can be useful to screen several plant species:

1. Germinate seeds of each genotype in vermiculite or Petri dishes containing a wetted disk of filter paper and incubate at 23–25 ° C in the dark for about 3–5 days. Seeds of wild species may require longer time to germinate.
2. Treat seeds of legume plant with the specific root nodulation bacterium [5].
3. Sow germinated seeds in pots of 10–15 cm in diam. Containing 800–1000 g of soil. To screen a large number of plants, and to save space and time, small pots, such as Cone-Tainer SC10, 3.5 cm in diam. and 20 cm long (Stuewe and Sons, Tangent, OR, USA), containing about 150–300 g soil can be used [56, 57].
4. Irrigate soon with a fertilizing solution up to wet all the soil of the pot and add the root nodulation bacterium if necessary. Seven to ten days after plant emergence, inoculate each pot with five to ten nematodes (mixed stages)/cm^3 soil in 1–3 mL water suspension into two to four holes 2 cm deep, around the plant stem.
5. Have some extra pots inoculate, of which some will be planted with the susceptible control, to follow the development of the nematodes and understand whether or not the nematode development proceeds as expected, and others left fallow to follow the decline of the nematode throughout the cycle of the susceptible control, to estimate the proportion of the Pi that may have survived until harvest.
6. One week later, do a second similar inoculation.
7. Thereafter, irrigate as required and fertilize every other week.

8. Four weeks later, if small pots were used, or otherwise 6–10 weeks later, discontinue the test and do the following evaluations.
9. Uproot plants, mix thoroughly the soil of the pots, take a sample, and extract and count nematode in it using one of the extraction methods available to you.
10. Discard the top plant and wash the root free of adhering soil, rate them in % of necrotic root, stain the nematode egg masses by immerging them in a 0.15 g/l phloxine B solution for about 15 min or better in the 20% McCormick Schilling red food color for 15 min [58], wash away the excess stain, and count the number of egg masses and nematode females colored in red; express them as % of those on the susceptible control cultivars.
11. Extract and count also eggs on the roots. Therefore, cut them in 1–2 cm long pieces and shake in a 1% hypochlorite solution for 3–4 min; the hypochlorite will dissolve the egg masses and separate eggs from each other.
12. Sieve all the egg–water suspension through a 100 μm aperture sieve nested on a 20 μm sieve (better if 5–10 μm aperture) and gently spray water on the larger sieve. With a spraying bottle, collect the eggs on the small sieve in a corner and then in a beaker, dilute and count eggs in three separate mL under a stereoscope, and refer the average to all suspension and to total per root and per g of roots.
13. Add total nematodes per root to total nematodes in the soil of each pot to obtain total nematodes per pot (Pf). Finally, calculate reproduction factor (Rf) per pot dividing Pf with the inoculated nematodes per pot (Pi).
14. If information on the effect of the nematode on plant growth is desired, record also weight and height of the top plant. However, the inoculum used could not be sufficiently high to cause significant growth reduction.
15. When screening large numbers of plants, as a preliminary test, it is enough to count just females/egg masses on the roots and based on the % of these per plant, compared to the susceptible control cultivar, and classify genotypes as <10% resistant, 10–30% moderately resistant, 31–60% moderately susceptible, and >60% susceptible. Resistant plants must be retested for a longer period to confirm their resistance. Some researchers consider a plant susceptible or resistant if nematode variables are significantly more or not, respectively, compared to the resistant control.

This method of screening has been demonstrated to be useful also for cotton [59], ananas [60], pigeon pea [61] and probably with other plant species. Moreover, to collect eggs from infested roots with the sodium hypochlorite method, sieve the eggs through a 74 μm aperture sieve nested on a 20 μm sieve, centrifuge in magnesium sulfate solution at 1.16–1.18 sp. g. (specific gravity), add kaolin (amount depends on the size of the centrifuge tube), inoculate plants with 6000 nematodes per pot of 20 cm in diam., and grow plants at 28–32 °C in a growing chamber, rather than in a greenhouse without good air-conditioning, for 60–70 days, thereby giving the greatest reproduction of the nematodes and the highest egg collections [62] (Fig. 19).

When the plants derive from breeding crosses, a preliminary nonplant destructive method is necessary to save resistant plants. Therefore, collect three to four soil cores per pot [63] using a 1 cm diam. Soil core, for a total of 40–200 cm^3 soil (depending on pot size) and extract nematode from them. Plants evaluated as resistant must be transferred to larger pots (20–30 cm in diam.) containing non-infested soil, to obtain self-pollinated seeds or for further immediate crosses (organize screening timing and procedure in accord with a breeder) (*see* **Note 9**).

3.8 Screening for Resistance of Citrus to Tylenchulus semipenetrans

Tylenchulus semipenetrans is mostly associated with citrus all over the world wherever citrus is cultivated and is known as the causal agent of the slow decline of citrus. It attacks a few other crop plants such as kaki, olive, and grape. Severe damages are caused also to grape in California (USA) [64] and Chile [65]. So far, three biotypes of the nematode are known. Control of the nematode with nematicides is possible but not suggested as treatment should be done every year and this would cause concerns for living being and the environment. The use of resistant rootstocks suitable to different agroclimatic conditions appears as the most feasible control option and is an encouraging breeding program of citrus and grapes for resistance to the nematode. Sources of resistance to the nematode do exist in *Poncirus trifoliata* and several hybrids of it are used since a long time as rootstocks for citrus [66]. Resistance has also been identified in wild grape now introgressed into grape rootstocks [67].

1. Collect seeds of the genotypes from fruits, wash in running water several times until pulp and juice residues are eliminated, put in a thermostatic bath at 52 °C for 10 min, submerge in a 1% hydroxyquinoline sulfate solution for 5 min, dry in the shade, keep in polythene bags (100–200 cm^3 volume), and store in a refrigerator at 5–6 °C until use (better if until after 5–6 months). Realize that 1 kg contains about 5000 seeds.

Fig. 19 Main steps of the screening procedure to evaluate cotton for resistance to *R. reniformis*: (**a**) Cotton seedlings planted in 5 L pots filled with a 3:2:1 (v:v:v) mixture of compost, sand, and peat; (**b**) Inoculation with 6000 *R. reniformis* eggs by injection of the appropriate nematode egg suspension; (**c**) Roots with soil, after 60 days of infestation; (**d**) Soaking of the roots with the soil in a bucket of water; (**e**) Harvest of the roots; (**f**) Blinding of the roots in 0.25% NaOCl solution; (**g–h**), Sieving of the blending with 75 and 20 μm mesh sieves; I, Collection of residue retained by the 20 μm mesh sieve; (**j**) A centrifuge tube after centrifugation; (**k**) After centrifugal flotation, putting of a 20 mL egg suspension in a Petri dish (especially divided) for counting; (**l**) Counting of eggs under a stereomicroscope; (**m**) Some *R. reniformis* eggs seen with the stereomicroscope. (From Ref. [62] with kind permission of Elsevier)

2. When desired, fill polypots (each tray of 30–40 pots) with a mixture of river sand, 25% peat, and 25% vermiculite, in order to have a soft soil, and sow one seed per pot 2 cm deep, irrigate as required, and arrange polypots in a greenhouse at 25 °C.
3. At seed germination, start to fertilize with a soluble NPK + microelement fertilizer. When 8–10 cm tall; uproot the seedlings; wash them free of soil gently, avoiding damaging the roots; and transfer plants in larger pots 15–20 cm in diam., but now containing soil mixture free of peat. When plants are 20–25 cm tall and new root flushes exploring the soil, inoculate about eight to ten nematodes/cm^3 soil into three to five holes 3–4 cm deep or more to have nematodes distributed in the central part of the pot soil.
4. Eggs, J2, and males of the nematodes for inoculation can be extracted from infected citrus roots by the sodium hypochlorite method [27] using a 0.5% solution. Alternatively, J2 can be extracted putting infected roots of citrus on Baermann's funnels, and on funnels in a mistifier chamber, or incubating in jars according to Young [68]. However, the use of eggs is easier.
5. Continue to irrigate and fertilize for 4–6 months. Then check a few extra control plants to see whether nematodes are already developed enough or not. If females are well developed and producing egg masses, discontinue the test; otherwise, delay the evaluation of 1–2 months more.
6. To extract and count nematodes from soil and roots, uproot plants and cut at soil level. Clean the roots free of soil, cut in 1–2 cm long piece, mix in water, dry between laboratory cleaning papers, weigh, take subsamples of at least 5 g, and extract eggs from the roots according to the hypochlorite method in a 1% solution. This method does not extract females and other stages from inside the roots. Therefore, if numbers of nematode females are required, the method of Greco and D'Addabbo [69] is suggested (*see* Chap. 1). This method extracts swollen females, eggs, J2, and males. It is similar to the method of Coolen [15] but the nematode pellet of the first centrifugation is resuspended in a colloidal silica (Du Pont De Nemours International S.A.) at 1.16 sp. g. Silica does not plasmolize females with broken neck, which does not become void of its body content, remain swollen, and can be clearly distinguished from root debris when observed under a stereoscope.
7. Calculate total nematodes in the pot, total eggs, and J2 in the soil plus nematodes in the roots (including swollen females if desired). Now determine Rf and RI and design genotypes as immune (no nematodes in the roots) and resistant (Rf <1, RI $<15\%$) [70].

The same procedure can be followed with screening of grapes, except that small-rooted cuttings must be used, unless seeds from breeding programs are to be screened [65].

3.9 Assessment of Plant Tolerance to Nematodes

Tolerant cultivars [3, 4, 61] may play an important role in the farm rotation, especially if no resistant cultivars are available. Owen [71] reported that the use of wheat cultivars tolerant to *Pratylenchus thornei* has reduced yield loss of wheat in Australia from AU$ 104 million in 2009 to AU$ 31 million in 2018. Therefore, identification of tolerant crop plants and a measure of the degree of their tolerance is very important. To achieve this goal, the general suggestions given to assess resistance to nematodes in this chapter and those for assessing tolerance in Chap. 3 are useful. Mwaura et al. [72] used the same method to evaluate resistance of potato to *Ditylenchus dipsaci* and *D. destructor* and used only one inoculum level (2000 nematode specimens/700 cm^3 soil). A number of genotypes were compared including cv. Desirée, very susceptible, and cv. Spunta, resistant.

To have a measure of the tolerance, near the end of the crop cycle, about 14–16 weeks after planting, it is necessary to count and weigh single and total potato tubers, to assess the grade of tubers according to market, and to determine the degree of damage visually after cleaning (based on extension of skin cracking) and internally (based on the extent of brownishment), after cutting tubers longitudinally, according to the evaluation scale by Mwaura et al. [40] in Fig. 13. Externally, if the skin is completely cracked, the tuber is considered 100% damaged, but if a tuber has no skin cracking, it is considered to have 0% damage. Internally, a tuber is considered fully damaged if the skin is entirely damaged and the internal part of the cortex partly damaged. However, only one half of each tuber is evaluated. The average % of total external tuber damage is calculated according to the following formula:

$$P = \sum (n1 + n2 + n3)/N$$

where P is the % tuber damage per replicate, N is the total number of tubers, and n is the % potato tuber showing lesions caused by either nematode. The same formula can be used for the internal tuber damage. To determine the average yield loss, Mwaura et al. [72] used the following formula:

$$\text{Loss (\%tuber weight reduction)} = \frac{\text{Control treatment tubertuberweight} - \text{Treatment tuber weight}}{\text{Control tuber weight}} \times 100$$

Based on tuber loss %, cultivars were considered as follows: 0–25% tolerant, 25.1–50% moderately resistant, 50.1–75% sensitive, and > 75.1% highly sensitive.

For potato affected by the mentioned *Ditylenchus* species, to be marketable, tubers must appear healthy from both outside and inside. Be aware that these and other nematodes, such as *Pratylenchus penetrans* and several *Meloidogyne* spp., may continue to reproduce and increase their damage during storage.

When assessing tolerance, nematodes in plant tissues and soil must be determined to measure reproduction rate and, therefore, if the test was reliable or not.

The data are then statistically analyzed and compared with the susceptible control and with non-infested plots. Based on this, a cultivar can be considered tolerant or not. The best situation would occur when the tolerant cultivar, despite being infected, presents neither tuber damage nor yield reduction, but this is the exception and not the rule.

According to Mwaura et al. [40], testing each nematode at the inoculum levels of one, four, and eight specimens per g or cm^3 soil, in rather large pots or microplots, is the best option. However, this author is of the opinion that in small pots (1 dm^3), as those used by Mwaura et al. [72], tubers would not develop well and plants would not fully express their yield potential. This author, while assessing the effect of initial nematode populations on the yield of different crop plants to various nematode species, obtained always good results using microplots of about 30 cm × 30 cm square section and 40–50 cm long sunk into the soil up to 5 cm of the ridge, containing about 35–40 dm^3 soil. In such microplots, a single plant of potato, sugar beet, tomato, pepper, broad bean, and eggplants or several plants of chickpea, peas, and wheat grew satisfactorily.

To assess the tolerance of carrot to *P. penetrans*, it is suggested to use a non-inoculated treatment and treatments with 2, 7, and 19 nematode specimens per cm^3 soil.

When a tolerant cv. has been identified, in greenhouse or microplot conditions, a final confirmation under field conditions, in naturally infested soil, will be necessary before suggesting to farmers the use of that cultivar. Therefore, trials must be arranged under different agroecological conditions and of soil infestation levels occurring more frequently in the area of interest. This requires designing tests in rather large plots (minimum of 100 m^2 each) and with at least six replicates. In such trials, the yield of the cv. evaluated as tolerant and nematode reproduction must be compared with a susceptible one and, if available, tolerant and resistant ones to decide if the supposed tolerant cv. can be useful or not in the area under study.

3.10 Assessment of Nematode Virulent Populations

On the base of presence or absence of reproduction of nematode populations of the same species on international standards of host plants, races have been identified in several nematodes described in the previous sections. Therefore, the required material is the same as that necessary for screening for resistance or tolerance. In general, the procedures are also the same. The only difference is that, in this case, not different genotypes, but the suggested different host plants and different nematode populations of the same species must be tested to characterize the virulence of a specific nematode population. To identify the nematode virulence group, it is suggested to extract and count nematodes in different differential plants, do comparisons, and, based on this, assign to each nematode population the virulence group to which it belongs.

The identification of virulence group is rather complicated in *Ditylenchus dipsaci* of which about 30 races have been reported, each named on the base of the most important host plant, such as onion race, oat race, alfalfa race, and so on. Moreover, *D. dipsaci* is believed to include several sibling species. Therefore, rather than the exact identification of the race, it is suggested to investigate which of the crop plants common in the area of study can be parasitized and damaged by the local nematode population, in comparison with host plants typical of some races, such as onion, garlic, oat, alfalfa, clover, carrot, sugar beet, broad bean, tulip, hyacinth, and narcissus (Table 2). Seinhorst [73], based on information from the Netherlands and elsewhere, prepared a table (Table 2) to differentiate 11 races of *D. dipsaci* based on their behavior on mine plant species. Since then, Seinhorst continued his investigation on *D. dipsaci* and collected more data never published (W.T. Runia, personal communication, 2002, the Netherlands). These data, and those of two Italian populations studied by the author, are presented in Table 3 and may serve as a guide.

Besides, *D. dipsaci* and *T. semipenetrans*, of which schemes are reported to identify races (Tables 2, 3 and 4), evidence of the existence of differences in virulence is reported for many other nematodes.

Abbad-Andaloussi [39] reported that a Moroccan population of the giant race of *D. dipsaci* (now *D. gigas*) reproduced on the faba bean line INRA 29H is reported earlier as resistant.

The root-lesion nematode *Radopholus similis* affects severely bananas and causes the quick decline of citrus in Florida (USA). This nematode has two races, the citrus races attacking both citrus and banana and the banana race attacking only banana. Therefore, banana and citrus can be used to characterize the nematode race.

Differences in virulence have been observed in species of *Pratylenchus* by France and Brodie [10] in *P. penetrans* from New York and Long Island states (USA) and in populations of *P. neglectus* from Montana (USA) by Al-Khafaji et al. [11].

In the literature, there are reports about the occurrence of races in *Nacobbus aberrans*. However, recently, these different populations have been re-examined and some have been described as a new species [3, 48]. Therefore, the status of races of several new species is to be reassessed.

Dasgupta and Seshadri [74] classified populations of *R. reniformis* in two races based on their ability to reproduce on cowpea, castor, or cotton, with race A reproducing on all these

Table 2
The distinction of 11 biological races of the bulb and stem nematode, *Ditylenchus dipsaci*, based on their behavior on nine differential plant species

Differential host plant	Host plant race										
	Teazel	Red clover	White clover	Lucerne	Rye	Onion	Potato 1	Potato 2	Hyacinth	Daffodil	Tulip
Teazel (*Dipsacus fullonum*)	+++	–			–	–	–	–		–	–
Red clover (*Trifolium pratense*)	–	+++	–	–	–	–	–	–		–	–
White clover (*T. Repens*)			–		–	–	–	–			
Lucerne (*Medicago sativa*)		–	–	+++	–	–	–	–			
Peas (*Pisum sativum*)	–	++			–	+++	+++	+++			
Potato (*Solanum tuberosum*)		–			+	+	+++	++		–	
Daffodil (*narcissus* sp.)		–			–	–	–	–	–	+++	+++
Hyacinth (*Hyacinthus orientalis*)											
Tulip (*Tulipa gesneriana*)		–			–	–	–	–	–	–	+++

Adapted from Ref. [73]

Table 3
Host plants for ten populations of *Ditylenchus dipsaci* from the Netherlands (From Seinhorst, unpublished data) and two Italian populations [91, 92] from onion (Apulia) and carrot (Sicily)

	Netherlands populations from										Italian populations from	
Host	Rye	Onion Flakkee	Onion N-Holl.	Potato	Red clover	Alfalfa	Daffodil	Hyacinth	Tulip	Veaver Card	Onion	Carrot
Rye	**			**	0	0					**	**
Oat	**	**		**	0	0	0			**	*	**[a]
Wheat	0	0		**	0	0						
Barley	0	0		0	0							
Maize	**	**		0	0							
Potato	*	0 of *	*	**	0							
Sugar beet	**	**		**	0							
Onion	**	**	**	**	0		**	0	**	**	**	*
Red clover	~S	~S	~S	~S	**	0						
White clover	~S	~S	~S	~S	~S	0						
Alfalfa	*S			~S	~S	**					**[b]	
French bean	**	**	**	**	**							
Broad bean	**	**	**	**	**		**	0		**	**	**
Pea	0	**	**	**	**						**	
Carrot	0	0	**	~S	**						**	**
Flax	~S	~S	~S	~S	~S							
Daffodil	0		0	0			**	0 ~ S	**			
Hyacinth	0		0	0			0	**	**			

Tulip	0		0	0		0	0	**		0	0
Veaver card	0	0	0	0					**		
Strawberry	~S	~S	~S	~S	*					**	*
Celery											**
Garlic										0	0[c]
Annual ryegrass										*	**[d]
Iris										0	

[a]Wild oats very susceptible in February
[b]Susceptible only soon after plant emergence under field conditions
[c]Garlic is very susceptible to populations from Trapani (Sicily) and Piacenza (Emilia-Romagna), Italy
[d]Annual ryegrass very susceptible in February; ** = very susceptible, * = susceptible, 0 = not susceptible, ~S = symptoms, no reproduction, 0 ~ S = not susceptible, symptoms, no reproduction

Table 4
Identification of the biotypes of populations of the citrus slow decline nematode, *Tylenchulus semipenetrans*, based on their reproduction on given differential host plants, according to [81]

	Nematode biotypes[a]		
Differential host plants	*Citrus*	*Mediterranean*	*Poncirus*
Citrus (*Citrus* sp.)	+	+	+
Kaki (*Diospyros kaki*)[b]	+	+	?
Olive (*Olea europaea*)	+	§	§
Poncirus trifoliata	–	–	+
Grape (*Vitis vinifera*)	+	+	+

§Indicates absence of nematode infection
?Indicates no data available
[a]Symbols + and – indicate good and poor nematode reproduction, respectively
[b]Different species of *Diospyros* and *Vitis* may differ in their reaction to *T. semipenetrans*

three hosts and race B reproducing only on cowpea. Afterward, Rao and Ganguly [75] added millet and mustard as differential host plants and demonstrated that among six Indian populations of the nematode, distinguishing race of the nematode was more complex. More recently, Agudelo et al. [76] found differences in the reproduction of US populations of *Rotylenchulus reniformis* on selected soybean and cotton genotypes. Genetic diversity has been reported in populations of *R. reniformis* by Arias et al. [77]. McGawley et al. [78] observed variation in reproduction and pathogenicity of populations of *R. reniformis* from six US states on soybean, while Khanal et al. [79] observed variability in reproduction of different populations that was confirmed by genetic variability observed later [80]. These findings support the hypothesis that virulent phenotypes and genotypes do exist in this nematode.

3.10.1 Schemes to Determine Different Virulence Groups Within Some of the Mentioned Nematodes

The only nematodes for which protocols to determine virulence groups are generally accepted are *Tylenchulus semipenetrans* [81] (Table 4) and *Radopholus similis*. For other nematodes, you may consider the information given in the previous subsections.

While handling these nematodes, realize that *R. similis* and *Ditylenchus* spp. may be listed as quarantine organisms for your country.

Plant species, cultivars, and plant introductions to use are listed in Tables 2, 3 and 4 or mentioned earlier. Calculation of the reproduction rate is necessary.

1. Basically, the methodology is the same described for the screening of plant resistance to different nematode species. The identification of virulent groups is based on the reproduction of the

nematodes on differential plants, such as reproduction rate or relative infection rate (% nematodes compared to susceptible plants). If the reproduction of the nematode on the differential plant is <1 or the relative infection rate is <5% or also >3%, the plant is rated resistant (nonhost); if these variables are larger, the plant is considered host (susceptible). Based on the host or nonhost status of the differential plants for a nematode population and on the schemes presented in Tables 2, 3 and 4, the nematode population can be assigned to a known race/pathotype or designated as a new virulence group.

2. For *Rotylenchulus reniformis* and *Tylenchulus semipenetrans*, nematode eggs can be extracted with the Hussey and Barker method [27], while for endoparasitic migratory nematodes (*Ditylenchus* spp., the root-lesion nematodes *Pratylenchus* spp. and *Radopholus similis*), the extraction of the nematodes from roots or plant tissues is necessary.
3. For details on extraction methods, see Chap. 1 in this book or as suggested by authors cited in the schemes.

3.11 Final Considerations

Protocols have herein been described for the most important nematodes, such as endoparasitic sedentary nematodes (Chap. 3) and migratory and semi-endoparasitic nematodes (this chapter). Another group of nematodes, considered as ectoparasites based on their feeding habit, may also attack and damage plants. For instance, longidorid and trichodorid nematodes, besides being vectors of plant viruses, may cause direct damage to many crop plants because of their feeding activity on plant tissues, such as roots and leaves, causing necrosis, root tip galls, plant deformations, and decolorations. Resistance to these nematodes is very limited and, therefore, no protocol has been designed. However, to screen for resistance in grape to *Xiphinema index*, Ferris et al. [64] considered the reproduction rate of the nematode in the soil and the typical symptoms of the nematode on grape roots, the apical galls.

In the UK, severe damage to sugar beet is caused by the docking disorder by *Trichodorus* spp., *Paratrichodorus* spp., and *Longidorus* spp., especially in light sandy soils and wet period of the years [84]. In several US states and Ontario (Canada), the ectoparasitic nematodes *Belonolaimus* sp., *Helicotylenchus* sp., and *Tylenchorhynchus* sp., along with other nematodes, are among the constraints of maize [85]. *Hoplolaimus* spp., and in particular *H. columbus*, known as lance nematodes damaging cotton, soybean, corn, and other crops, are spread throughout southern and mid-western USA [86]. *Helicotylenchus multicinctus* is considered as the second most important nematode of bananas and plantains [87]; however, this nematode, contrary to other species of the genus, is regarded as an endoparasite. Many ectoparasitic nematodes are known to affect grape production all over the world [88] and others may have a rather local importance. In early June 1995, the author (N. Greco) observed a positive correlation between the severity of stunting and yellowing of wheat and soil populations of a *Tylenchorhynchus* sp. in the transitional area of Turkey, between Ankara and Eskisehir.

Should it be necessary, to screen for resistance to these nematodes, the reproduction rate of the nematodes and the severity of the symptoms they cause to the host plants must be considered to discriminate between the tested genotypes, both under protected and field conditions. The required inoculum can be collected from infested fields near or soon after harvest time. However, several of the mentioned nematodes can also be reproduced in greenhouse on one of the best host plants.

4 Notes

1. It is important that, at inoculation time, nematodes find enough roots to infect for their reproduction; otherwise, many will die of starvation, leading to underestimation of Rf. To evaluate the genotypes according to both extension of root necrosis and reproduction rate, to discriminate better between susceptible control and different genotypes, it is suggested to do the evaluation when plants are still actively growing with no more than 60–70% of the roots necrotized. When plants start to senesce or roots are much damaged, nematodes in both soil and roots start to decline, and it may happen that nematodes in the susceptible control are less than in tested genotypes, thus leading to wrong evaluation. To avoid this problem, it is suggested to do some preliminary observations on the extra susceptible plants and do the evaluation at the right time. Toktay et al. [82], studying factors that may affect the results of screening spring wheat for reaction to *P. thornei*, concluded that inoculating each plant with 400 specimens from carrot disk culture 1 week after sowing well pre-germinated seeds (having three 1 cm long radicles), use of sandy soil (70:29:1 sand, field soil, organic matter), small size tubes (15 × 100 mm), 9-week harvest time, and bottom perlite watering system are the best conditions for screening.
2. Screening of *Musa* for resistance to nematodes can be conducted both under field conditions and in pot in greenhouse.
3. As numbers of gall on the roots are a few, for a higher statistical significance, it is suggested to increase the size of the pots to 10–15 cm in diam. and inoculate two to four nematodes per cm^3 soil. Discontinuing the test 90 days after inoculation would mean that plants have remained in small pots (200 g soil) and become pot bounds, which in turn reduces plant growth, nematode development, and Rf.

4. The method used by Dijkstra [31] has been and is still used, with some adjustment, with many seeds and especially with clovers and alfalfa. However, as a special filter paper, several investigators use chromatographic paper cut at the required size. Instead of pots, especially for alfalfa, clover, and sugar beet, aluminum foil or celluloid tubes are used, which are about 20–25 cm long and 3 cm in diameter. Screening of onion to *D. dipsaci* can also be conducted in small pots as for other crop plants [83].
5. To judge the results of the screening, check the Rf in the two controls. If Rf in the susceptible control is rather low, most probably many specimens of the inoculum died before infecting. Check how environment conditions in the greenhouse proceeded especially soon after inoculation. If Rf of the resistant control is >1, probably this cv. was not really resistant or the nematode population was too aggressive compared to other populations occurring in the area. However, if looking outside and inside the tubers appear very damaged, this may have reduced also Rf. Therefore, in future tests, reduce the numbers of inoculated nematodes and/or anticipate the final observation of 4–8 weeks. To follow infection and damage progress, you may wish to plant extra control plants to be checked after 8–10 weeks.
6. The use of small pots (15–20 cm in diam) in greenhouse could be used but it is not suggested. Besides the difficulty of pod setting, in greenhouse there could be severe problems caused by mites, white fly, and other pests. Also, it is important to decide when to do the final observation. In the soil, when plants start to senesce, the nematode starts its natural decline. Therefore, to avoid misinterpretation of the Rf, soil sampling and nematode extraction must be done just before the plants start to senesce. If the evaluation will be done on fruit quality, waiting until plant senescence is suggested.
7. This last method is more appropriate when screening several thousands of seeds as when dealing with progenies from breeding programs because every seed is equally important and must be scored to decide to or not save it. Resistant plants can be transplanted singly in larger pots and grown to produce seeds or to be crossed.
8. The methods to extract nematodes from rice plants are destructive. Therefore, to figure the presence of infections, the development of the symptoms must be observed. Unfortunately, not all infected plants show symptoms of the nematode infections. Therefore, when screening progenies from crosses of a breeding program, we suggest eliminating all plants that show symptoms, while all the remaining plants must be grown to

produce mature seeds. Then part of the seed of each plant must be used to extract nematodes and only plants whose seeds are free of nematodes can be designated as not infected.

9. Some researchers use as inoculum eggs and mixed stages of the nematodes extracted from the roots ground in sodium hypochlorite solution followed by centrifugation in magnesium solution, while others prefer mixed nematode stages extracted from soil from a nematode culture. We suggest using the second method because it consists of nematode not damaged. Nematodes extracted with hypochlorite followed by centrifugation in magnesium sulfate or different chemical solutions may have reduced infectivity because of the damaging effect of the chemicals.

Acknowledgments

I wish to thank all journals and colleagues who have granted permissions to reuse their figures; Dr. Graham Stirling, Australia, for the useful suggestions; and Dr. Alberto Troccoli, Italy, for the great work done with figures.

References

1. Peng Y, Moens M (2003) Host resistance and tolerance to migratory plant-parasitic nematodes. Nematology 5:145–177
2. Roberts PA (1990) Resistance to nematodes: definitions, concepts, and consequences. In: Starr JL (ed) Methods for evaluating plant species for resistance to plant-parasitic nematodes. Society of Nematology Inc., Hyattsville, pp 1–15
3. Roberts PA (2002) In: Starr JL, Bridge J, Cook R (eds) Concept and consequence of resistance. Plant resistance to parasitic nematodes. CABI Publishing, Wallingford, pp 23–41
4. Trudgill DL (1991) Resistance to and tolerance of plant parasitic nematodes in plants. Annu Rev Phytopathol 29:167–192
5. Andrews M, Andrews ME (2017) Specificity in legume-rhizobia symbioses. Int J Mol Sci 18:705. https://doi.org/10.3390/ijms18040705
6. Reen RA, Mumford MH, Thomson JP (2019) Novel sources of resistance to root-lesion nematode (*Pratylenchus thornei*) in a new collection of wild *Cicer* species (*C. reticulatum* and *C. echinospermum*) to improve resistance in cultivated chickpea (*C. arietinum*). Phytopathology 109:1270–1279. https://doi.org/10.1094/PHYTO-02-19-0047-R
7. Thomson JP, Sheedy JG, Robinson NA (2020) Resistance of wheat genotypes to root-lesion nematode (*Pratylenchus thornei*) can be used to predict final nematode population densities, crop greenness, and grain yield in the field. Phytopathology 110:505–516. https://doi.org/10.1094/PHYTO-06-19-0203-R
8. Rostad HE, Reen RA, Munford MH, Zuart RS, Thompson JP (2022) Resistance to root-lesion nematode *Pratylenchus neglectus* identified in a new collection of two wild chickpea species (*Cicer reticulatum* and *C. echinospermum*) from Turkey. Plant Pathol 71:1205–1219. https://doi.org/10.1111/ppa.13544
9. Castillo P, Vovlas N (2007) In: Hunt DJ, Perry RN (eds) *Pratylenchus* (Nematoda: Pratylenchidae): diagnosis, biology, pathogenicity and management, Nematology monographs and perspectives No. 6. Brill, Leiden, p 529
10. France RA, Brodie BB (1995) Differentiation of two New York isolates of *Pratylenchus penetrans* based on their reaction on potato. J Nematol 27:339–345
11. Al-Khafaji RT, Troth EEG, Lambert KN, Jhonston JA, Dyer AT (2019) Pathotypes detected among populations of *Pratylenchus neglectus* collected from Montana. Plant Dis

103:3259–3264. https://doi.org/10.1094/PDIS-12-18-2234-RE

12. Thompson AL, Smiley RW, Paulitz TC, Garland-Campbell K (2016) Identification of resistance to *Pratylenchus neglectus* and *Pratylenchus thornei* in Iranian landrace accessions of wheat. Crop Sci 56:654–672. https://doi.org/10.2135/cropsci2015.07.0438
13. Hollaway K, Knights S (2022) NVT harvest report – Southern Queensland. GRDC-National Variety Trials. ISSN: 2652-5771 (online)
14. Figueiredo J, Vieira P, Abrantes I, Esteves I (2022) Potato commercial cultivars exhibit distinct susceptibility to root lesion nematode *Pratylenchus penetrans.* Horticulturae 8:244. https://doi.org/10.3390/horticulturae8030244
15. Colen WA (1979) Methods for the extraction of *Meloidogyne* spp. and other nematodes from roots and soil. In: Lamberti F, Taylor CE (eds) Root-knot nematodes (*Meloidogyne* spp.) – systematics, biology and control. Academic, London, pp 317–329
16. Marul J, Pinochet J (1991) Host suitability of *Prunus* rootstocks to four *Meloidogyne* species and *Pratylenchus vulnus* in Spain. Nematropica 21:185–195
17. Stalin N, Salesses G, Pinochet J, Minot JC, Voisin R, Esmenjaud D (1998) Comparative host suitability to *Meloidogyne* spp. and *Pratylenchus vulnus* in Myrobalan plums (*Prunus cerasifera*). Plant Path 47:211–215
18. Nyczepir AP, Pinochet J (2001) Assessment of Guardian peach rootstock for resistance to two isolates of *Pratylenchus vulnus.* J Nematol 33: 302–305
19. Bhuiyan SA, Croft BJ, Stirling GR, Wong E, Jackson P, Cox M (2016) Assessment of resistance to root-lesion and root-knot nematodes in Australian hybrid clones of sugarcane and its wild relatives. Australas Plant Pathol 45:166–173
20. Dos Santos PR, Viana AP, Gomes VM, da Preisigke SC, Santos EA, Cavalcante NR, De Almeida OF, Walker MA (2018) Clonal selection in interspecific *Vitis* spp. hybrids resistant to the root-lesion nematode *Pratylenchus brachyurus* by REML/BLUP. Fruits 73:191–197. https://doi.org/10.17660/th2018/73.3.6
21. Elsen A, Stoffelen R, Tuyet NT, Bamey H, de Boulois HD, De Waele D (2002) *In vitro* screening for resistance to *Radopholus similis* in *Musa* spp. Plant Sci 163:407–416
22. Dochez C, Dusabe J, Tenkouano A, Ortiz R, White J, De Waele D (2013) Screening *Musa* germplasm for resistance to borrowing nematode populations from Uganda. Genet Resour Crop Evol 60:367–337. https://doi.org/10.1007/s10722-012-9841-7
23. Sarah JL (1996) A laboratory method for early varietal screening of banana for resistance to nematodes. In: New frontiers in resistance breeding for nematode, Fusarium and Sigatoka. IPGRI, Rome, pp 58–61
24. Stoffelen R, Verlinden R, Pinochet J, Swennen RL, De Waele D (2000) Host plant response of Fusarium wilt resistant *Musa* genotype to *Radopholus similis* and *Pratylenchus coffeae.* Int J Pest Manag 46:289–293
25. Manzanilla-Lopez RH, Costilla MA, Doucet M, Franco J, Inserra RN, Lehman PS, Cid Del Prado-Vera I, Souza RM, Evans K (2002) The genus *Nacobbus* Thorne et Allen, 1944 (Nematoda: Pratylenchidae) systematics, distribution, biology and management. Nematropica 32:149–227
26. Lax P, Dueñas JCR, Coronel NB, Cardenal CG, Bima P, Doucet ME (2011) Host range of Argentine *Nacobbus aberrans sensu* Sher populations and comments on the differential host test. Crop Prot 30:1414–1420
27. Hussey RS, Barker KR (1973) A comparison of methods of collecting inocula of *Meloidogyne* spp., including a new technique. Plant Dis Rep 57:1025–1028
28. Suárez SI, Chavez EJ, Clausen AM, Franco J (2009) *Solanum* tuber-bearing species behavior against *Nacobbus aberrans.* J Nematol 41:5–10
29. Cazarez-Álvarez AA, Mendéz-Inocenzio Zepeda-Jazo I, Fernández-Herrera E, Medina-Medrano JR, Villar-Luna E (2019) Host suitability of five populations of wild tomato (*Solanum lycopersicum* var. *cerasiforme*) for nematode *Nacobbus aberrans sensu lato.* Rev Mex Fitopatol 37:10.18781/r.mex.fit.1905-1
30. Greco N, Vovlas N (2007) The stem and bulb nematode *Ditylenchus dipsaci*: a worldwide severe parasite of many crop plants. Fitopatologia Venezuelana 20:2–14
31. Dijkstra J (1956) Experience with breeding of red clover resistant to the stem eelworm. Euphytica 5:298–307
32. Williams WM (1972) Laboratory screening of white clover for resistance to stem nematode. N Z J Agric 15:363–370
33. Whithehead AG, Fraser JE, Nichols AJF (1987) Variation in the development of stem nematode, *Ditylenchus dipsaci*, in susceptible and resistant crop plants. Ann Appl Biol 111: 373–383
34. Cook R, Evans DR (1988) Observation on resistance in white clover (*Trifolium repens* L.)

to the stem nematode (*Ditylenchus dipsaci* (Kühn) Filipjev). J Agric Sci 110:145–154
35. Storelli A, Kiewnick S, Daub M, Malhein A-K, Schumann M, Beyer W, Keiser A (2021) Virulence and pathogenicity of four *Ditylenchus dipsaci* populations on sugar beet. Eur J Plant Pathol 161:63–71. https://doi.org/10.1007/s10658-021-22304-w
36. Vovlas N, Troccoli A, Palomares-Rius JE, De Luca F, Liébanas G, Landa BB, Subbotin SA, Castillo P (2011) *Ditylenchus gigas* n. sp. parasitizing broad bean: a new stem nematode singled out from the *Ditylenchus dipsaci* species complex using a polyphasic approach with molecular phylogeny. Plant Pathol 60:762–775. https://doi.org/10.1111/j.1365-3059.2011.02430.x
37. Caubel G, Leclercq D (1989) Estimation de la resistance à la race geant de *Ditylenchus dipsaci* par les symptoms chez la feverole (*Vicia Faba* L.). Nematologica 35:216–224
38. Plowright RA, Caubel G, Mizen KA (2002) *Ditylenchus* species. In: Starr JL, Cook R, Bridge J (eds) Plant resistance to parasitic nematodes. CAB International, Wallingford, pp 107–139
39. Abbad Andaloussi F (2001) Screening of *Vicia faba* for resistance to the "giant race" of *Ditylenchus dipsaci* in Morocco. Nematol Mediterr 29:29–33
40. Mwaura P, Niere B, Vidal S (2015) Effect of initial population densities of *Ditylenchus destructor* and *D. dipsaci* on potato tuber damage and nematode reproduction. Nematology 17:193–202
41. Smaliev H, Markova D (2015) Resistance of potato cultivars to *Ditylenchus dipsaci* and *Ditylenchus destructor*. Sci Technol 5:1–7
42. De Waele D, Jones BL, Bolton C, van den Berg E (1989) *Ditylenchus destructor* in hulls and seeds of peanut. J Nematol 21:10–15
43. Steenkamp S, McDonald AH, De Waele D (2010) Resistance to *Ditylenchus africanus* present in peanut breeding lines. J Nematol 42:159–165
44. Haque A, Latif MA (2011) Screening of ufra (*Ditylenchus angustus*) resistant rice genotypes. Eco-Friendly Agric J 4:527–530
45. Khanam S, Akanda AM, Ali MA, Kyndt T, Gheisan G (2015) Identification of Bangladeshi rice varieties resistant to ufra disease caused by the nematode *Ditylenchus angustus*. Crop Prot 79:162–169. https://doi.org/10.1016/j.cropro.2015.09.009
46. Hui F, Li-hui W, Mao-song L, Yi-jung Z (2014) Assessment of rice cultivars in China for resistance to *Aphelenchoides besseyi*. J Integr Agric 13:2221–2228
47. Azizah N, Indarti S, Widiastuti A, Trisyomo YA (2019) Detection and development rate of *Aphelenchoides besseyi* on various rice seed varieties. Jurnal Perlindungan Tanaman Indonesia 23:305–310
48. Subbotin SA, Oliveira CJ, Alvarez-Ortega S, Desaeger JA, Row WC, Oversyreet C, Leahy R, Vau S, Inserra RN (2021) The taxonomic status of *Aphelenchoides besseyi* Christie, 1942 (Nematoda: Aphelenchoididae) populations from the southeastern USA, and description of *Aphelenchoides pseudobesseyi* sp. n. Nematology 23:381–413
49. Latif MA, Rahman ML, Bakr MA (1997) Evaluation of inoculation method for white tip disease of rice. J Bangladesh Agric Univ 7:15–19
50. Xie J, Yang F, Wang Y, Peng Y, Ji H (2019) Studies on the efficiency of different inoculation methods of rice white-tip nematode, *Aphelenchoides besseyi*. Nematology 21:673–678
51. Godoy FMC, Overstreet C, McGawley EC, Hollier CA, Kularathna MT, Khanal C, McInnes B (2019) Incidence of *Aphelenchoides besseyi* in rice in Louisiana and host status of the most widely planted cultivars. Nematropica 49: 107–123
52. Jamali S, Mousanejad S (2011) Resistance of rice cultivars to white tip disease caused by *Aphelenchoides besseyi* Christie. J Agric Technol 7:441–447
53. Parveen R, Khan AA, Imran M, Ansari AA (2003) Response of wheat varieties to the seed gall nematode, *Anguina tritici*. Nematol Mediterr 31:103–104
54. Qassem NE, Al-Taae HHW, Thanoon (2021) Screening of varieties of wheat for infestation by the seed gall nematode *Anguina tritici*. Plant Cell Biotechnol Mol Biol 22:94–105
55. Tülek A, Kepenekci I, Tülek B, Sakin A (2021) Determination of the reaction of some bread and durum wheat varieties to the wheat seed gall nematode [*Anguina tritici* (Steinbuch) Filipjev]. Plant Prot Bull 61:13–18. https://doi.org/10.16955/bitkorb.822783
56. Stetina SR, Smith JR, Ray JD (2014) Identification of *Rotylenchulus reniformis* resistant *Glycine* lines. J Nematol 46:1–7
57. Han J, Locke SP, Herman TK, Schroeder NE, Hartman GL (2022) Evaluation of perennial *Glycines* species for response to *Meloidogyne incognita*, *Rotylenchulus reniformis*, and *Pratylenchus penetrans*. J Nematol 54. https://doi.org/10.21307/jofnem-2022-001
58. Thies JA, Merril SB, Corley EL (2002) Red food coloring stain: new safer procedures for

staining nematodes in roots and egg masses on root surface. J Nematol 34:179–181

59. Robinson AF, Bridges AC, Oercival AE (2004) New sources of resistance to reniform (*Rotylenchulus reniformis* Linford and Oliveira) and root-knot (*Meloidogyne incognita* Kofoid &White) Chitwood in upland (*Gossypium hirsutum* L) and Sea Island (*G. barbadense* L.) cotton. J Cotton Sci 8:191–197
60. Sipes BS, Schmitt DP (1994) Evaluation of pineapple (*Ananas comosus*) for host plant resistance and tolerance to *Rotylenchulus reniformis* and *Meloidogyne javanica*. Nematropica 24:113–121
61. Sharma SB, Jain KC, Lingaraju S (2000) Tolerance to reniform nematode (*Rotylenchulus reniformis*) race A in pigeonpea (*Cajanus cajan*). Ann Appl Biol 136:247–252
62. Konan NO, De Proft M, Ruano O, Baduoin J-P, Mergeai G (2014) A screening procedure for evaluating cotton for *Rotylenchulus reniformis* resistance in controlled conditions. Tropicoltura 32:3–9
63. Sacks EJ, Robinson AF (2009) Introgression of resistance to reniform nematode (*Rotylenchulus reniformis*) into upland cotton (*Gossypium hirsutum*) from *Gossypium arboreum* and a *Gossypium/Gossypium aridum* bridging line. Field Crop Res 112:1–6
64. Ferris H, Zheng L, Walker MA (2012) Resistance of grape rootstock to plant-parasitic nematodes. J Nematol 44:377–386
65. Aballay E, Navarro A (2005) Tolerance of some grapevine root-stock to *Tylenchulus semipenetrans* Cobb 1914 in Chile. Agricultura Técnica (Chile) 65:319–322
66. Inserra RN, Duncan LW, O'Bannon JH, Fuller SA (1994) Citrus nematode biotypes and resistant citrus rootstocks in Florida. Nematology Circular 205. Gaineswille, FL, USA, Florida Department of Agriculture and Consumer Service, Division of Plant Industry, 4 pp
67. Walker A (2006) New nematode-resistant rootstocks are nearing release. Foundation Plant Service, FPS Grape Programme Newsletter, November: 8–10
68. Young TW (1954) An incubation method for collecting endo-parasitic migratory nematodes. Plant Dis Rep 38:794–795
69. Greco N, D'Addabbo T (1990) Efficient procedure for extracting *Tylenchulus semipenetrans* from citrus roots. J Nematol 22:590–593
70. Galeano M, Verdejo-Lucas S, Sorribas FJ, Ornat C, Forner JB, Alcade A (2003) New citrus selection from Cleopatra mandarin × *Poncirus trifoliata* with resistance to *Tylenchulus semipenetrans* Cobb. Nematology 5:227–234
71. Owen K (2022) A triumph of tolerance: managing the threat to wheat production by the root lesion nematode *Pratylenchus thornei* in the subtropical grain region of eastern Australia. In: Sikora RA, Desaeger J, Molendijk LPG (eds) Integrated nematode management: state-of-the-art and vision for the future. CABI, Wallingford, pp 13–19
72. Mwaura P, Niere B, Vidal S (2015) Resistance and tolerance of potato varieties to potato rot nematode (*Ditylenchus destructor*) and stem nematode (*Ditylenchus dipsaci*). Ann Appl Biol 166:257–270
73. Seinhorst JW (1957) Some aspect of the biology and ecology of stem eelworm. Nematologica 2(Suppl):355–361
74. Dasgupta DR, Seshadri AR (1971) Races of the reniform nematode, *Rotylenchulus reniformis* Linford and Oliveira, 1940. Indian J Nematol 1:21–24
75. Rao GMVP, Ganguly S (1996) Host preference of six geographical isolates of reniform nematode, *Rotylenchulus reniformis*. Indian J Nematol 26:19–22
76. Agudelo P, Robbins RT, Steward JMCD, Szalanski AL (2005) Intraspecific variability of *Rotylenchulus reniformis* from cotton-growing regions in the United States. J Nematol 37: 105–114
77. Arias RS, Stetina SR, Tonos JL, Scheffler JA, Scheffler BE (2009) Microsatellites reveal genetic diversity in *Rotylenchulus reniformis* populations. J Nematol 41:146–156
78. McGawley ED, Overstreet C, Pontif MJ (2011) Variation in reproduction and pathogenicity of geographic isolates of *Rotylenchulus reniformis* on soybean. Nematropica 41:12–22
79. Khanal C, McGawley EC, Overstreet C, Stetina SR, Myers GO, Kularathna MT, McInnes B, Godoy FMC (2018) Reproduction and pathogenicity of endemic populations of *Rotylenchulus reniformis* in cotton. Nematropica 48:68–81
80. Khanal C, Khularathna MT, Ray GD, Stetina SR, McGawley EC, Overstreet C (2019) Single nucleotide polymorphism analysis using KASP reveals genetic variability in *Rotylenchulus reniformis*. Plant Dis 103:1835–1842. https://doi.org/10.1094/PDIS-11-18-1975
81. Inserra RN, Vovlas N, O'Bannon JH (1980) A classification of *Tylenchulus semipenetrans* biotypes. J Nematol 12:283–287
82. Toktay H, Imren M, Nicol JM, Dadabat A, Elekçioğlu IH (2012) Improved methodology for resistance screening in spring wheat against

the root lesion nematode, *Pratylenchus thornei* (Sher *et* Allen) (Tylenchida: Pratylenchidae). Türk Entomol Derg 36:533–540

83. Yavuzaslanoglu E (2019) Resistance and tolerance of commercial onion cultivars to stem and bulb nematode, *Ditylenchus dipsaci*. J Agric Sci 25:409–416. https://doi.org/10.15832/ankutbd.440179
84. Cook DA (1989) Damage to sugar-beet crops by ectoparasitic nematodes, and its control by soil-applied granular pesticides. Crop Prot 8: 63–70
85. Simon ACM, Lopez-Nicora HD, Niblack TL (2022) Impact of plant parasitic nematodes on maize in mid-western USA: an unrecognized or ignored treat to production. In: Sikora RA, Desaeger J, Molendijk LPG (eds) Integrated nematode management: state-of-the art and vision for the future. CABI, Wallingford, pp 34–40
86. Mueller J (2022) *Hoplolaimus columbus*: a prime candidate for site-specific management in cotton and soybean production. In: Sikora RA, Desaeger J, Molendijk LPG (eds) Integrated nematode management: state-of-the art and vision for the future. CABI, Wallingford, pp 80–86
87. Gowen SR, Quénéhervé P, Fogain R (2005) Nematode parasites of bananas and plantains. In: Luc M, Sikora RA, Bridge J (eds) Plant parasitic nematodes in subtropical and tropical agriculture, 2nd edn. CABI Publishing, Wallingford, pp 611–643
88. Zasada I, Forge T (2022) Ectoparasitic nematodes: emerging challenge to wine grape production in the Pacific northwest of North America. In: Sikora RA, Desaeger J, Molendijk LPG (eds) Integrated nematode management: state-of-the art and vision for the future. CABI, Wallingford, pp 192–198
89. Holgado R, Oppen Skau KA, Magnusson C (2009) Field damage in potato by lesion nematode *Pratylenchus penetrans*, its association with tuber symptoms and its survival in storage. Nematol Medit 37:25–29
90. EFSA Panel on Plant Health (PLH) (2014) Scientific opinion on the pest categorization of Ditylenchus destructor Thorne. EFSA J 12: 3834
91. Lamberti F, Greco N (1974) Piante coltivate ospiti di *Ditylenchus dipsaci* (Kühn) Filipjev. Nematol Meditrr 2:159–164
92. Greco N, Brandonisio A, Boncoraglio P (2002) Investigations on *Ditylenchus dipsaci* damaging carrot in Italy. Nematol Meditrr 30: 139–146
93. Hanna MR, Hawn EJ (1965) Seedling inoculation studies with alfalfa stem nematode. Can J Plant Sci 45:357–363

Chapter 5

Molecular Techniques for Root-Knot Nematode Identification

Regina M. D. G. Carneiro, Caio F. B. Souza, Vanessa S. Mattos, and Valdir R. Correia

Abstract

Among plant-parasitic nematodes, root-knot nematodes (RKN), *Meloidogyne* spp., are the most important parasite infecting economically important crops globally and causing severe losses in crop production. The use of efficient nematode control methods against these parasites depends upon their correct detection in roots and soil samples. Currently, the use of integrated identification methods, including biochemical, molecular, and morphological-based characters, is preferred. But the techniques using morphology and phylogenetic analysis are time-consuming and not suitable for routine analysis. They have only been used for studies of cryptic species, which were identified using integrative taxonomy. Here we describe the enzymatic and molecular-based methods that have successfully been used in Brazil for more than 25 years in the Nematology Lab at Embrapa Genetic Resources and Biotechnology for routine analysis. This technique is a combination of isozyme esterase profiling and molecular markers, with the aim of having a rapid and correct diagnosis of *Meloidogyne* spp. populations from field and greenhouse.

Key words Diagnosis, Esterase phenotypes, Multiplex PCR, PCR, RKN molecular identification, SCAR markers

1 Introduction

Nematodes are microscopic multicellular animals comprising free-living and plant-parasitic species. Plant-parasitic nematodes parasitize important cultivated crops, including monocots and dicots, and are considered the main constraint on crop production, with an estimated annual crop loss of $ 78 billion worldwide and an average crop yield loss of 10–15% [1–3].

Root-knot nematodes (RKN), *Meloidogyne* spp., are the most aggressive, damaging, and economically important group of plant-parasitic nematodes infecting several plant species, including important crops globally [2, 3]. Among the 21 species detected in Brazil, the most widespread and major RKN species include

Sergio Molinari (ed.), *Plant-Nematode Interactions*, Methods in Molecular Biology, vol. 2756,
https://doi.org/10.1007/978-1-0716-3638-1_5,

Meloidogyne javanica and *M. incognita*; likewise, other RKN species are widespread and detected in important crops in certain regions, e.g., *M. exigua*, *M. paranaensis*, *M. graminicola*, and *M. enterolobii*, and have become a constraint on crop production in Brazil [4, 5].

Some chemical nematicides have been removed from the market due to their toxicity and harmful effects. Alternatively, other methods in controlling nematodes in the field have been applied. These include the use of resistant cultivars, crop rotation, crop succession, antagonistic plants, fallow, new nematicides, and biological control. To efficiently use these control methods, a correct and precise diagnosis of nematode species detected in plants and soil samples is needed. However, because these nematodes are diverse, polyphagous, with different parasitism strategies and with varying morphological characters, the identification of species relying solely on morphology is extremely difficult, even for nematode taxonomists [6–10].

Moreover, nematodes are difficult to identify due to their small size and lack of unique features observed under light microscope. *Meloidogyne* spp. diagnosis has traditionally relied on the report of female perineal patterns, morphology of adults, and morphometrics. Nevertheless, these characters are uneven and overlap among RKN species, making them non-diagnostic for most species (e.g., *M. incognita*, *M. paranaensis*, *M. luci*, *M. ethiopica*, *M. izalcoensis*, and *M. enterolobii*) [9]. This has led to misidentification of species when using morphology as a single criterion [10, 11]. Alternatively, the use of biochemical and molecular methods, such as isozyme esterase profiling and DNA-based markers, has proved reliable and allowed diagnosis in most cases, reviewed in [4, 8–10, 12–17].

The first conclusive study of the identification of *Meloidogyne* species through esterases, using individualized females, was carried out by Janati et al. [18], who studied 86 populations, having identified 75 at a specific level. The remaining 11 populations were atypical phenotypes and later characterized and described in studies using integrative taxonomy, as is the case of *M. paranaensis*, which appears in this work and was described 14 years later [19, 20]. Subsequently, Esbenshade and Triantaphyllou [21], already in collaboration with the International *Meloidogyne* Project (IMP), studied about 300 populations from 65 countries and several continents and reported the esterase patterns of several *Meloidogyne* species, the most common phenotypes being Est A2 and Est A3 (*M. arenaria*), Est H1 (*M. hapla*), Est I1 (*M. incognita*), and Est J3 (*M. javanica*). Some cryptic species were characterized in this work and later described. Carneiro et al. [10, 22] found 18 esterase phenotypes in more than 100 populations of *Meloidogyne* spp. from Brazil and other countries in the Americas, characterizing three cryptic species in Brazil. Among

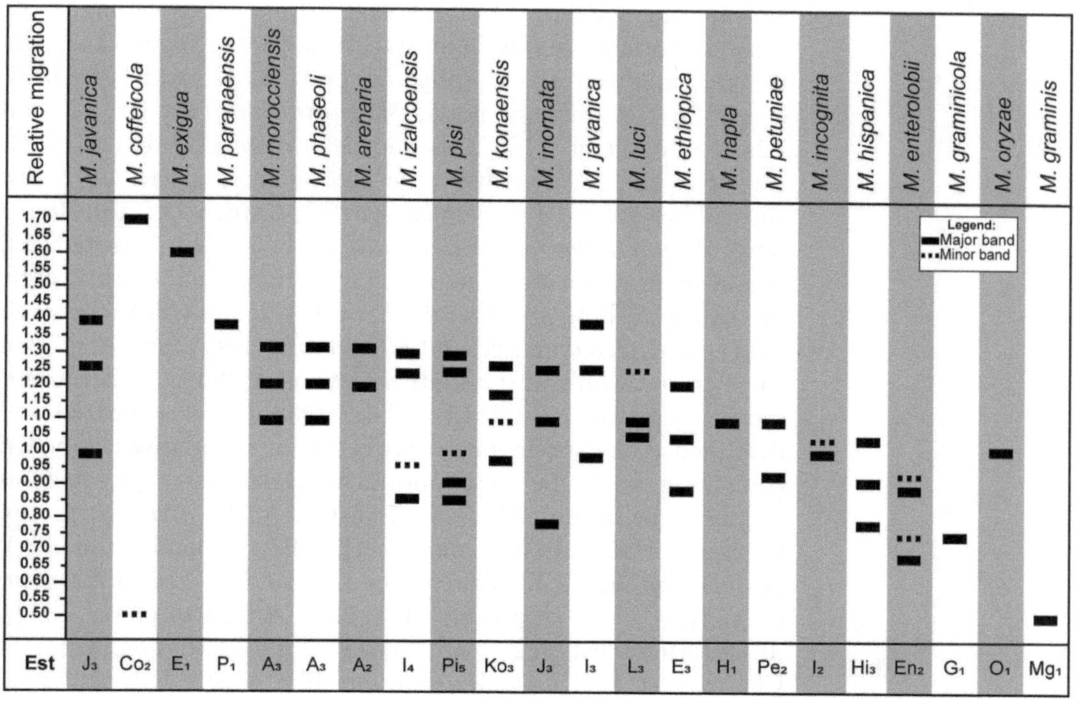

Fig. 1 Esterase (Est) phenotypes of 21 *Meloidogyne* species detected in Brazil. Relative migration (Rm), migration ratio in relation to the slowest band of *M. javanica* (Rm = 1.0)

these 18 phenotypes studied, *M. enterolobii* stands out, appearing soon after, in Brazil, decimating guava trees in Petrolina and other localities. After 2000, isoenzymes became more widely used in prospecting studies of *Meloidogyne* spp., in different crops and different regions in Brazil, including private laboratories. These findings were reviewed by Carneiro et al. [4]. Although a recent review of the use of enzymes in the identification of *Meloidogyne* species has been done, no schematic diagram has been provided [23]. In the diagram provided in this chapter (*see* Fig. 1), the enzymatic phenotypes were designated using the first letter as the species name and the number of bands of that species (e.g., *M. javanica*, Est J3). Species with the same initials and number of bands were distinguished by small letters (e.g., *M. enterolobii*, Est En2; *M. hispanica*, Est Hi3) [10, 21, 22, 24]. Enzyme patterns are generally compared with a known pattern of *M. javanica* (Est J3), which is included in electrophoresis gels to determine the relative migration distances.

Esterase isozyme patterning (Fig. 1) has been used for diagnosing *Meloidogyne* spp. from a wide range of samples and has been proved to be species-specific for almost all species detected in Brazil [4, 10, 21], except for the *M. arenaria* group, including *M. morocciensis* and *M. phaseoli*, which are swarm species. After 25 years of species identification in different crops, *M. javanica*

(Est J3, J2a, and J2b, Santos et al. [25]) proved to be one of the most important species in hosts like soybean, sugar cane, and tomato, followed by *M. incognita* (Est I1, I2, and S2, Santos et al. [26]) in different hosts; *M. paranaensis* (Est P1, P2, and P2a, Santos et al. [20, 27]) and *M. exigua* (Est E1, E2, E3, Muniz et al. [28, 29]) were the predominant species on coffee and rubber trees; *M. coffeicola* (Est C2, Carneiro et al. [10]) is rare. *M. enterolobii* (Est En2, Carneiro et al. [4]) was detected mainly in the guava crop and on some vegetables and more recently on cotton [30]. *M. graminicola* (Est G1, G2, and G3, Soares et al. [31]) was detected on rice, and *M. arenaria* (Est A2 and A1 [7]) in some crops, mainly in the southern region of Brazil. Other minor species were also identified based on atypical esterase profiles and described as new species or redescribed using integrative taxonomy, comparing with the descriptions, in different crops, such as *M. morocciensis* and *M. phaseoli* (Est A 3, *M. arenaria* group, Carneiro et al. [7]), *M. ethiopica* (Est E3, Carneiro et al. [6]), *M. inornata* (Est I3, Carneiro et al. [32]), *M. luci* (Est L3, Carneiro et al. [33]), *M. hispanica* (Est Hi3, Carneiro et al. [34]), *M. petuniae* (Est Pe2, Carneiro et al. [4]), *M. pisi* (Est Pi5, Charchar et al. [35]), *M. graminis* (Est Mgr1, Oliveira et al. [36]), *M. oryzae* (Est O1, Mattos et al. [37]), and *M. ottersoni* without any esterase band (Est Ot0, Leite et al. [38]). All these esterase profiles are presented in Fig. 1. Unfortunately, with the advent of molecular techniques, the use of esterase was abandoned in many laboratories around the world, and the interpretation of enzyme phenotypes became difficult and little used, leading to misidentifications [39]. This enzymatic technique also allowed the synonymization of two new Brazilian RKN species (*M. brasiliensis* = *M. ethiopica*, Monteiro et al. [40]) and *M. polycephannulata* = *M. incognita*, Monteiro et al. [41]) that presented the same esterase profiles.

Alternatively, DNA-based detection tools make excellent methods of *Meloidogyne* diagnosis since they are independent of phenotypic variation and interpretation, simple, accurate, and fast and can be used in a high-throughput manner [8, 9, 14, 42].

Although molecular-based identification methods have been proved reliable and simple, there are some obstacles when identifying RKN species using only molecular tools. Problems associated with molecular RKN diagnosis include the following: (i) the optimization and validation of DNA extraction protocols that are reliable for routine use; (ii) the amount and quality of DNA may be insufficient for downstream analyses, such as PCR analysis and DNA sequencing, especially if the extraction is from samples that are not in good shape, such as some quarantine materials; (iii) species-specific primers have been developed for detection of several RKN species, but they are not available for all RKN species, and some sets of primers may show problems related to specificity, sensitivity, false-positive and false-negative results; and (iv) when developing a species-specific primer, not several isolates of a RKN species may be tested and validated due to limitations of available isolates [23, 43]. Nevertheless, even with these challenges and limitations, molecular tools have been applied to characterize different species and populations, and it is already part of an integrative diagnostic in our and many laboratories. Indeed, when DNA-based markers are used in previous combination with esterase patterning, there is more chance of correctly diagnosing RKN populations, because the species is identified first by the esterase profile, making it easier to choose the specific primers to confirm it [30]. The presence of more than one species or atypical populations always appears when esterases are used and are not detected by specific primers [4].

Specific primers have been developed based on target regions of the ribosomal DNA (rDNA) cistron, e.g., the external transcribed spacer (ETS), internal transcribed spacers 1 and 2 (ITS1 and ITS2), intergenic space regions 1 and 2 (IGS1 and IGS2), and D2-D3 segment [8, 9, 44] (Fig. 2). Specific primers have also been developed based on SCAR markers, which are sequences drawn from randomly scattered differential sequences along RKN genomes, e.g., the DNA band obtained from RAPD (random amplified polymorphic DNA) gels or AFLP (amplified fragment length polymorphism), with subsequent cloning and sequencing of differential bands between related species. Additionally, specific primers have been developed based on differential sequences from mitochondrial genomes, satellite DNA, or another particular target region, e.g., specific effector proteins [42, 45]. These primers can be used in single or multiplex PCR reactions for RKN diagnosis. Thus, PCR-based RKN detection is simple, fast, cheap, and accurate and can be used in a high-throughput manner [9].

The importance of these molecular methods is well evidenced in samples from coffee-growing areas, for example (Fig. 3). Although coffee-related *Meloidogyne* species can be identified through esterase profiles, the root condition may not be ideal to obtain females for esterase, so for over 20 years, the primers established for *M. incognita*, *M. exigua*, *M. paranaensis*, *M. izalcoensis*, and *M. arabicida*, known as the SCAR coffee kit, have successfully been used in our laboratory, and below in this chapter the technique is described [46, 47].

Here we reported a review and describe protocols on current enzymatic and molecular-based methods that have been successfully used in combinations to diagnose and identify *Meloidogyne* spp. isolates from the field based on our experience in studying *Meloidogyne* spp. populations in Brazil.

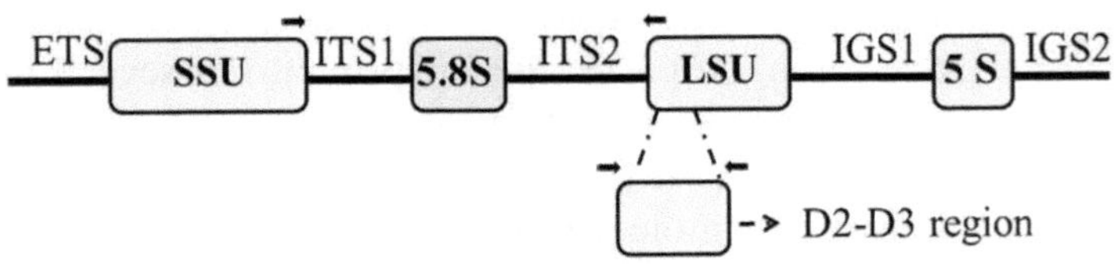

Fig. 2 Schematic representation of nuclear rRNA genes in eukaryotic cells. SSU, 18S small subunit; LSU, 28S large subunit; ETS, external transcribed spacer region; ITS1 and ITS2, internal transcribed spacers; IGS1 and IGS2, intergenic spacer region; arrows indicate possible starting point for primer amplification. Open box indicates D2–D3 expansion segments for the 28S rRNA [9]

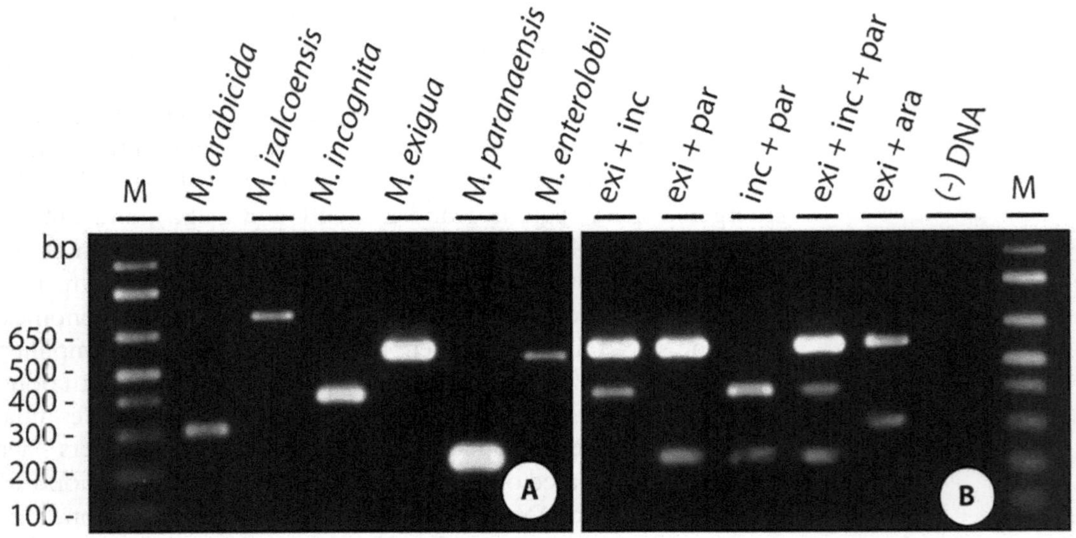

Fig. 3 SCAR-PCR amplification patterns for *Meloidogyne* spp. infecting coffee plants in the Americas. Individual (**a**) and multiplex PCR (**b**). *Meloidogyne arabicida* (ara-1), *M. izalcoensis* (iza-1), *M. incognita* (inc-1), *M. exigua* (exi-1), *M. paranaensis* (par), and *M. enterolobii* (ent). (–) DNA: negative control. M = 1 kb Plus DNA ladder [46]

2 Materials

2.1 Acrylamide Gel Preparation for α-Esterase Staining

Solution A, buffer for the incubator: Lithium hydroxide 1.20 g, boric acid 11.89 g, distilled water 1000 mL.

Solution B, buffer for the gel: Tris–HCL base 6.2 g, citric acid 1.6 g, distilled water 1000 mL.

Solution C, gel buffer (final): Solution A 100 mL, solution B 900 mL.

Solution D, migration gel: Acrylamide 6% (5.7 g) or 7% (6.65 g), bis-acrylamide 6% (0.3 g) or 7% (0.35 g), solution C 100 mL, TEMED 0.1 mL.

Solution E, ammonium persulfate (10%): 1 g in 10 mL of distilled water.

Solution F, bromophenol blue 0.1%: 0.1 g 100 mL of distilled water.

2.1.1 α-Esterase Staining

Solution G, phosphate buffer: NaHPO4 13.789 g/500 mL of water, NaH_2PO_4 7.162 g/100 mL of H_2O. Mix the two and make up the volume to 1 liter (store in a dark glass bottle in the refrigerator).

Solution H, α-naphthylacetate 0.05 g, acetone 1 mL (freshly prepared).

Solution I, staining solution: Solution G 100 mL, Fast Blue RR Salt 0.1 g, homogenize and filter, and add solution H 1 mL (freshly prepared).

Solution J, gel fixative solution: Methyl alcohol 50 mL, distilled water 50 mL, acetic acid 10 mL.

2.2 Detection of Meloidogyne spp. with SCAR-PCR

Reagents can be used for detection of a single species (individual PCR) or a combination of three species (multiplex PCR, e.g., *M. incognita*, *M. exigua*, and *M. paranaensis*) [47].

Reagents for individual PCR: 2.5 μL from 10x PCR buffer, 4 μL dNTPs from 1.25 mM solution, 1 μL from 10 μM forward primer, 1 μL from 10 μM reverse primer, 0.25 μL from 5 U Taq DNA polymerase, 2 μL from 3 ng DNA template, 14.25 μL deionized-autoclaved water.

Reagents for multiplex PCR: use 3 μL of primer solutions, 6 μL DNA template, 6.25 μL deionized-autoclaved water.

10X TBE**:** Tris base 108 g, boric acid 55 g, EDTA 7.45 g, deionized water 600 mL (dissolve reagents and fill up volume to 1 L with water).

NIB buffer: NaCl 5 M 1 mL, Tris base 1 M pH 8 1.5 mL, EDTA 0.5 M 1 mL, β-mercaptoethanol 35 μL, Triton 280 μL, deionized water 40 mL (mix reagents and fill up volume to 50 mL with water).

Homogenization buffer: NaCl 5 M 1 mL, sucrose 3.42, EDTA 0.5 M 1 mL, deionized water 40 mL (mix reagents and fill up volume to 50 mL with water).

Lysis buffer: Tris HCl 2 M pH 9.2 12.5 mL, EDTA 0.5 M 12.5 mL, SDS 10% 11.5 mL, deionized water 40 mL (mix reagents and fill up volume to 50 mL with water).

HLB buffer: Tris HCl 0.2 M pH 8 272 μL, NaCl 0.2 M 272 μL, β-mercaptoethanol 1% 10 μL, proteinase K 1 mg/mL 400 μL, deionized water 402.5 μL (HLB buffer is kept at 4 °C for about 3 months).

Primers adopted for SCAR-PCR are listed in Table 1.

Table 1
List of species-specific PCR primers developed for diagnosis of *Meloidogyne* spp.

Meloidogyne spp.	Target region	Primer sets (5′—3′)	Amplicon (bp)	References
M. arenaria	SCAR	TCGGCGATAGAGGTAAATGAC	420	[48]
		TCGGCGATAGACACTACAACT		
	SCAR	TCGAGGGCATCTAATAAAGG	950	[49]
		GGGCTGAATAATCAAAGGAA		
M. arabicida	SCAR	TCGGCGATAGTACGTATTTAGCG	300	[46]
		TAGTGATTTCGGCGATAGGC		
M. chitwoodi	SCAR	CCAATGATAGAGATAGGAAC	400	[50]
		CTGGCTTCCTCTTGTCCAAA		
	IGS	GATCTATGGCAGATGGTATGGA	900	[51]
		AGCCAAAACAGCGACCGTCTAC		
	SCAR	TGGAGAGCAGCAGGAGAAAGA	800	[52]
		GGTCTGAGTGAGGACAAGAGTA		
M. exigua	SCAR	CATCCGTGCTGTAGCTGCGAG	562	[47]
		CTCCGTGGGAAGAAAGACTG		
M. fallax	IGS	TGGGTAGTGGTCCCACTCTG	1100	[51]
		AGCCAAAACAGCGACCGTCTAC		
	SCAR	CCAAACTATCGTAATGCATTATT	515	[52]
		GGACACAGTAATTCATGAGCTAG		
M. hapla	SCAR	CAGGCCCTTCCAGCTAAAGA	960	[50]
		CTTCGTTGGGGAACTGAAGA		
	SCAR	TGACGGCGGTGAGTGCGA	610	[52]
		TGACGGCGGTACCTCATAG		
	SCAR	GGCTGAGCATAGTAGATGATGTT	1500	[53]
		ACCCATTAAAGAGGAGTTTTGC		
	IGS	GGATGGCGTGCTTTCAAC	440	[54]
		AAAAATCCCCTCGAAAAATCCACC		
M. incognita	SCAR	CTCTGCCCAATGAGCTGTCC	1200	[48]
		CTCTGCCCTCACATTAGG		
	SCAR	TAGGCAGTAGGTTGTCGGG	1350	[49]
		CAGATATCTCTGCATTGGTGC		
	SCAR	GGGATGTGTAAATGCTCCTG	399	[47]

(continued)

Table 1 (continued)

Meloidogyne spp.	Target region	Primer sets (5′—3′)	Amplicon (bp)	References
		CCCGCTACACCCTCAACTTC		
	SCAR	GTGAGGATTCAGCTCCCCAG	955	[55]
		ACGAGGAACATACTTCTCCGTCC		
M. izalcoensis	SCAR	GGAAACCCCTAATTAGGATACACT	670	[46]
		CGCTTGATTTGAGCAGTAGG		
M. javanica	SCAR	CCTTAATGTCAACACTAGAGCC	1650	[49]
		GGCCTTAACCGACAATTAGA		
	SCAR	GGTGCGCGATTGAACTGAGC	670	[48]
		CAGGCCCTTCAGTGGAACTATAC		
	SCAR	ACGCTAGAATTCGACCCTGG	517	[55]
		GGTACCAGAAGCAGCCATGC		
M. enterolobii	mtDNA	GAAATTGCTTTATTGTTACTAAG	322	[56]
		TAGCCACAGCAAAATAGTTTTC		
	SCAR	GATCAGAGGCGGGCGCATTGCGA	520	[11]
		CGAACTCGCTCGAACTCGAC		
M. naasi	ITS	CTCTTTTGGAGAATAATCGT	433	[57]
		CCTCCGCTTACTGATATG		
M. paranaensis	SCAR	GCCCGACTCCATTTGACGGA	208	[47]
		CCGTCCAGATCCATCGAAGTC		
M. graminicola	SCAR	TTCGACTCTGTACGAAAGC	230	[37]
		CAAAAGTAACCGGACACTCTTTT		
	SCAR	GGGGAAGACATTTAATTGATGA TCAAC	640	[58]
		GGTACCGAAACTTAGGGAAAG		
	ITS	TTATCGCATCATTTTATTTG	369	[59]
		CGCTTTGTTAGAAAATGACCCT		
M. oryzae	SCAR	CCAGCATCCGCTGTTGTAT	120	[37]
		AACAGGCTCCAGGTGAAAAG		
M. salasi	SCAR	CAAAGAACGGGGTTTATTCG	160	[37]
		GGTTATCCGAACTCCCCAAT		
M. luci	SCAR	ACTCCTGCGACCTCATGG	771	[60]
		ACTCCTGCGAACACAACA		

SCAR sequence characterized amplified region, *IGS* intergenic spacer region, *ITS* internal transcribed spacer, *mtDNA* mitochondrial DNA

3 Methods

3.1 Enzyme-Based Method: α-Esterase

3.1.1 Preparation of Extraction Tubes

Tubes must be constructed from non-heparinized micro hematocrit tubes 75 mm in length and 1.2 mm in internal diameter. The tubes are cut in half, thus obtaining two tubes that are closed at one end, by the flame of a lamp, rotating the tube between the thumb and forefinger, so that the bottom is perfectly flat.

3.1.2 Extraction Medium for Females

Extraction medium for the esterase enzyme should be prepared according to the recipe below and kept under refrigerated conditions in amber bottles (10 mL) for approximately 15 days to avoid fungal contamination. Recipe is sucrose (2 g) + Triton X 100 (200 microliters) + distilled water (to top up to 10 mL).

3.1.3 Obtaining and Selecting *Meloidogyne* Females

RKN populations obtained from naturally infected plants are maintained on susceptible tomato plants "Santa Cruz" in a greenhouse at 25–30 °C. Populations that do not infect tomato plants must be maintained in their respective hosts. To obtain good results, the females must be young and milky white, starting the egg-laying phase. Roots should be washed, and females extracted individually and transferred to extraction tubes containing 3–5 microliters of extraction buffer. These tubes must be kept in containers with ice during the entire female extraction process [61]. Females that are not used immediately after extraction must be frozen at −10 °C. If the objective is to proceed with the purification of mixed species, for each female removed from the root, the corresponding egg mass must be extracted and placed in 0.1 mL of 1% saline solution in "Eppendorf" tubes, which must be kept at room temperature [61].

3.1.4 Electrophoresis Apparatus

There are many electrophoresis incubators available on the market that allow the use of polyacrylamide gels from 0.6 to 1.0 mm in thickness and therefore are ideal for studies with *Meloidogyne* spp. females. The incubator used in this work was made by acrylate (Brasilia DF). The cassettes (20 × 13 cm) were made by glaziers, using glass plates, with spacers (glass strips) of 1 mm thickness glued on the glass plates with silicone [61].

3.1.5 Gel Preparation, Sample Application, and Migration

The gel composition described in this work has been used in several papers for *Meloidogyne* spp. identification [4, 10, 22]. Polyacrylamide gels are obtained by polymerization of acrylamide and bis-acrylamide (Bis) in the presence of ammonium persulfate and tetramethylethylenediamine (TEMED). The stock solutions (Subheading 2.1), after being prepared, must be filtered and stored in brown glass flasks in the refrigerator. All solutions tolerate a storage period of about 3 months, except the ammonium persulfate solution, which must be prepared more frequently, having a shelf life of approximately 15 days [61]. Preparation of the migration gel is done by mixing solutions D (30 mL) and E (0.3 mL) (Subheading 2.1.1). Next, the migration gel should be placed in rectangular plates with a 1-mm-thick spacer and 20 × 13 cm channels placed in molds [61]. When placing the cap over the plate, it is important not to allow bubbles to form, which could interfere with the migration of enzymes in the gel. These plates must remain for 30 min at room temperature for the gel to solidify and then be placed in the refrigerator at 4 °C to store the gel. When applying the samples, carefully remove the cover, and mark the wells with the metal comb, vertically about 3 cm from one end of the gel. Place a transparent PVC (plastic) film on top to protect the gel, and refrigerate (4 °C) before applying the samples. The females placed in the tubes must be macerated by means of rotating movements of a piston. The macerate must be applied, with the aid of a Hamilton syringe, on Whatman 3 mm papers, cut into pieces of 1.5 × 4.0 mm, and arranged on an acrylic plate [61]. Soon after that, the filter papers with the samples must be placed in the polyacrylamide gel wells. The first well must always contain the *M. javanica* pattern. Be careful to insert the papers well into the gel, and clean the forceps after each sample change. Bromophenol drops (Subheading 2.1, solution F) should be placed on the samples to monitor migration. When placing the cassette in the apparatus, care must be taken to ensure that the filter paper that comes into contact with the plug is approximately 1 cm away from the wells and that the distance between the two papers on the cassette is of 8 cm [61].

The electrophoresis apparatus must be connected to a power supply of approximately 80–120 V and placed on a refrigerated counter or in the refrigerator at a temperature of 4–8 °C. The voltage must be measured periodically with a voltmeter, placing the meters at the two opposite ends of the gel (positive and negative poles) [61]. Migration should be followed by displacement of the bromophenol blue trace. Take care that the migration is rectilinear: if the voltage is low, the migration line will be concave; if it is high, the migration line will be convex.

After a migration of approximately 5 cm (6% gel) and 6 cm (7% gel) from the wells (± 2 h), the apparatus must be turned off. The cassette is removed from the apparatus, and the filter papers, where the samples were deposited, are removed with the aid of tweezers. With the help of a spatula, the gel must be transferred directly from the glass plate to a mold with the staining solution (Subheading 2.1.1 [61]).

3.1.6 Esterase Staining and Fixation in Polyacrylamide Gel

Solutions for revealing enzymatic activity (Subheading 2.1.1) must be freshly prepared. Only buffer solutions, stored in dark glass in the refrigerator, can be organized in advance.

After the preparation of solution I (Subheading 2.1.1), at the moment when the staining will be done, solution I must be poured into a glass dish, and the gel must be gently placed inside this solution so that it is completely covered. After incubation in the dark, in an oven at 37 °C, for approximately 20–30 min, the regions of esterase activity appear dark on a light background.

After staining, the gels must be placed in a fixative solution (Subheading 2.1.1) and kept there for approximately half an hour. After washing the gels twice with tap water, the gels are placed between two wet cellophane papers and placed to dry [61] or immediately photographed.

3.2 Molecular-Based Methods

Nematode eggs can be extracted from infected roots as described by Carneiro et al. [6] and be used for DNA extraction according to Randig et al. [47]. Both protocols are described in details below. DNA can also be obtained from a single nematode (juvenile, female, or male) with protocols specifically designed for this purpose or using commercial kits (see below). Diagnosis using a single nematode has the advantage of direct sample processing, without prior culturing of nematode population. Several other protocols have been published in the literature, including DNA extraction from galls and plant roots infected with *Meloidogyne* spp. or from infested soil [14].

Reagents and some RKN-specific primers are described in Subheading 2.2 and listed in Table 1, respectively. Other materials include equipment normally used in a molecular biology laboratory, e.g., sets of pipettors, microcentrifuge tubes, vortexes, gel electrophoresis apparatus and gel imaging system.

3.2 1 Egg Extraction

1. Thoroughly wash infected roots under tap water to completely remove soil waste.
2. Cut infected roots into small pieces (1–2 cm) and place in a blender.
3. Add enough 1.25% NaOCl (sodium hypochlorite) to cover roots, and blend for 30 s.
4. Pour off suspension through a set of sieves (20-200-500 meshes), and wash thoroughly with tap water to remove bleach. Eggs are retained in the 500-mesh sieve.
5. Eggs are gently washed with tap water to remove any traces of hypochlorite.
6. Collect eggs from the 500-mesh sieve into a 50 mL Falcon tube.
7. Add one tablespoon of kaolin and centrifuge at 2000 g for 5 min.
8. Gently discard the supernatant (upper phase).

9. Fill up tubes with cold (4 °C) sugar solution (30% w:v), and stir with a spatula to break up any pellets (must be completely dissolved).
10. Centrifuge at 2000 g for 2 min, and pour off supernatant through a 500-mesh sieve.
11. Rinse gently with distilled water and transfer to a new 15 mL Falcon tube.
12. Centrifuge at 2000 g for 3 min.
13. Carefully eliminate supernatant with a pipette, transfer pellet to a new 1.5 mL microcentrifuge tube, and spin at 10,000 g for 2 min to eliminate excess of liquids.
14. Label tubes, freeze in liquid nitrogen, and store at −80 °C freezer for later use [6].

3.2.2 DNA Extraction from Eggs

1. Remove eggs from −80 °C freezer and keep on ice. Pre-chill mortars and pestles with liquid nitrogen, and transfer eggs (pellet) to a mortar.
2. Macerate eggs with a pestle until a fine powder is achieved.
3. Transfer powder with a spatula into 2 mL microcentrifuge tube.
4. Add 500 μL NIB buffer and vortex for 30 s.
5. Centrifuge at 14,000 rpm for 2 min and discard supernatant.
6. Add 500 μL NIB, dissolve pellet with a pipette tip, and centrifuge at 14,000 rpm for 2 min.
7. Discard the supernatant, add 800 μL homogenization buffer, disrupt the pellet, add 200 μL lysis buffer, and vortex for 30 s.
8. Incubate samples at 55 °C for 30 min, and allow samples to cool to room temperature for 9 min (or keep at 4 °C for 1 h).
9. To precipitate DNA, add 1 volume phenol solution to each tube and vortex.
10. Centrifuge at 14,000 for 3 min, carefully remove the upper phase, and transfer into a clean 1.5 mL microcentrifuge tube.
11. Add an equal volume of phenol:chloroform and vortex.
12. Centrifuge at 14,000 rpm for 3 min, carefully remove the upper phase, and transfer into a clean microcentrifuge tube.
13. Add 200 μL ether and vortex.
14. Centrifuge at 14,000 for 3 min and carefully discard the upper phase.
15. Precipitate the DNA with 1 mL 100% ethanol by inverting the tubes 5–10 times.
16. Incubate at −80 °C for 30 min (or −20 °C for 2 h).
17. Centrifuge at 14,000 rpm for 10 min.

18. Carefully pipette off ethanol and add 1 mL 70% ethanol.
19. Centrifuge at 14,000 rpm for 5 min and discard the ethanol.
20. Invert tubes on a clean paper towel, and let it dry for a few minutes, without over-drying.
21. Resuspend pellet in 20 μL autoclaved deionized water.
22. Store DNA at −20 °C until use.
23. Estimate DNA concentration by running an aliquot of DNA in 1% agarose gel comparing with lambda DNA [47]

3.2.3 DNA Extraction from Single J2, Male or Female

Protocol A (adapted from Zymo Research Instruction Manual). Reagents are from the kit.

1. Add 50 μL reagent 1 into 1.5 mL Eppendorf tube.
2. Add a single nematode (also works for 1–15 individuals) and vortex for 10 s.
3. Incubate at 25 °C for 10 min.
4. Transfer lysate into 2 mL spin-column (provided in the kit) and centrifuge at 12,000 rpm for 1 min.
5. Add 100 μL reagent 2 and centrifuge at 12,000 rpm for 1 min.
6. Add 100 μL reagent 3, and centrifuge at 12,000 rpm for 1 min.
7. Transfer the spin-column into a clean 2 mL tube, add 50 μL reagent four, incubate at 25 °C for 10 min, and centrifuge at 12,000 rpm for 1 min.
8. Transfer DNA into a clean 1.5 mL Eppendorf tube, and store at −20 °C until use. Use 2–4 μL of DNA for PCR reaction.

Protocol B

1. Add 25 μL of autoclaved deionized water +25 μL HLB buffer into 0.2 mL PCR tubes, and keep on ice.
2. Add a single nematode (J2, male or female) in each tube.
3. Add a drop of mineral oil in each tube (depending on type of thermo cycler), and incubate tubes in the thermo cycler (PTC-100, MJ Research) for 2 h at 65 °C and then at 95 °C for 5 min.
4. Cool down to 4 °C and centrifuge at 3000 rpm for 2 min.
5. Transfer lysate (upper phase) into a new 1.5 mL Eppendorf tube, and store at −20 °C until use.
6. Use 2–3 μL for PCR reaction.

Adapted from Holterman et al. [62]

3.2.4 Diagnosis of RKN Nematodes Using PCR Analysis

In the Nematology Lab (Embrapa, Brazil), diagnosis of samples is mostly done using esterase isozyme analysis (Est) combined with PCR analysis when needed. Other characters are also taken into account: crop, typical symptoms, and host range (e.g., coffee-associated RKN species). For instance, samples coming from coffee fields may arrive in poor conditions, not allowing the removal of females in good shape for the Est analysis. In this case, second-stage juveniles are used for PCR analysis using specific primers developed for most RKN species infecting coffee [46, 47]. Adopted primers are shown in Table 1.

3.2.5 A PCR Protocol for Diagnosis of Coffee Infecting RKN Species

1. Thaw all PCR components including the DNA target on ice and vortex briefly.
2. Prepare the master mix according to the amounts shown in Subheading 2.2, plus at least one or two extra reactions.
3. Keep the Taq polymerase in the freezer at −20 °C, and add it last in the master mix.
4. Add 23 μL of PCR master mix into each 0.2 mL PCR tube.
5. Lastly, add 2 μL of DNA template or water (negative control) into each tube.
6. Make sure to include a positive control DNA as well.
7. Incubate samples in the thermo cycler, e.g., PTC 100 (MJ Research), using the following cycling settings [28]: 5 min at 94 °C; 30 cycles of 30 s at 94 °C; 45 s at 64 °C; 1 min at 70 °C; and a final incubation of 8 min at 70 °C. This PCR program was optimized to run an individual PCR (one primer set at once) or a multiplex PCR (a combination of three primer sets at once).
8. To analyze the PCR amplification results, add 2–3 μL of 6X loading dye to each sample, and run 10–15 μL of sample in a 1.2% agarose gel. Gels are run in 0.5X TBE buffer (Tris–borate–EDTA) for about 60–80 min with ca. 70 V. Stain the gel in an ethidium bromide solution (0.3 μg/mL) for 30–50 min, destain in 0.5X TBE for the same time, and photograph the gel under UV light, according to gel imaging setup.

As an example, an electrophoresis gel image of conventional PCR (single and multiplex) using SCAR markers for major RKN species (*M. arabicida*, *M. izalcoensis*, *M. incognita*, *M. exigua*, *M. paranaensis*, and *M. enterolobii*) infecting coffee plants in the Americas is shown in Fig. 3 [46].

Acknowledgments

This work was supported by Embrapa Recursos Genéticos e Biotecnologia and Conselho Nacional de Desenvolvimento Científico e Tecnológico (CNPq), Brazil.

References

1. Maggenti A (1981) General nematology, 1st edn. Springer, New York, p 372. https://doi.org/10.1007/978-1-4612-5938-1
2. Hunt DJ, Handoo ZA (2009) Taxonomy, identification and principal species. In: Perry RN, Moens M, Starr JL (eds) Root-knot nematodes. CABI Publishing, pp 55–97
3. Lopes EA, Ferraz S (2016) Importância dos nematoides na agricultura. In: Oliveira CMG, Santos MA, Castro LHS (eds) Diagnose de fitonematoides. Millenium, Campinas, pp 1–10
4. Carneiro RMDG, Monteiro JMS, Silva UC, Gomes G (2016) Gênero *Meloidogyne*: diagnose através de eletroforese de isoenzimas e marcadores SCAR. In: Oliveira CM, Santos MA, Castro LHS (eds) Diagnose de fitonematoides. Millenium, Campinas, pp 47–70
5. Castro JMC (2019) *Meloidogyne enterolobii* e sua evolução nos cultivos brasileiros. Inf Agropecuário 40:41–48
6. Carneiro R, Randig O, Almeida MR, Gomes AC (2004) Additional information on *Meloidogyne ethiopica* Whitehead, 1968 (Tylenchida: Meloidogynidae), a root-knot nematode parasitising kiwi fruit and grape-vine from Brazil and Chile. Nematology 6:109–123. https://doi.org/10.1163/156854104323072982
7. Carneiro RMDG, Santos MFA, Almeida MRA, Mota FC, Gomes ACMM, Tigano MS (2008) Diversity of *Meloidogyne arenaria* using morphological, cytological and molecular approaches. Nematology 10:819–834. https://doi.org/10.1163/156854108786161526
8. Castagnone-Sereno P, Skantar A, Robertson L (2011) Molecular tools for diagnostics. In: Jones J, Gheysen G, Fenoll C (eds) Genomics and molecular genetics of plant-nematode interactions. Springer, Dordrecht, pp 443–464
9. Carneiro RMDG, Lima FSO, Correia VR (2017) Methods and tools currently used for the identification of plant parasitic nematodes. In: Shah MM (ed) Nematology – concepts, diagnosis and control. InTech
10. Carneiro RMDG, Quénéhervé P, Almeida MR (2000) Enzyme phenotypes of *Meloidogyne* spp. populations. Nematology 2:645–654. https://doi.org/10.1163/156854100509510
11. Tigano M, De Siqueira K, Castagnone-Sereno P, Mulet K, Queiroz P, Santos M et al (2010) Genetic diversity of the root-knot nematode *Meloidogyne enterolobii* and development of a SCAR marker for this guava-damaging species. Plant Pathol 59:1054–1061. https://doi.org/10.1111/j.1365-3059.2010.02350.x
12. Powers T (2004) Nematode molecular diagnostics: from bands to barcodes. Annu Rev Phytopathol 42:367–383. https://doi.org/10.1146/annurev.phyto.42.040803.140348
13. Blok VC (2005) Achievements in and future prospects for molecular diagnostics of plant-parasitic nematodes. Can J Plant Pathol 27:176–185. https://doi.org/10.1080/07060660509507214
14. Blok VC, Powers TO (2009) Biochemical and molecular identification. In: Root-knot nematodes. CABI Publishing, pp 98–118
15. Perry RN, Subbotin SA, Moens M (2007) Molecular diagnostics of plant-parasitic nematodes. In: Punja Z, De Boer SH, Sanfaçon H (eds) Biotechnology and plant disease management. CAB International, pp 195–226
16. Oliveira CMG, Monteiro AR, Blok VC (2011) Morphological and molecular diagnostics for plant-parasitic nematodes: working together to get the identification done. Trop Plant Pathol 36:65–73. https://doi.org/10.1590/S1982-56762011000200001
17. Cunha TG, Visôtto LE, Lopes EA, Oliveira CMG, God PIVG (2018) Diagnostic methods for identification of root-knot nematodes species from Brazil. Ciência Rural 48. https://doi.org/10.1590/0103-8478cr20170449
18. Janati A, Bergé JB, Triantaphylou AC, Dalmasso A (1982) Nouvelles données sur l'utilisation des isoesterases pour l'identification des *Meloidogyne*. Rev Nematol 5:147–154
19. Carneiro RMDG, Carneiro RG, Abrantes IM, Santos MS, Almeida MR (1996) *Meloidogyne paranaensis* n. sp. (Nemata: Meloidogynidae), a root-knot nematode parasitizing coffee in Brazil. J Nematol 28:177–189
20. Santos MFA, Mattos VS, Gomes ACMM, Monteiro JMS, Souza DA, Torres CAR et al (2020) Integrative taxonomy of *Meloidogyne paranaensis* populations belonging to different esterase phenotypes. Nematology 22:453–468. https://doi.org/10.1163/15685411-00003316
21. Esbenshade PR, Triantaphyllou AC (1985) Use of enzyme phenotypes for identification of *Meloidogyne* species. J Nematol 17:6–20
22. Carneiro RMDG, Almeida MRA, Carneiro RG (1996) Enzyme phenotypes of Brazilian populations of *Meloidogyne* spp. Fundam Appl Nematol 19:555–560
23. Subottin SA, Rius JEP, Castillo PC (2021) Methods of biochemical and molecular

diagnostics. In: Hunt DJ, Perry RN (eds) Systematics of root-knot nematodes (Nematoda: Meloidogynidae). Brill, Leiden-Boston, pp 63–83

24. Esbenshade PR, Triantaphyllou AC (1990) Isozyme phenotypes for the identification of Meloidogyne species. J Nematol 22:10–15
25. Santos MFA, Mattos VS, Monteiro JMS, Almeida MRA, Jorge AS, Cares JE et al (2019) Diversity of *Meloidogyne* spp. from peri-urban areas of sub-Saharan Africa and their genetic similarity with populations from the Latin America. Physiol Mol Plant Pathol 105:110–118. https://doi.org/10.1016/j.pmpp.2018.08.004
26. Santos MFA, Furlanetto C, Almeida MRA, Carneiro MDG, Mota FC, Gomes ACMM et al (2012) Biometrical, biological, biochemical and molecular characteristics of *Meloidogyne incognita* isolates and related species. Eur J Plant Pathol 134:671–684. https://doi.org/10.1007/s10658-012-0018-1
27. Santos MFA, Correa VR, Peixoto JR, Mattos VS, Silva JGP, Moita AW et al (2018) Genetic variability of *Meloidogyne paranaensis* populations and their aggressiveness to susceptible coffee genotypes. Plant Pathol 67:193–201. https://doi.org/10.1111/ppa.12718
28. Muniz MF, Carneiro RMDG, Almeida MR, Campos VP, Castagnone-Sereno P et al (2008) Diversity of *Meloidogyne exigua* (Tylenchida: Meloidogynidae) populations from coffee and rubber tree. Nematology 10: 897–910. https://doi.org/10.1163/156854108786161418
29. Muniz MF, Campos VP, Almeida MR, Gomes ACMM, Santos MF, Mota FC et al (2009) Additional information on an atypical population of *Meloidogyne exigua* Göldi, 1887 (Tylenchida: Meloidogynidae) parasitising rubber tree in Brazil. Nematology 11:95–106. https://doi.org/10.1163/156854108X398444
30. Souza CFB, Galbieri R, Belot J-L, Negri BF, Perina FJ, Cares JE et al (2022) Occurrence of a new race of *Meloidogyne enterolobii* and avirulent *M. incognita* populations parasitizing cotton in western Bahia state, Brazil. Physiol Mol Plant Pathol 121:101874. https://doi.org/10.1016/j.pmpp.2022.101874
31. Soares MRC, Mattos VS, Leite RR, Gomes ACMM, Gomes CB, Castagnone-Sereno P et al (2020) Integrative taxonomy of *Meloidogyne graminicola* populations with different esterase phenotypes parasitising rice in Brazil. Nematology 23:627–643. https://doi.org/10.1163/15685411-bja10065
32. Carneiro RMDG, Mendes MM, Almeida MRA, Santos MFA, Gomes ACMM, Karssen G (2008) Additional information on *Meloidogyne inornata* Lordello, 1956 (Tylenchida: Meloidogynidae) and its characterisation as a valid species. Nematology 10:123–136. https://doi.org/10.1163/156854108783360131
33. Carneiro RMDG, Correa VR, Almeida MRA, Gomes ACMM, Mohammad Deimi A, Castagnone-Sereno P et al (2014) *Meloidogyne luci* n. sp. (Nematoda: Meloidogynidae), a root-knot nematode parasitising different crops in Brazil, Chile and Iran. Nematology 16:289–301. https://doi.org/10.1163/15685411-00002765
34. Carneiro RMDG, Almeida MRA, Gomes ACMM (2004) First record of *Meloidogyne hispanica* Hirschmann, 1986 on squash in State of Bahia,. Nematol Bras 28:215–218
35. Charchar J, Charchar MJ, Eisenback J, Boiteux ME (2008) *Meloidogyne pisi* n. sp. (Nematoda: Meloidogynidae), a root-knot nematode parasitising pea in Brazil. Nematology 10:479–493. https://doi.org/10.1163/156854108784513905
36. Oliveira SA, Oliveira CMG, Maleita CMN, Silva MFA, Abrantes IMO, Wilcken SRS (2018) First report of *Meloidogyne graminis* on golf courses turfgrass in Brazil. PLoS One 13:e0192397. https://doi.org/10.1371/journal.pone.0192397
37. Mattos VS, Mulet K, Cares JE, Gomes CB, Fernandez D, Grossi-de-Sá MF et al (2019) Development of diagnostic SCAR markers for *Meloidogyne graminicola*, *M. oryzae*, and *M. salasi* associated with irrigated rice fields in Americas. Plant Dis 103:83–88. https://doi.org/10.1094/PDIS-12-17-2015-RE
38. Leite RR, Mattos VS, Gomes ACMM, Py LG, Souza DA, Castagnone-Sereno P et al (2020) Integrative taxonomy of *Meloidogyne ottersoni* (Thorne, 1969) Franklin, 1971 (Nematoda: Meloidogynidae) parasitizing flooded rice in Brazil. Eur J Plant Pathol 157:943–959. https://doi.org/10.1007/s10658-020-02049-y
39. Stare BG, Strajnar P, Susič N, Urek G, Širca S (2017) Reported populations of *Meloidogyne ethiopica* in Europe identified as *Meloidogyne luci*. Plant Dis 101:1627–1632. https://doi.org/10.1094/PDIS-02-17-0220-RE
40. Monteiro JMS, Cares JE, Correa VR, Pinheiro JB, Mattos VS, Silva JGP et al (2017) *Meloidogyne brasiliensis* Charchar & Eisenback, 2002 is a junior synonym of *M. ethiopica* Whitehead, 1968. Nematology 19:655–669. https://doi.org/10.1163/15685411-00003078

41. Monteiro JMS, Mattos VS, Santos MFA, Gomes ACMM, Correa VR, Sousa DA et al (2019) Additional information on *Meloidogyne polycephannulata* and its proposal as a junior synonym of *M. incognita*. Nematology 21: 129–146. https://doi.org/10.1163/15685411-00003202
42. Castagnone-Sereno P, Danchin EGJ, Perfus-Barbeoch L, Abad P (2013) Diversity and evolution of root-knot nematodes, genus *Meloidogyne*: new insights from the genomic era. Annu Rev Phytopathol 51:203–220. https://doi.org/10.1146/annurev-phyto-082712-102300
43. Nega A (2014) Review on nematode molecular diagnostics: from bands to barcode. J Biol Agric Healthc 4:129–153
44. Castagnone-Sereno P, Esparrago G, Abad P, Leroy F, Bongiovanni M (1995) Satellite DNA as a target for PCR-specific detection of the plant-parasitic nematode Meloidogyne hapla. Curr Genet 28:566–570. https://doi.org/10.1007/BF00518170
45. Randig O, Deau F, Santos MFA, Tigano MS, Carneiro RMDG, Castagnone-Sereno P (2009) A novel species-specific satellite DNA family in the invasive root-knot nematode *Meloidogyne mayaguensis* and its potential use for diagnostics. Eur J Plant Pathol 125:485–495. https://doi.org/10.1007/s10658-009-9497-0
46. Correa VR, Santos MFA, Almeida MRA, Peixoto JR, Castagnone-Sereno P, Carneiro RMDG (2013) Species-specific DNA markers for identification of two root-knot nematodes of coffee: *Meloidogyne arabicida* and *M. izalcoensis*. Eur J Plant Pathol 137:305–313. https://doi.org/10.1007/s10658-013-0242-3
47. Randig O, Bongiovanni M, Carneiro RMD, Castagnone-Sereno P (2002) Genetic diversity of root-knot nematodes from Brazil and development of SCAR markers specific for the coffee-damaging species. Genome 45:862–870. https://doi.org/10.1139/g02-054
48. Zijlstra C, Donkers-Venne DTHM, Fargette M (2000) Identification of *Meloidogyne incognita, M. javanica* and *M. arenaria* using sequence characterised amplified region (SCAR) based PCR assays. Nematology 2:847–853. https://doi.org/10.1163/156854100750112798
49. Dong K, Dean RA, Fortnum BA, Lewis SA (2001) Development of PCR primers to identify species of root-knot nematodes: *Meloidogyne arenaria, M. hapla, M. incognita* and *M. javanica*. Nematropica 31:271–280
50. Williamson VM, Caswell-Chen EP, Westerdahl BB, Wu FF, Caryl G (1997) A PCR assay to identify and distinguish single juveniles of *Meloidogyne hapla* and *M. chitwoodi*. J Nematol 29:9–15
51. Petersen DJ, Zijlstra C, Wishart J, Blok V, Vrain TC (1997) Specific probes efficiently distinguish root-knot nematode species using signature sequences in the ribosomal intergenic spacer. Fundam Appl Nematol 20:619–626
52. Zijlstra C (2000) Identification of *Meloidogyne chitwoodi, M. fallax* and *M. hapla* based on SCAR-PCR: a powerful way of enabling reliable identification of populations or individuals that share common traits. Eur J Plant Pathol 106:283–290. https://doi.org/10.1023/A:1008765303364
53. Dong K, Dean RA, Fortnum BA, Lewis SA (2001) A species-specific DNA probe for the identification of *Meloidogyne hapla*. Nematropica 31:17–23
54. Wishart J, Phillips MS, Blok VC (2002) Ribosomal intergenic spacer: a polymerase chain reaction diagnostic for *Meloidogyne chitwoodi, M. fallax*, and *M. hapla*. Phytopathology 92: 884–892. https://doi.org/10.1094/PHYTO.2002.92.8.884
55. Meng QP, Long H, Xu JH (2004) PCR assays for rapid and sensitive identification of three major root-knot nematodes, *Meloidogyne incognita, M. javanica* and *M. arenaria*. Acta Phytopath Sin 34:204–210
56. Blok V, Fargette M, Wishart J, Phillips M, Berthier K (2002) Mitochondrial DNA differences distinguishing *Meloidogyne mayaguensis* from the major species of tropical root-knot nematodes. Nematology 4:773–781. https://doi.org/10.1163/15685410276040255a9
57. Zijlstra C, Van Hoof R, Donkers-venne D (2004) A PCR test to detect the cereal root-knot nematode *Meloidogyne naasi*. Eur J Plant Pathol 110:855–860. https://doi.org/10.1007/s10658-004-2492-6
58. Bellafiore S, Jougla C, Chapuis É, Besnard G, Suong M, Vu PN et al (2015) Intraspecific variability of the facultative meiotic parthenogenetic root-knot nematode (*Meloidogyne graminicola*) from rice fields in Vietnam. C R Biol 338:471–483. https://doi.org/10.1016/j.crvi.2015.04.002
59. Htay C, Peng H, Huang W, Kong L, He W, Holgado R et al (2016) The development and molecular characterization of a rapid detection method for Rice root-knot nematode (*Meloidogyne graminicola*). Eur J Plant Pathol 146: 281–291. https://doi.org/10.1007/s10658-016-0913-y
60. Maleita C, Cardoso JMS, Rusinque L, Esteves I, Abrantes I (2021) Species-specific

molecular detection of the root knot nematode *Meloidogyne luci*. Biology (Basel) 10:775. https://doi.org/10.3390/biology10080775

61. Carneiro RMDG, Almeida MRA (2001) Técnica de eletroforese usada no estudo de enzimas de nematóides de galhas para identificação de espécies. Nematol Bras 25:35–44

62. Holterman M, Wurff A, Elsen S, Megen H, Bongers T, Holovachov O et al (2006) Phylum-wide analysis of SSU rDNA reveals deep phylogenetic relationships among nematodes and accelerated evolution toward crown clades. Mol Biol Evol 23:1792–1800. https://doi.org/10.1093/molbev/msl044

Chapter 6

Use of PCR-DGGE-Based Molecular Methods to Analyze Nematode Community Diversity

Margherita Festa, Pamela Abbruscato, and Barbara Manachini

Abstract

DGGE (denaturing gradient gel electrophoresis) is a nucleic acid separation technique applied to the evaluation of microbial biodiversity. This technique is quite rapid and cheap compared to other types of analysis. Here we describe the comparison of nematode communities inhabiting different ecosystems. After an ecologically representative sampling collection and the nematode extraction from soil, nematodes are centrifuged in Eppendorf tubes to facilitate DNA extraction. DNA from the whole community of each type of soil is extracted, amplified with primers for 18 S rDNA and used in DGGE analysis. The profiles of DGGE can be analyzed with appropriate software, and biodiversity indices can be estimated.

Key words Nematofauna, Denaturing gradient gel electrophoresis, 18S rDNA, Biodiversity indices, Ecological indices

1 Introduction

The soil biodiversity is an important characteristic for all men agricultural activities. A recently considered important indicator of soil biodiversity and health is the nematode community. Nematodes are abundant and ubiquitous in the soil, representing a wide group that affects directly the crop growth (feeding on the plants) or indirectly as consumers of bacteria, fungi, and other organisms, regulating decomposition and balances among different levels of food webs [1, 2]. The study of the nematode community structure offers a powerful ecological tool for assessing soil disturbances [3–6] and the effects of crop management on soil biodiversity and proprieties [7–9]. Classification of nematodes is time-consuming and needs uncommon expertise. In the last decades, a lot of molecular techniques, fast and reliable, have been developed to evaluate the biodiversity of different soil communities. Techniques such as 16S or 18S rDNA amplification, DGGE and TGGE [10, 11], various types of restriction analysis, cloning, and sequencing have

Sergio Molinari (ed.), *Plant-Nematode Interactions*, Methods in Molecular Biology, vol. 2756,
https://doi.org/10.1007/978-1-0716-3638-1_6,

been applied to nematode biodiversity assessment. In particular three molecular methods have been mainly used: denaturing gradient gel electrophoresis (DGGE), marker gene DNA sequences, and pyrosequencing [12].

DGGE technique allows DNA fragments of the same length but different sequence to be separated on a denaturing gradient acrylamide gel. As a result of the sequence diversity, sequences melt at different temperatures; thus molecules migrate to different positions in the denaturing gel. This technique is applied to study nematode community mainly. Amplification and sequencing of 18 S rDNA gene is applied [13, 14] to identify dangerous or quarantinable species [15] or to assess the variation within and among phytophagous pest species [16, 17]. Different studies have applied DGGE technique to determine nematode species biodiversity [12, 18–22] or even to understand the relationship between soil bacteria diversity and bacteria feeder nematodes [23].

The aim of this study is to compare different nematode communities from very different types of soils (chestnut forest, maize monoculture, and grassland) with a molecular tool (DGGE) and describe their level of biodiversity.

2 Materials

2.1 Nematode Sampling

In the field nematode distribution is always patchy because nematode population density is often affected by the crop grown, presence of weed hosts, chemicals used, and environmental factors. Thus, it is very important that the soil sample is truly representative of the area/crop sampled. To ensure a representative sampling, it is necessary to collect the soil from many spots around the field rather than from only one or two spots. The samples should be collected by a systemic, zigzag pattern using a core sampler making subsamples. As nematodes may not survive in upper 1–2 inches of soil, due to extreme environmental conditions, the upper soil layer is removed, and 20–30 soil cores are taken in a zigzag pattern from regularly spaced locations. The number of soil cores depends on the crop, size of crop, and locations.

2.2 Nematode Extraction from Soil

- Baermann funnels equipped with silicon hose and squeeze clip.
- Eppendorf tube, plastic containers (50 mL).
- Square plates to count the nematodes.
- Centrifuge.
- Stereomicroscope.

2.3 DNA Extraction and Precipitation Solutions

- 1.5 mL microtubes (sterilized).
- Genomic Prep™ Cells and Tissue DNA Isolation Kit (Amersham-Biosciences).
- Cell lysis solution: Tris–HCl 200 mM pH8.5, 250 mM NaCl, EDTA 250 mM, 0.5% SDS.
- Proteinase K (ThermoFisher Scientific) (20 mg/mL).
- RNAse A solution (10 mg/mL).
- Phenol–chloroform–isoamyl alcohol (25:24:1) v/v solution.
- Sodium acetate 3 M in water pH 5.2 with acetic acid.
- Isopropanol 100 %.
- Ethanol 70 %.
- TAE buffer Tris–acetate–EDTA (Tris 40 mM, acetic acid 20 mM, EDTA 1 mM).
- All the solutions are prepared with analytical grade reagents using ultrapure water at 25 °C.

2.4 PCR Condition and Primers

Perform polymerase chain reaction (PCR) using the following primer set:

- **Forward primer F1**-GGAGAAT<u>W</u>AGGGTTCGACTC (where W=A or T)
- **Reverse primer R1**-AGTCAAATTAAGCCGCAG
- **GC Clamp F1**-CGCCCGCCGCGCGCGGCGGGCGGGG CGGGGGCACGGGGAGGAGAAT<u>W</u>AGGGTTCGACTC
- **Reverse prime R2**-TCCAAGAATTTCACCTCT<u>M</u>ACG (where M=A or C)
- Taq polymerase (Sigma-Aldrich)

2.5 DGGE: Gel Preparation

The DCode™ Universal Mutation Detection System BIORAD for DGGE (including electrophoresis tank, electrophoresis, temperature control module, core, casting stand, sandwich clamps, comb gasket holder, combs, and glass plates (Bio-Rad 16×16).

475 Gradient Delivery System®.

Bio-Rad PowerPac® 1000.

- **40% acrylamide/bis-acrylamide** at the ratio (37.5:1) prepared as follows:

Acrylamide	38.93 g
Bis-acrylamide	1.07 g
dH_2O	to 100.0 mL

Filter through a 0.45 μ filter and store at 4 °C.

– **Denaturing acrylamide/bis-acrylamide 20%**

40% acrylamide/bis-acrylamide	1.8 mL
TAE 50 X*	2.0 mL
Urea 7 M	8.4 g
Formamide	8.0 mL
H_2O	71.20 mL

– **Denaturing acrylamide/bis-acrylamide 60%**

40% acrylamide/bis-acrylamide	18.8 mL
TAE 50 X*	2.0 mL
Urea 7 M	25.2 g
Formamide	24.0 mL
H_2O	71.2 mL

– **10% ammonium persulfate (Biorad, 1610700EDU)**

Ammonium persulfate	0.1 g
dH_2O	1.0 mL

Store at –20 °C for about a week.

Add a final concentration of 0.09% (v/v) each of ammonium persulfate and TEMED (Bio-Rad, 1610800) solutions. Few minutes before casting, add APS and TEMED to allow the gel polymerization.

3 Methods

3.1 Soil Sampling

Collected soil cores are thoroughly mixed into one composite sample. From each composite sample, about 500 g of soil are transferred into a labeled plastic bag for the assay. Nematodes are extracted from 100 g of soil (fresh weight) by Baermann funnel modified method (other extraction methods can be performed according to the equipment available) (*see* **Note 1**).

3.2 Nematode Enrichment with Centrifugations

After extraction from the soil, the nematodes are counted under a stereomicroscope and centrifuged. The nematodes extracted from the soil with Baermann funnel method are concentrated alive (to extract good quality DNA) in a centrifuge at 13000 g; the upper aqueous part is discarded and kept to control the number of nematodes lost. This procedure is repeated different times, to finish the volume of water obtained from Baermann funnels. The nematodes are collected finally in a volume of 100 μL in Eppendorf tubes of 1.5 mL to facilitate DNA extraction (*see* **Note 2**).

3.3 DNA Extraction

DNA is extracted from the nematodes concentrated in Eppendorf in a volume of 100 μL. The tubes are kept in the freezer (−20 °C) before extraction, and then put in boiling water for 10 min. Then, an equal volume of cell lysis solution from Genomic Prep™ Cells and Tissue DNA Isolation Kit (Amersham Biosciences) is added with 1 μL of K proteinase solution. After overnight incubation at 55 °C, 1 μL of RNAse A solution is added and incubated for 1 hour at 37 °C. Then the phenol–chloroform–isoamyl alcohol solution (25:24:1) is added, and the tubes inverted several times, vortexed, and then centrifuged for 10 min at 13000 g. The upper phase is removed, and another 100 μL of phenol–chloroform–isoamyl alcohol solution is added. The tubes are inverted, vortexed, and centrifuged for 10 min at 13000 g. The upper aqueous phase is removed, and a solution of 3 M sodium acetate (50 uL) and isopropanol (100 uL) is added. Then the tubes are inverted, vortexed, and placed in the freezer overnight. After another centrifugation (10 min at 13000 g), the supernatant is discarded. The pellet is washed with a 70% ethanol solution, centrifuged again, then the obtained pellets are air-dried, and the DNA is re-suspended in 100 μL of Milli-Q water.

DNA concentration is determined with spectrophotometer absorbance at 260 nm (Eppendorf spectrophotometer), and concentrations are adjusted to reach a value of 1 μg μL^{-1} which is used for PCR amplification (*see* **Note 3**).

3.4 DNA Amplification

- DNA is amplified using two primers for 18S rDNA: the forward primer F1 (5′- GGAGAATWAGGGTTCGACTC-3′) and the reverse primer R1 (5′-AGTCAAATTAAGCCGCAG-3′) [e.g., 19] which give a product of 820 bp.
- To profile communities an internal primer R2 5′ TCCAAG AATTTCACCTCTMACG 3′ [19] is used with the same primer F1 but with an addition of a 39GC- clamp at 5′ end [10]. Moreover, degenerated nucleotides (underlined) are introduced to amplify a wider number of genera.
- The PCR is performed in a 25 μL volume using 10 μL of 50 mM $mgCl_2$ solution, 2.5 μL of each primer (50 μM, Invitrogen), 2.5 μL of dNTPs (20 μM, Sigma-Aldrich), and 0.04 units of Taq polymerase (Sigma-Aldrich) for every μL of master mix.
- Thermal conditions in the reaction are 95 °C 1 min; 52 °C 1 min; and 72 °C 1 min for 30 cycles using the thermal cycler PCR Express Hybaid (Franklin, MA, USA).

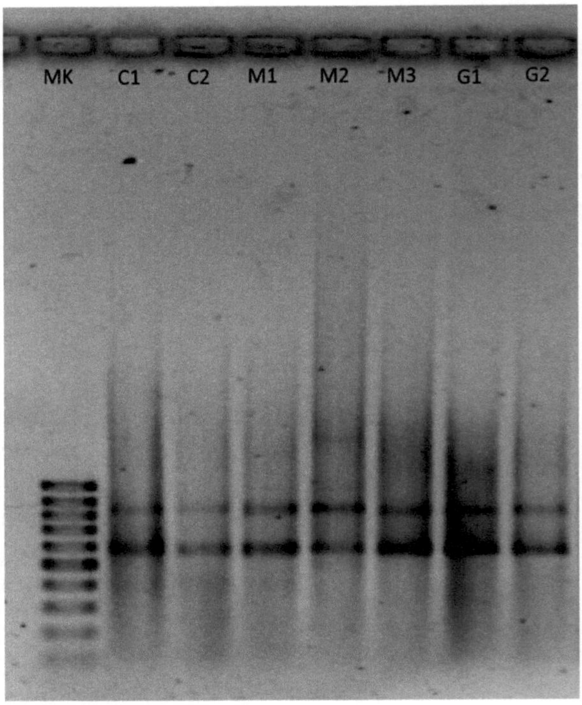

Fig. 1 Results of agarose gel 1% TAE with the hemi-nested PCR products of 820 bp and 560 bp from different agroecosystems (MK, Marker 100 bp Promega first lane, and different amplified samples C1, C2, M1, M2, M3, G1, G2)

- Amplification products are checked for length on agarose 1% TAE gel electrophoresis, stained with ethidium bromide and viewed with a Bio-Rad transilluminator.
- The hemi-nested PCR is carried out at 95 °C for 1 min, 60 °C for 45 s, and 72 °C for 45 s reducing the annealing temperature of 0.5° at every cycle for a total of 12 cycles and then 95 °C for 1 min, 54 °C for 45 s, 72 °C for 45 s for 18 cycles, and 72 °C for 5 min to reduce unspecific amplification (*see* Fig. 1, **Note 4**).

3.5 DGGE and Analysis

- The gel is prepared with a concentration of 8% with a first denaturing gradient 35–55%, and another gel is prepared using a denaturing gradient 20–50%. The gel is left to polymerize overnight.
- First of all the code foresees a pre-heating run at 58 °C to warm up the buffer and a pre-running at 70 V for 1 h which allow to empty the wells from acrylamide. Afterward the PCR products are loaded after mixing with loading solution *DCode dye solution* (1:1).
- Gels are run with DCode® for 6 h at 220 V in TAE 1× at 60 °C, stained with SYBR® Green 1× Stock 10000X in dimethyl sulfoxide (*see* Fig. 2, **Note 5**).

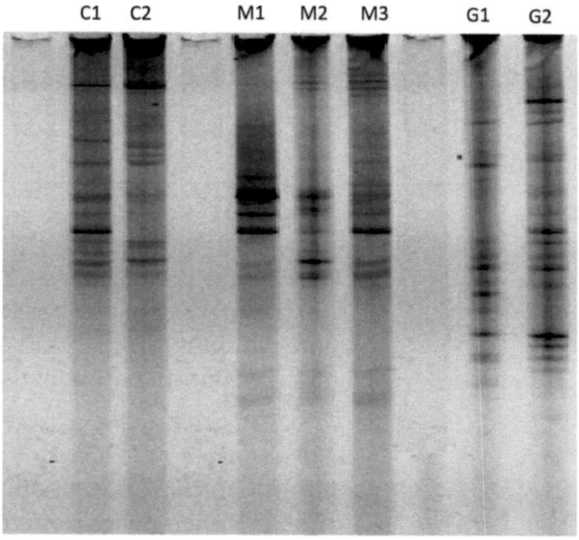

Fig. 2 DGGE profiles with gradient 35–55% of amplified 18S rDNA fragments from nematode soil samples from different habitats: C chestnut forest, M maize monoculture, G grassland

Table 1
Examples of biodiversity indices, of nematode communities from different soil habitats, calculated through classical analysis and through the DGGE

	Classical analyses			DGGE		
Indices	Chestnut forest	Maize monoculture	Grassland	Chestnut forest	Maize monoculture	Grassland
H′	1.08	0.70	1.20	0.98	0.65	0.99
N	34	11	27	29	7	15
D	0.13	0.27	0.11	0.17	0.32	0.16
J	0.31	0.30	0.36	0.39	0.33	0.37

3.6 Calculation of Nematode Diversity Indices

- The banding patterns of DGGE profiles are analyzed by Quantity One 4.2.3 software (Bio-Rad). Each band is quantified as volume (INT*mm^2).
- Data are exported in Excel or other software to calculate the indices using the intensity of the band and the number of bands.
- The nematode diversity index is expressed in terms of Shannon-Weaver diversity index and calculated using the formula $H = -\Sigma$ pi ln pi where for PCR-DGGE analysis, pi is the importance probability of the bands in a gel lane and measured as pi = ni/N, where ni is the intensity of a band and N is the sum of all band intensities in the densitometry profile (*see* Table 1).

– Other indices are applied to DGGE analyses such as Simpson index or dominance index calculated as $D = \Sigma (pi)^2$ and the Pielou index or evenness index as $J' = H'/\text{Log } N0$ where N0 is the number of bands and H′ is Shannon-Weaver Index.

4 Notes

1. The extraction efficiency decreases when heavy clay soil, such as soil from a paddy field, is used. In this case, use more trays (i.e., more amount of soil) for each sample, or use the density gradient centrifugation method.
2. To be sure that DNA extraction will be successful, it is important to ensure that nematodes are no lost during the concentration process. Collect some part of the waste solution, and check whether a large number of nematodes are present by using a stereomicroscope.
3. Perform agarose gel electrophoresis using a 5 μL aliquot of the PCR product to confirm that a product of the expected size (approximately 590 bp) has been obtained.
4. If DGGE is not performed immediately, store the purified product at −20 °C or lower.
5. If the movement of the marker bands is irregular, the bands from the top show irregular movement. Care must be taken when comparing different gels.

Acknowledgments

This work was financially supported by MUR Fondi PNR D.M. 737/2021 (PRJ 0823) project CEISA. The authors want to thank the University of Milano because this team worked for the application of DGGE to nematode biodiversity, which started years ago, when all authors had as affiliation the University of Milano.

References

1. Neher DA (1999) Soil community composition and ecosystem processes: comparing agricultural ecosystems with natural ecosystems. Agrofor Syst 45:159–185
2. Ferris H, Bongers T, de Goede RGM (2001) A framework for soil food web diagnostics: extension of the nematode faunal analysis concept. Appl Soil Ecol 18:13–29
3. Bongers T (1990) The maturity index: an ecological measure of environmental disturbance based on nematode species composition. Oecologia 83:14–19
4. Yeates GW, Bongers T (1999) Nematode diversity in agroecosystems. Agric Ecosyst Environ 74:113–135
5. Ritz K, Trudgill DL (1999) Utility of nematode community analysis as an integrated measure of the functional state of soils: perspective and challenges. Plant Soil 212:1–11

6. Corsini A, Zaccheo P, Crippa L, Manachini B (2020) Nematode colonization of pyrite cinder-polluted soil. Redia 103:137–145
7. Manachini B, Corsini A, Bocchi S (2009) Soil quality indicators as affected by a long term barley-maize and maize cropping systems. Ital J Agron 1:15–22
8. Manachini B (2008) Biodiversity of nematofauna associated with the rice field agroecosystem in north Italy. Redia XCI:151–153
9. Freckman DW, Ettema CH (1993) Assessing nematode communities in agroecosystems of varying human intervention. Elsiever Science Publishers BV
10. Muyzer G, De Waal EC, Uitterlinden AG (1993) Profiling of complex microbial populations by denaturing gradient gel electrophoresis analysis of polymerase chain reaction-amplified genes coding for 16S rRNA. Appl Environ Microbiol 59:695–700
11. Muyzer G (1998) Smalla, K (1998) Application of denaturing gradient gel electrophoresis (DGGE) and temperature gradient gel electrophoresis (TGGE) in microbial ecology. Antonie Van Leeuwenhoek 73:127–141. https://doi.org/10.1023/A:1000669317571
12. Nisa RU, Tantray AY, Shah AA (2022) Shift from morphological to recent advanced molecular approaches for the identification of nematodes. Genomics 114:110295
13. Blaxter ML, de LeyP GJR, Liu LX, Scheldeman P, Visraete A, Vanfleteren JR, Mackey LY, Dorris M, Frisse LM, Vida JT, Thomas WK (1998) A molecular evolutionary framework for the phylum Nematoda. Nature 392:71–75
14. Floyd R, Abebe E, Papert A, Blaxter M (2002) Molecular barcodes for soil nematode identification. Mol Ecol 11:839–850
15. Vejl P, Skupinová S, Sedlák S, Domkářová J (2002) Identification of PCN species (*Globodera rostochiensis*, *G. pallida*) by using of ITS-1 region's polymorphism. Rostl Výroba 48:486–489
16. Nuaima RH, Roeb J, Hallmann J, Daub M, Heuer H (2020) Significant genetic differences among *Heterodera schachtii* populations within and among sugar beet production areas. Nematology 22:165–177
17. Nuaima RH, Roeb J, Hallmann J, Daub M, Otte S, Heuer H (2018) Effector gene based DGGE fingerprinting to assess variation within and among populations. J Nematol 50:517–528
18. Foucher AFL, Wilson MJ (2002) Development of a polymerase chain reaction-based denaturing gradient gel electrophoresis technique to study nematode species biodiversity using the 18S rDNA gene. Mol Ecol Notes 11:45–48
19. Waite IS, O'Donnell AG, Harrison A, Davie JT, Colvan SR, Ekschmitt K, Dgan H, Wolters V, Bongers T, Bongers M, Gakonyi G, Nagy P, Papatheodorou EM, Stamou GP, Boström S (2003) Design and evaluation of nematode 18SrDNA primers for PCR and denaturing gradient gel electrophoresis (DGGE) of soil community DNA. Soil Biol Biochem 35:1165–1173
20. Foucher ALJL, Bongers T, Noble LR, Wilson MJ (2004) Assessment of nematode biodiversity using DGGE of 18S rDNA following extraction of nematodes from soil. Soil Biol Biochem 36:2027–2032
21. Okada H, Oba H (2008) Comparison of nematode community similarities assessed by polymerase chain reaction-denaturing gradient gel electrophoresis (DGGE) and by morphological identification. Nematology 10:689–700
22. Ikoyi I, Egeter B, Chaves C (2020) Responses of soil microbiota and nematodes to application of organic and inorganic fertilizers in grassland columns. Biol Fertil Soils 56:647–662
23. Jiang Y, Wang Z, Wu Y, Li H, Xue X (2023) The effect of auxin status driven by bacterivorous nematodes on root growth of *Arabidopsis thaliana*. Appl Soil Ecol 182:104730

Chapter 7

Investigations on Microbes Attached to the Cuticle of Phytonematodes

Olivera Topalović, Zoran S. Radaković, Ahmed Elhady, Sandra Bredenbruch, and Holger Heuer

Abstract

Nematodes form various associations with soil microbiome. Experimental studies on nematode-attached microbes can improve mechanistic understanding of these associations and lead to new discoveries relevant for the field of nematode biocontrol. Microbial attachment to the surface of phytonematodes is very specific and influenced by a multitude of factors, including the designation of nematodes and microbes, environmental and biological factors in soil, time of incubation, and the ratio and evolutionary trajectories between nematodes and microbes. Here, we describe how the classical nematological and microbiological techniques can be coupled with the advanced molecular tools to study the microbial attachment to phytonematodes in soil. We focus on the characterization of nematode-attached microbes using classical microbiological approaches and high-throughput amplicon sequencing and on the effects of nematode-attached microbes on plant defense responses.

Key words Nematodes, Microbiome, Microbial attachment, Soil, Amplicon sequencing, Plant defense

1 Introduction

Some microorganisms specifically attach to phytonematodes in soil [1–3] and can antagonize nematodes [4, 5]. However, until recently, microbial attachment to the surface of phytonematodes has been almost exclusively studied for the gram-positive bacterium *Pasteuria* spp. [6–9]. The infective stages of endoparasitic phytonematodes cannot ingest microorganisms because of a very narrow stylet; thus, their associations with microbes occur mostly via the body surface [8]. Although the studies on *Pasteuria* spp. have led to significant advances in understanding plant–nematode–microbial interactions, the associations of nematodes with microbial communities in soil have only recently received more attention [1, 3]. This has been possible especially by the advancement of

Sergio Molinari (ed.), *Plant-Nematode Interactions*, Methods in Molecular Biology, vol. 2756,
https://doi.org/10.1007/978-1-0716-3638-1_7,

next-generation sequencing (NGS) technologies that allow the in-depth analyses of microbial communities that associate with nematodes [2, 3]. We established protocols that rely on classical nematological and microbiological techniques and molecular tools to characterize nematode-attached microbes and unravel their potential to suppress nematodes [2, 3, 5, 10]. We have modified some of the previous techniques to reduce the time and effort necessary for an efficient surface disinfection of nematodes [11, 12] and removal of loosely attached microbes from nematode surface [1]. We also modified the techniques for studying the attachment of bacterial cells to nematode surface [13] and invasion of nematodes into the roots, with and without attached microbes [14]. For the characterization of nematode-attached microbes, we combined the culture-dependent methods, such as isolation of nematode-attached bacteria on growth media and DNA identification of the bacterial isolates [5]. We also established the methods for characterizing the whole microbial communities attached to the nematode surface using high-throughput amplicon sequencing [2, 3]. Furthermore, we established an assay for quantifying early plant defense responses to the invading nematodes with and without attached microbes, including the expressions of plant defense genes and oxidative burst [10].

Our article contains descriptions of assays to analyze the attachment of single bacterial isolates and microbial community to the cuticle of phytonematodes, as exemplified by the soil-dwelling stages of root-knot nematodes (*Meloidogyne* spp.) and root-lesion nematodes (*Pratylenchus* spp.). It includes extraction and surface disinfection of nematodes, preparation of microbial suspensions, attachment assays, characterization of nematode-attached microbiome, root invasion of nematodes with and without the attached microbiome, and studying the relative expressions of plant defense genes and oxidative burst in response to the invading nematodes.

2 Materials

2.1 Nematode Extraction Tools

1. *Meloidogyne* spp.: Plant culture with *Meloidogyne* spp., scissors, blender, 1% sodium hypochlorite, set of sieves (from top to bottom, 1 mm, 250 μm, 100 μm, and 20 μm), beaker for collecting eggs, 35% sucrose for purification of eggs, tap water, 50 mL Falcon tubes, pipette with 5 mL pipette tips, centrifuge for 50 mL Falcon tubes, Baermann tray (glass bowl, plastic rings, cheesecloth, paper tissue), bottle for collecting the nematodes.
2. *Pratylenchus* spp.: 2-month axenic carrot disc cultures with nematodes, forceps, scalpel, Baermann funnel, coarse sieve,

paper tissue, tap water, 50 mL Falcon tubes, bottle for collecting the nematodes.

3. Surface sterilization of nematodes: 4 mg/mL streptomycin–sulfate, 25 mg/mL rifampicin, CellCulture Guard, Falcon tubes, autoclaved tap water, centrifuge, pipette, and pipette tips.

2.2 Attachment Assays in Soil Suspensions

1. Soil, 1 mm sieve, autoclaved tap water, Stomacher, 50 mL Falcon tubes, 20 μm sieve, glass bottle, glass jars with lids, sodium hypochlorite, handmade 5 μm sieves, commercial 5 μm sieves (e.g., CellTrics filters, Sysmex, Norderstedt, Germany, or handmade, *see* **Note 1**).
2. Centrifuge, R2A and PDA plates with antibiotics (*see* Subheading 2.5), sterilization flame, sterile spatula for bacterial plating, pipette and pipette tips, incubator at 28 °C.

2.3 Characterization of Nematode-Attached Bacteria

1. Petri dishes with R2A and PDA media (*see* Subheading 2.4 for materials), pipette and pipette tips, inoculation loops, 1.5 and 2 mL Eppendorf tubes, TE buffer—50 mM Tris–HCl pH 8.0, 50 mM EDTA, 0.5% Tween 20, 0.5% Triton X-100, containing 200 μg lysozyme, 90 μg proteinase K, and 20 μg RNase A.
2. Centrifuge, incubator at 37 °C and 50 °C, 3 M guanidine hydrochloride, 20% Tween 20, silica matrix, washing solution (100 mM NaCl, 1 mM EDTA, 10 mM Tris–HCl, pH 7.5, 50% ethanol), 10 mM Tris–HCl, 0.1 mM EDTA pH 8.0.
3. BOX-PCR fingerprinting solution: GoTaq Flexi Buffer, $MgCl_2$, dNTP, DMSO, primer BOXA1R (5′-CTA CGG CAA GGC GAC GCT GAC TGA CG-3′), GoTaq Flexi DNA polymerase, thermocycler, electrophoresis device, 1.5% agarose gel, staining dye, GelCompar II (Applied Maths, Sint-Martens-Latem, Belgium) (*see* **Note 2**).
4. DNA extraction and amplicon sequencing: DNA extraction kit (FastDNA Spin Kit for Soil, MP Bio, Heidelberg, Germany), bead-beating tubes, FastPrep FP120 bead beating system, bacterial primers 341F (5'-CCTAYGGGRBGCASCAG-3') and 806R (5'-GGACTACNNGGGTATCTAAT-3'), fungal primers gITS7 (GTGARTCATCGARTCTTTG) and ITS4 (TCCTCCGCTTATTGATATGC).

2.4 Re-attachment Assay

Falcon tubes, liquid LB media, autoclaved distilled water, 1.5 and 2 mL Eppendorf sterile tubes, centrifuge for 50 mL tubes and for Eppendorf tubes, inoculation loop, sterilization flame, small spoon, glass beads, pipette and pipette tips, plates with R2A media with antibiotics (*see* Subheading 3.5).

2.5 Antibiotics and Growing Media

1. Nematode surface sterilization: 4 mg/mL streptomycin–sulfate, 25 mg/mL rifampicin, CellCulture Guard (PanReac, AppliChem, Darmstadt, Germany).
2. Media: R2A media, 10 mg/mL cycloheximide (*see* **Note 3**), PDA media, 30 mg/mL penicillin, 25 mg/mL streptomycin, 15 mg/mL chloramphenicol (*see* **Note 4**).

2.6 Nematode Invasion Assay

Petri dishes, Murashige and Skoog (MS) media, sucrose, Gelrite, distilled water, autoclaved distilled water, 1 L bottle, Falcon tubes, tomato seeds (cv. Moneymaker), 70% ethanol, 2.5% sodium hypochlorite, forceps, sterilization flame, climate chamber at 20 °C.

1. 800 mL MS media: weigh 3.76 g MS media and 16 g sucrose, and transfer them to a 1 L bottle containing the stirring magnet, add 800 mL of distilled water, adjust pH to 5.5–5.8, add 4 g Gelrite, and autoclave the media at 121 °C for 20 min.

2.7 Nematode Staining

Small beakers, forceps, scissors, tap water, 3% sodium hypochlorite, acid fuchsin solution, microwave, Petri dishes, glycerol.

2.8 RNA Extraction from the Roots and RT-q-PCR

2.8.1 RNA Extraction

Aluminum foil, container with liquid nitrogen, forceps, paper tissue, scissors, 2 mL Eppendorf tubes, 5 mm metal beads, TissueLyser II, FastRNA Pro Green Kit, DNA free DNA Removal Kit, Superscript IV reverse transcriptase, oligo dT(23), 1 μL of 10 mM dNTP, nuclease-free water, PCR reaction tubes, SSIV buffer, 100 mM DTT, 1 μL RNaseOUT, ice, incubators.

2.8.2 RT-q-PCR (Expressions of Defense Genes in Plants)

PCR tubes or plate, 0.25 mM $MgCl_2$, 0.2 mM of each dNTP, EvaGreen (Biotium), gene specific primers (or reference gene, respectively), 0.5 U Hot Start Taq DNA Polymerase, cDNA, thermocycler.

2.8.3 RT-q-PCR (Relative Quantification of Nematodes)

PCR tubes or plate, Standard Taq Reaction Buffer, 0.25 mM $MgCl_2$, 0.2 mM of each dNTP, EvaGreen, 1 μM of nematode-specific primers (*Meloidogyne* spp.-specific primers Melo1f 5′-AAGATATCTGGTTGATCCTGCCTGA-3′ and Melo723r 5′-GTTCAAAGTAAACTTGCAAYGMCTG-3′), 0.625 U Hot Start Taq DNA Polymerase, cDNA, thermocycler.

2.9 Oxidative Burst Assay

12-day-old tomato plants, approx. 10,000 surface-sterilized nematodes, bacterial suspensions (as described in Subheading 3.5), autoclaved tap water, 1.5 mL tubes, centrifugation, sterile filters, brush, 96-well plate, 0.1 μM luminol derivate 8-amino-5-chloro-2,3-dihydro-7-phenyl-pyrido[3,4-d] pyridazine sodium salt, 20 μg/mL horseradish peroxidase, 1 μM flagellin peptide (Flg22), luminescence 96-well microplate reader instrument software, Microsoft Office Excel. A measurement period of 120 min is recommended.

3 Methods

3.1 Nematode Extraction and Surface Disinfection of the Cuticle of Nematodes

1. Extraction of *Meloidogyne* spp.: Wash the roots infested with nematodes extensively (*see* Fig. 1). Chop the infested roots into small pieces (1–2 cm), and transfer them to the blender. Add 1% sodium hypochlorite to the blender with roots (*see* **Note 5**). Blend the roots for 15 s at a low speed and 15 s at a high speed (*see* **Note 6**). Pour the blended roots with nematode eggs over the set of sieves (the eggs are collected on a 20 μm sieve). Wash the eggs extensively (*see* **Note 7**). Transfer eggs to the beaker.
2. Centrifugation method to purify the eggs (optional): Transfer egg suspensions to 50 mL Falcon tubes, and centrifuge at 1000 x*g* for 1 min. Remove the supernatant. Add 30 mL of 35% sucrose to the tube. Invert the tube several times to resuspend the eggs. Using a 5 mL pipette tip, slowly add 5 mL of tap water on top of the sucrose suspension with eggs. Without stirring, centrifuge the tube for 3 min at 1000 x*g*. Carefully, collect eggs separated on the top using a 5 mL pipette tip. Wash the eggs on a 20 μm sieve using tap water (*see* **Note 8**).
3. Extraction of *Pratylenchus* spp.: Set up the Baermann funnel. Place the sieve with a paper tissue inside the funnel. Cut the carrot disks into small pieces using scalpel, and distribute them over the sieve [15]. Collect nematodes daily (*see* **Note 9**).
4. Nematode surface sterilization: Add 2 mL of CellCulture Guard, 100 μL of 4 mg/mL streptomycin, and 10 μL of 25 mg/mL rifampicin to a 15 mL Falcon tube, and top it up with hatched nematodes to 4 mL. Mix the content by inverting the tube, and incubate it with a slow shaking (approx. 30 rpm) for 4 h at room temperature.
5. Washing of the surface-sterilized nematodes using the sieve: Transfer the nematodes to a sterile 5 μm sieve (*see* **Note 10**). Wash the nematodes extensively with autoclaved tap water (*see* Fig. 2).
6. Washing of the surface-sterilized nematodes using a centrifugation method: Transfer the nematodes to a 50 mL Falcon tube. Add approx. 20 mL of an autoclaved tap water, and centrifuge the tube for 1 min at 1000 × *g*. Remove the supernatant and repeat the washing three times.

3.2 Extraction of Microbial Suspensions from Soil

1. Extraction using Stomacher (*see* **Note 11**): Sieve the soil using a 1 mm sieve. Weigh 10 g of sieved soil in a Stomacher bag, and add 20 mL of sterile tap water. Blend the soil for 1 min at a high speed (300 rpm). Transfer the soil suspension to a 50 mL Falcon tube. Repeat the extraction with another 20 mL of sterile tap water. Centrifuge the combined suspensions for 5 min at 500 × *g*. Transfer the supernatant (microbial suspension) into a

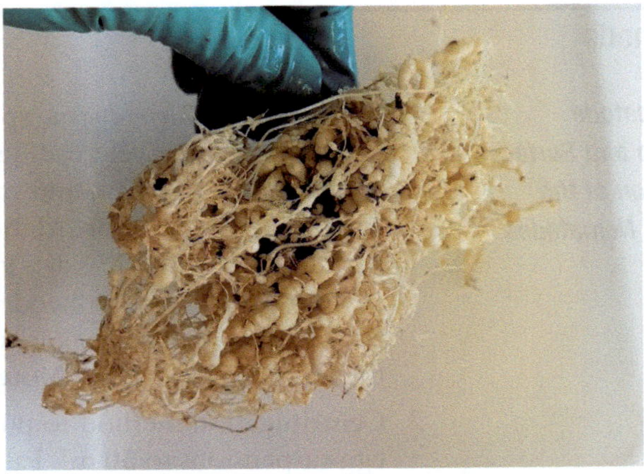

Fig. 1 Washed tomato roots infested with *Meloidogyne incognita*

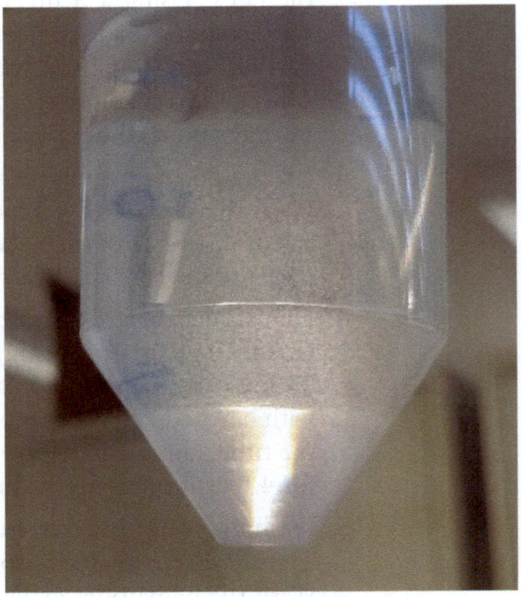

Fig. 2 Surface-sterilized J2s of *Meloidogyne* sp.

new tube. Sieve the microbial suspension through a 10 μm sieve (*see* **Note 12**). Centrifuge the suspension for 5 min at 5000 × *g*, and resuspend the pellet in 25 mL of sterile tap water.

2. Extraction using shaking: Weigh 100 g of soil to a 2 L glass bottle (optionally, with pre-added sterilized metal beads). Add 900 mL of autoclaved tap water. Place the bottle on a shaker horizontally and shake it for 1 h at 100 rpm. After shaking, let the coarse soil particles sediment for 10 min, and collect the supernatant into Falcon tubes. Proceed with centrifugation as previously described.

Fig. 3 Nematode incubation in soil suspensions using sieves

3.3 Attachment Assays

1. Microbial attachment to nematodes in soil suspensions (Method 1): Sterilize the glass jars in the oven for 2 h at 180 °C. Sterilize the handmade 5 μm sieves by incubation in 6% sodium hypochlorite for 10 min, and wash them using an autoclaved water. Add 25–30 mL of soil suspensions (*see* Subheading 3.2) to the glass jars. Insert the sterilized sieves in the glass jars, and add surface-sterilized nematodes on top of the sieves using the pipette (*see* Fig. 3). Close the jars with the lids, and incubate them on a shaker (approx. 30 rpm) at room temperature overnight (*see* **Note 13**).
2. Microbial attachment to nematodes in soil suspensions (Method 2): Incubate the nematodes in soil suspensions as previously described, but without the use of sieves.
3. Washing of nematodes: Wash the nematodes on the sieves with 50 mL autoclaved tap water. Collect the washed nematodes in 50 mL Falcon tubes, and centrifuge at 1000 × *g* for 2 min (*see* **Note 14**). Remove the supernatant, and transfer 500 μL of nematodes with attached microbes to an Eppendorf tube.

Prepare serial dilutions using autoclaved tap water, and plate the nematodes on R2A and PDA plates that are supplemented with antibiotics as described in Subheading 2.3. Incubate the plates at 28 °C for several days.

3.4 Characterization of Nematode-Attached Bacteria

1. Culture-dependent methods and BOX-PCR fingerprinting: Collect single bacterial colonies, and purify them using the inoculation loop and a streak plate method.
2. Resuspend three loops with bacterial lawn in 1 mL of liquid LB media containing 20% glycerol in cryogenic tubes. Store the tubes at −20 °C as freeze stock cultures.
3. Resuspend 10 μL of a bacterial lawn produced from the single isolates in 100 μL TE buffer in 1.5 mL tubes.

4. Lyse the cells by adding 100 μL of TE buffer (*see* Subheading 2.3). Incubate the tubes at 37 °C for at least 30 min. Add 3 M guanidine hydrochloride and 20% Tween 20, and incubate at 50 °C for at least 30 min.
5. Capture the DNA from the lysate for 5 min with 200 μL GeneClean Spin Glassmilk. Wash the pelleted Glassmilk twice with 500 μL washing solution and air-dry for 10 min.
6. Elute the DNA with 100 μL 10 mM Tris–HCl and 0.1 mM EDTA pH 8.0, and separate it from the Glassmilk by centrifugation at maximum speed for 2 min (12,000 × *g*).
7. Store the supernatant containing the bacterial DNA at −20 °C until use.
8. For BOX-PCR fingerprinting, add approx. 20 ng of template DNA to a 25 μL PCR reaction containing GoTaq Flexi Buffer, 3.75 mM $MgCl_2$, 0.2 mM of each dNTP, 5% w/v DMSO, 0.2 mM primer BOXA1R (5′-CTA CGG CAA GGC GAC GCT GAC TGA CG-3′), and 1 U GoTaq Flexi DNA polymerase; amplify the DNA in a thermocycler using the following conditions: denaturation step for 7 min at 94 °C, 35 cycles of 1 min at 94 °C, 1 min at 53 °C, and 8 min at 65 °C; perform the final extension step for 16 min at 65 °C.
9. Visualize the PCR products using electrophoresis in 1.5% agarose gel for 3 h at 80 V and UV transillumination (254 nm) after staining with ethidium bromide; analyze the band patterns by GelCompar II.
10. DNA extraction and amplicon sequencing of nematode-attached microbes: Transfer the washed nematodes with the attached microbes to the bead-beating tubes from the kit used for DNA extraction.
11. Lyse the cells using the FastPrep FP120 bead beating system for 30 s at a high speed.
12. Proceed with the DNA extraction according to the manufacturer's instructions.
13. Amplify the 460 bp variable V3–V4 region of the 16S rRNA gene from the total DNA using the bacterial primers 341F and 806R.
14. Amplify the fungal ITS fragments with primers gITS7 and ITS4.
15. Consult with the sequencing company on the most suitable library preparation for the amplicon sequencing of nematode-attached microbes.

3.5 Re-attachment Assay

3.5.1 Preparation of Bacterial Cells

1. Grow the isolate from the freeze stock culture on R2A overnight. Purify a single colony by additional streaking on R2A. Use the sterile toothpick or inoculation loop to transfer a purified isolate in 4 mL of liquid LB in a glass test tube for 24 h. Resuspend 8 μL of overnight culture in 3992 μL of LB and grow it for 24 h.

2. Pellet the bacteria for 5 min at 5000 × *g* and wash twice with autoclaved tap water. Measure the bacterial optical density at 600 nm (OD_{600}) using a spectrophotometer. Prepare 240 μL of bacterial suspensions with OD_{600} 0.2 in sterilized glass tubes for nematode incubation. Serially dilute the bacterial suspensions, and plate aliquots on R2A media to determine the number of colony forming units (CFU) in the suspensions (CFU/mL).

3.5.2 Attachment Studies

1. Add approx. 500 surface-sterilized nematodes (10 μL) to the glass tube with bacterial suspension. Close the tube and gently stir. Incubate the tube on a shaker (30 rpm) at room temperature overnight.
2. Transfer the suspension to the 1.5 mL tubes and centrifuge at 1000 × *g* for 1 min.
3. Replace the supernatant with 500 μL autoclaved tap water, and repeat the washing three times. Add equal amount of glass beads (*see* **Note 15**) to each tube and vortex for 10 s.
4. Serially dilute the suspensions down to 1:1000 using autoclaved tap water.
5. Plate aliquots of 50 μL on R2A media and store the plates at 28 °C for 2 days.
6. Determine the number of CFU per mL and calculate the number of CFU per nematode.

3.6 Nematode Invasion Assay

3.6.1 Surface Sterilization of Tomato Seeds

1. Distribute MS media equally to Petri dishes.
2. Transfer tomato seeds to a 50 mL Falcon tube, and wash them three times with autoclaved distilled water (by inverting the tube and replacing the water). Remove the water.
3. Add 20–30 mL of 70% ethanol to the tube with seeds, and invert the tube for 1 min, remove the ethanol, add 20–30 mL of 2.5% sodium hypochlorite, and invert the tube for 3 min, and remove the sodium hypochlorite.
4. Wash the seeds four times with autoclaved distilled water. Transfer the seeds to the plates with MS media (1–2 seed per plate).
5. Germinate the seeds in a climate chamber at 20 °C for 2–4 days (*see* Fig. 4, **Note 16**).

3.6.2 Nematode Invasion into the Roots

1. Surface-sterilize the nematodes as described in Subheading 3.1, and incubate them in soil suspensions as described in Subheadings 3.3 and 3.5 or in autoclaved tap water overnight. Collect the washed nematodes in a 2 mL tube. Place five 10 μL drops with approx. 100 nematodes around the root tips of a 2-week-old tomato seedling in a Petri dish. Use autoclaved tap

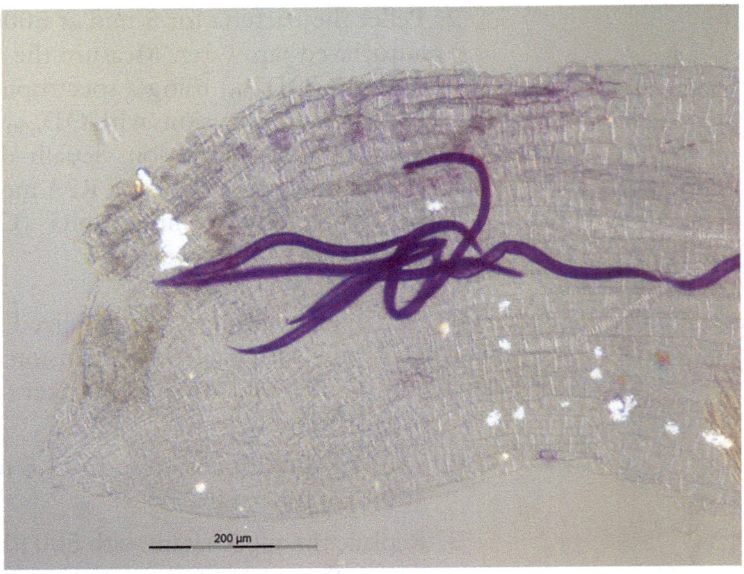

Fig. 4 Tomato roots invaded with the J2s of *M. hapla*

water to inoculate the control plants. Close the Petri dishes with parafilm, and store them back in the climate chamber (*see* **Note 17**).

2. To stain nematodes inside the roots (*see* Fig. 4), remove the plants from the Petri dishes, excise the roots using scissors, and wash them using tap water. Incubate the roots in 10–20 mL of 3% NaOCl for 4 min (agitate occasionally). Rinse the roots with running tap water, and incubate them in water for 15 to 30 min. Boil the roots in 30–50 mL of tap water supplemented with 1 mL of stock acid–fuchsin solution for 30 s. Dry the roots slightly by placing them on a paper tissue, and transfer to the designated Petri dish. Cover the roots with glycerol, and store the Petri dishes at +4 °C (*see* **Note 18**). Count the numbers of invaded nematodes using a stereomicroscope.

3.7 RNA Extraction from the Roots

1. Sample collection: Use the forceps to remove the plants from the Petri dish and a paper tissue to remove the attached media if needed.
2. Separate the leaves and roots with scissors.
3. Transfer the plant material to the labeled bags made of aluminum foil, place the bags with plant material in a container with liquid nitrogen, and store the samples at –80 °C until use.
4. RNA extraction: Transfer the plant material to a 2 mL tube containing a 5 mm metal bead. Lyse the material in a Tissue-Lyser II twice for 20 s at the highest speed (30 Hz), keeping samples frozen in a pre-cooled block.

5. Transfer the frozen root powder with lysing solution to bead beating tubes of the FastRNA Pro Green Kit to extract the total RNA from the homogenized plant material according to the manufacturer's instructions.
6. Remove the traces of DNA from aliquots of 1 μg extracted RNA by DNase I digestion, DNase inactivation, and removal, using the DNA free DNA Removal Kit.
7. For reverse transcription of a purified RNA, use Superscript IV reverse transcriptase according to the manufacturer's instructions: incubate 8 μL RNA (ca. 1 μg) with 1 μL 50 μM oligo dT (23), 1 μL dNTP (10 mM each), and 3 μL nuclease-free water in the PCR tubes at 65 °C for 5 min. Place the tubes on ice for at least 1 min, add Superscript IV reverse transcriptase (200 U in 1 μL), 4 μL 5× SSIV Buffer, 1 μL 100 mM DTT, and 1 μL RNaseOUT to each reaction tube. Incubate the tubes at 55 °C for 10 min, followed by 80 °C for 10 min. Store cDNA at − 20 °C.

3.8 RT-q-PCR

This technique is used for studying the expressions of plant defense genes in response to nematodes and their attached microbes.

1. Add 0.25 mM $MgCl_2$, 0.2 mM of each dNTP, EvaGreen, 1 μM of plant defense gene specific primers (or reference gene, respectively), 1 μL of cDNA, and 0.5 U Hot Start Taq DNA Polymerase to each reaction tube with two technical replicates.
2. Place the tubes to a thermocycler and use the following conditions: initial denaturation at 94 °C for 5 min, 40 cycles of a denaturation step at 94 °C for 20 s, annealing step at 60 °C for 30 s, and an extension step at 68 °C for 30 s and 80 °C for 30 s.
3. Read the fluorescence at the 80 °C step of each cycle, generate the melting curve between 65° and 95 °C, and use the reference gene to normalize the Ct values of the expressed defense genes.
4. Determine the relative gene expressions in each sample against gene expressions in control plants relative to expression of a housekeeping gene (e.g., ubiquitin) using the $2^{-\Delta\Delta Ct}$ method [16].

3.9 Relative Quantification of the Nematode 18S rRNA in Roots

1. Prepare 25 μL PCR reactions containing Standard Taq Reaction Buffer (*see* Subheading 2.8.3).
2. Place the tubes in a thermocycler and use the following conditions: initial denaturation at 94 °C for 2 min, 40 cycles with a denaturation at 94 °C for 20 s, annealing at 57 °C for 20 s, extension at 72 °C for 45 s, and fluorescence detection at 80 °C for 30 s.

3. Generate the melting curves between 65° and 95 °C, and normalize the Ct values against the ubiquitin gene expression. Use the $2^{-\Delta\Delta Ct}$ method [16] to determine the relative 18S rRNA of the nematodes with attached microbes against the nematodes without attached microbes.

3.10 ROS Burst Assay

1. Preparation of the elicitor samples: Incubate approx. 10,000 surface-sterilized nematodes in 200 μL bacterial suspensions (as described in Subheading 3.5) or for the control in 200 μL of autoclaved tap water in 1.5 mL microtubes at room temperature overnight without shaking.
2. The following day, centrifuge the tubes at 1000 x*g* for 2 min, and separate the pellet and supernatant (*see* **Note 19**). Sterile-filter the latter, and resuspend nematode pellets or nematode–bacteria pellets in 200 μL of autoclaved tap water to obtain the elicitor samples (*see* **Note 20**).
3. To obtain plant tissue, excise the well-developed cotyledon leaves from 12-day-old tomato plants in distilled water in the dark for 12 h, carefully transfer the leaf samples with a brush to a 96-well plate containing 80 μL of 0.1 μM luminol derivate 8-amino-5-chloro-2,3-dihydro-7-phenyl-pyrido[3,4-d] pyridazine sodium salt, and incubate for 1 h in the dark.
4. Add 20 μL of 20 μg mL^{-1} horseradish peroxidase to each well. Add 50 μL elicitor, with 1 μM flagellin peptide (Flg22) as positive control and autoclaved tap water as negative controls.
5. Immediately start the measurement using the luminescence program of a 96-well microplate reader over 120 min.
6. Analyze the data using instrument software and Microsoft Office Excel. A measurement period of 120 min is recommended, but an individual adaption might be required. The program for the analysis is optional (*see* **Note 21**). Analyze the data using instrument software and Microsoft Office Excel.

4 Notes

1. Sieves can be handmade from the small plastic pots and the filter cloth. Use the hot scalpel to remove the bottom of the pot. Use the high temperature on a heating plate to bind the cloth to the smooth edges of the pot.
2. GelCompar II allows the discrimination of duplicate isolates.
3. When the autoclaved media cools down to approx. 60 °C, add 1 mL of cycloheximide stock solution per 100 mL R2A media.
4. Add 2 mL of the stock solution of penicillin–streptomycin and 1.25 mL of chloramphenicol per 250 mL of liquid PDA media (<60 °C).

5. The chopped roots should be completely soaked in sodium hypochlorite.
6. Optionally, the roots can be left in sodium hypochlorite for 1–2 min after blending to increase the release of eggs from the roots.
7. Wash the content on top of every sieve extensively (the eggs on the 20 μm sieve should lose the chlorine smell).
8. To increase the success of the nematode surface sterilization, purify the eggs under the laminar flow. In that case, sterilize the sieve using 6% sodium hypochlorite, and use autoclaved tap water during every step of purification.
9. After 48 h only the J2s of *Pratylenchus* spp. will be collected as they will hatch from the eggs within the carrot tissues.
10. You can obtain the sterile sieves from Sysmex (sysmex-europe.com).
11. Use Stomacher (Stomacher®80, Seward, London, United Kingdom) for small amounts of soil (e.g., 10 g).
12. Additional sieving is done to remove native soil nematodes.
13. Using sieves eases the washing of nematodes.
14. Additional centrifugation will reduce the volume of water. The disadvantage of using sieves is that larger soil microbes as fungal mycelium may fail to reach the nematodes.
15. Use the tip of the smallest measuring spoon to transfer the glass beads (0.1 mm) to the tubes with nematodes and the attached bacteria. Glass beads will allow nematodes to break and homogenize within the suspension to suspend attached microbes.
16. Optionally, the seeds can be left in dark to germinate.
17. Depending on the purpose of your experiment, you can collect the roots at different time points. Keep in mind that the longer you grow the roots, the more humid it gets inside the Petri dish, and the more stressed the plants are.
18. Acid fuchsin stock solution: Dissolve 3.5 g acid fuchsin in 250 mL acetic acid, and add 750 mL distilled water.
19. The sterile-filtered supernatant represents the surface-released compounds from the nematodes with and without bacterial cell-free supernatants.
20. These are the suspensions of live nematodes with and without the living bacteria.
21. Recommendation: To compare the data as curve graph as well as based on the total luminescence over a defined period.

References

1. Adam M, Westphal A, Hallmann J, Heuer H (2014) Specific microbial attachment to root knot nematodes in suppressive soil. Appl Environ Microbiol 80:2679–2686
2. Elhady A, Gine A, Topalovic O, Jacquiod S, Sørensen SJ, Sorribas FJ, Heuer H (2017) Microbiomes associated with infective stages of root-knot and lesion nematodes in soil. PLoS One 12:e0177145
3. TopalovićO SSS, Heuer H, Nesme J, Kanfra X, Sørensen SJ, Vestergård M (2022) Deciphering bacteria associated with a pre-parasitic stage of the root-knot nematode *Meloidogyne hapla* in nemato-suppressive and nemato-conducive soils. Appl Soil Ecol 172:104344
4. Elhady A, Topalovic O, Heuer H (2021) Plants specifically modulate the microbiome of root-lesion nematodes in the rhizosphere, affecting their fitness. Microorganisms 9:679
5. Topalović O, Elhady A, Hallmann J, Richert-Pöggeler K, Heuer H (2019) Bacteria isolated from the cuticle of plant-parasitic nematodes attached to and antagonized the root-knot nematode *Meloidogyne hapla*. Sci Rep 9:11477
6. Danks C, Davies KG (1993) Carbohydrate/protein interactions between the cuticle of infective juveniles of *Meloidogyne incognita* and spores of the obligate hyperparasite *Pasteuria penetrans*. Nematologica 39:53–64
7. Davies KG (2009) Understanding the interaction between an obligate hyperparasitic bacterium, *Pasteuria penetrans* and its obligate plant-parasitic nematode host, *Meloidogyne* spp. In: Webster JP (ed) Natural history of host-parasite interactions. Academic Press, pp 211–245
8. Davies KG, Curtis RHC (2011) Cuticle surface coat of plant-parasitic nematodes. Annu Rev Phytopathol 49:135–156
9. Orr JN, Neilson R, Freitag TE, Roberts DM, Davies KG, Blok VC, Cock PJA (2020) Parallel microbial ecology of *Pasteuria* and nematode species in Scottish soils. Front Plant Sci 10:1763
10. Topalović O, Bredenbruch S, Schleker S, Heuer H (2020) Microbes attaching to endoparasitic phytonematodes in soil trigger plant defense upon root penetration by the nematode. Front Plant Sci 11:138
11. Tsai AY-L, Iwamoto Y, Tsumuraya Y, Oota M, Konishi T, Ito S, Kotake T, Ishikawa H, Sawa S (2021) Root-knot nematode chemotaxis is positively regulated by l-galactose sidechains of mucilage carbohydrate rhamnogalacturonan-I. Sci Adv 7:eabh4182
12. van Bezooijen J (2006) Methods and techniques for nematology. Wageningen University, Wageningen
13. Maghodia AB, Spiegel Y, Sela S (2008) Interactions between *Escherichia coli* and the plant-parasitic nematode *Meloidogynejavanica*. J Appl Microbiol 105:1810–1816
14. Mendy B, Wang'ombe MW, Radakovic ZS, Holbein J, Ilyas M, Chopra D, Holton N, Zipfel C, Grundler FMW, Siddique S (2017) Arabidopsis leucine-rich repeat receptor-like kinase NILR1 is required for induction of innate immunity to parasitic nematodes. PLoS Pathog 13:e1006284
15. European and Mediterranean Plant Protection Organization (2013) PM 7/119 (1) Nematode extraction. EPPO Bull 43:471–495
16. Pfaffl MW (2001) A new mathematical model for relative quantification in real-time RT–PCR. Nucleic Acids Res 29:e45

Chapter 8

Detection of the Effects of Root Exudates (Diffusates) on Nematode Hatching and Attraction

Roland N. Perry

Abstract

Plant-parasitic nematodes have enormous economic and social impacts. The majority of plant-parasitic nematodes are soil dwelling and feed on plant roots. Exudates from actively growing roots initiate hatch of some nematode species, thus ensuring infective juveniles emerge in close proximity to host plant roots. Several gradients of volatile and non-volatile compounds are established around plant roots, at least some of which are used by nematodes to orientate toward the roots. Plant-parasitic nematodes are microscopic in size (less than 1 mm in length and between 15 and 20 μm in diameter), so investigations into behavior are challenging. Various in vitro techniques have been used to evaluate the effects of root exudates. The techniques can also be used to evaluate the comparative attractiveness of different plants or cultivars of the same plant species. This chapter describes some examples of different types of basic in vitro assays.

Key words Agar plate assay, Chemosensory responses, Cyst nematode, Hatching factor, Nematode behavior, Nematode invasion, Plant-parasitic nematode, Pluronic gel

1 Introduction

For soil-dwelling plant-parasitic nematodes, cues from host plant roots are vital to enable the nematodes to locate a food source. Chemicals in root exudates (also termed root diffusates or root leachates) from host roots include signals that elicit a hatch response in many plant-parasitic nematodes, especially the cyst nematodes such as the potato cyst nematodes, *Globodera rostochiensis* and *G. pallida*, where the host–parasite relationship is so close that substantial hatch of the second-stage juveniles (J2) occurs only in response to exudates from host plants. Once free in the soil, the nematodes need to locate a host root, and volatile and non-volatile components from roots are central in enabling the nematode to orientate to roots and, for endoparasitic nematodes, to locate the preferred invasion sites. Investigation of interactions between plant hosts and plant-parasitic nematodes has included examination of

Sergio Molinari (ed.), *Plant-Nematode Interactions*, Methods in Molecular Biology, vol. 2756,
https://doi.org/10.1007/978-1-0716-3638-1_8,

the effects of chemicals emanating from host roots on hatching and orientation of nematodes, and some of the basic techniques will be outlined here.

2 Materials

Plant-parasitic nematodes can be counted using tally counter and a stereomicroscope at ×25 magnification. All experiments are conducted at room temperature, unless otherwise stated.

2.1 Hatching

2.1.1 Collection of Root Exudates

1. Grow potato plants from chits in pots (15 cm diameter) containing soil, potting compost, or growth medium, under standard glasshouse conditions.
2. Distilled water (dH_2O) to saturate pots.
3. Whatman No. 1 filter paper.
4. A 0.2 μm cellulose acetate to remove fungal and bacterial contaminants.
5. If using seeds, sterilize seeds with 70% ethanol and 1% sodium hypochlorite followed by dH_2O.
6. Sterilized 3:1 soil and sand mixture.
7. Grow seedlings at 26 ± 4 °C and 12 h/12 h dark/light.
8. Use 500 mL glass beakers containing 400 mL deionized water of HPLC quality.

2.1.2 Hatching Assays

1. Cylinders of glass tube (approx. 2.5 cm long) with nylon mesh (aperture 60 μm) fixed at the end with a thin plastic ring.
2. 20 mL capacity vial.
3. An alternative unit is a 96-well microtiter plate with wire mesh "thimbles" fitted into each well.
4. For egg masses of *Meloidogyne*, use a 48 μm sieve.
5. A microscope is used to visualize and count hatched juveniles.
6. Use 10% NaOCl and homogenize unhatched eggs to obtain numbers of unhatched juveniles.
7. A concentration of 0.001% acridine orange dissolved in PRD can visualize hatchable juveniles under a fluorescent microscope.

2.1.3 Adenosine Triphosphate Hatching Assay

1. Use bioluminescent photometry to determine adenosine triphosphate (ATP) in nematodes.
2. Use 0.1 M arsenate buffer at pH4 and 4 °C to extract ATP.

3. Measure the ATP content of the cysts using a photometer to determine bioluminescence of a luciferin–luciferase extract from fireflies.

2.2 Attraction

2.2.1 Agar Plate Attraction Assays

1. The standard attraction assay uses Petri dishes (5 cm diameter) with 5 mL water agar (0.4%).
2. A simpler form uses Petri dishes (10 cm diameter) containing 1.7% distilled water agar (15 mL dish).
3. Ethanol (70%).

2.2.2 Pluronic Gel Attraction Assays

1. Pluronic F-127 (PF-127) (NF Prill Poloxamer 407, BASF).
2. Petri dishes (35 or 60 mm diameter).
3. Tissue culture plates (6- or 12-well).

2.2.3 Movement Assays for Root Exudates

1. Polythene tubes (usually, height 1.0 cm); nylon mesh (53 μm aperture).
2. Sand (particle size 250–600 μm).
3. Small tube or sample jar.

2.2.4 Attraction Choice Assay

1. Four hydrophobically lined PVC Pasteur pipettes.
2. Parafilm.
3. Sterilized sand (630 g).
4. Seeding compost.
5. Three 50 mL Falcon tubes, each filled with 40 mL of 1:1 (w/v) concentrated sucrose solution.

3 Methods

3.1 Hatching

3.1.1 Collection of Root Exudates

The common method for obtaining exudate is based on collection of root exudates from potato plants, grown from potato chips, for hatching potato cyst nematodes. For cyst nematodes in particular, the activity of exudate varies with the age of the host plant with cyst nematodes, such as *Heterodera goettingiana*, only hatching in exudates from plants of a specific age [1]. Several artificial hatching agents have been used to induce hatching of species of *Globodera* and *Heterodera* (Table 1).

1. Plants can be grown in pots (15 cm diameter) containing soil, potting compost, or growth medium that ensures good root growth.
2. When plants have grown either to a set leaf stage or time (e.g., from planting or shoot emergence), exudates can be obtained (*see* **Note 1**).
3. Add sufficient dH_2O to saturate the pots.

Table 1
Some artificial hatching agents and the suggested concentrations for common cyst nematodes

Species	Compound	Concentration (mM)
Globodera rostochiensis	Sodium metavanadate Picrolonic acid	0.6 0.3
G. tabacum	Zinc sulfate	2.0
Heterodera schachtii	Picric acid Zinc chloride	3.0 3.0
H. cruciferae	Flavianic acid	0.6
H. glycines	Flavianic acid Zinc chloride	3.0 3.0
H. trifolii	Zinc chloride	3.0
H. carotae	Zinc chloride	10.0

From Ref. [46]

4. Add a further volume of dH_2O (usually 100 mL), and collect the runoff from the pot.
5. Collection can be repeated every other day, depending on the required volume.
6. Filter solution through Whatman No. 1 filter paper.
7. For sterile root exudates, pass the solution through a 0.2 μm cellulose acetate filter (0.2-FW) to retain bacterial and fungal spores.
8. This is the stock solution, which for potato root exudates is usually diluted 1 in 4 with dH_2O to remove the effect of hatch inhibitors.
9. Store the stock solution in glass bottles at about 5 °C.

Exudates can be collected from roots of plants grown from seed, such as cowpea and tomato, as described in [2] for obtaining exudates for GC-MS analysis.

1. Sterilize seeds with 70% ethanol for 30 s, followed by 1% sodium hypochlorite for 1 min, and finally wash three times with dH_2O.
2. Sow seeds in sterilized 3:1 soil and sand mixture.
3. Grow emerged seedlings for 4 weeks in a glasshouse at 26 ± 4 ° C and 12 h/12 h dark/light.
4. Water plants with sterilized water every second day, after which they are uprooted and washed gently in sterile deionized water.

5. Submerge roots of intact plants in 500 mL glass beakers containing 400 mL deionized water of HPLC quality, and leave for 10 h at 25 °C to collect exudates.
6. The exudates from five beakers, each containing five plants, can then be pooled to make 2 L of final sample which was stored at 4 °C.
7. These exudates were used for GC-MS analysis for profiling the root exudates.

3.1.2 Standard Hatching Assays

Determination of the hatching of a nematode population is often referred to as a viability assay. However, hatching assays may not accurately reflect viability as some unhatched nematodes may be in diapause [1], thus being viable but unresponsive to root exudates. Diapause of *G. rostochiensis* and *G. pallida* can be avoided by preventing desiccation or can be broken by storing at 5 °C for 2–3 weeks. Hatching assays can be done using watch glasses containing the test solution and the biological material. Other methods use hatching units.

1. There are numerous descriptions of hatching units that can be used with cysts or free eggs. A basic unit (ideally made from glass; *see* **Note 2**) should consist of two compartments, one containing water or the hatching agent and the other containing the test nematodes, separated from each other by a material permeable for water or the hatching agent and the hatched nematodes. For hatch determination of free eggs of cyst nematodes, cut cysts open to release the eggs.
2. Use cylinders of glass tube approximately 2.5 cm long with nylon mesh (aperture 60 μm) fixed at the end with a thin plastic ring. The mesh size is sufficient to retain eggs while allowing passage of hatched nematodes. Larger tubes can be used for cysts (*see* **Note 3**) or *Meloidogyne* egg masses (see below).
3. Place cylinders in a 20 mL capacity vial containing the root exudates (or test solution).
4. Variation of hatching data is reduced the more cysts are used, but this depends on availability and source of biological material (*see* **Note 4**). Ideally, 100 cysts per run, split as four batches of 25 cysts or five batches of 20 cysts is sufficient for biological replication; technical replication is always advisable.
5. The number of juveniles that have hatched can be counted at set intervals, usually weekly, for a 3–5-week period. Replace with fresh exudate at the set intervals. Hatching tests can also be continued until hatching has ceased or, for example, <10 J2 hatched; this may take 6–8 weeks or longer.
6. At the end of the period, the number of hatched juveniles can be counted under a microscope or by photographing the

number in a set volume and counting from the photograph. The result can be expressed as a percentage of the total number of hatched and unhatched juveniles.

7. Microtiter plates have been used as hatching systems for cysts [3, 4]. Use a 96-well microtiter plate with wire mesh "thimbles" fitted into each well as a hatching assay unit for *G. rostochiensis* cysts [3].
8. Each thimble, containing five cysts, contains the root exudates (or test solution). Hatched J2 moved through the mesh into the well and can be counted.
9. To reduce inter-replicate variability, cysts can be graded through sieves into three size classes: <300 μm diam., 300–500 μm, and >500 μm [3]. The juvenile content of cysts in the three classes can be determined by crushing a representative sample of each size cyst.

   Meloidogyne hatching assays can utilize free eggs or egg masses (*see* **Note 5**). Free eggs are obtained by releasing eggs from egg masses using 0.53 or 1.05% sodium hypochlorite (NaOCl) for 4 min [5] or by blending egg masses in 100 mL tap water plus 500 mL 1% NaOCl at high speed for 40 s, followed by sieving and washing [6].
10. Egg masses are often difficult to obtain intact and root pieces containing one female, or one egg mass may be used.
11. Wesemael et al. [7] used 20 root pieces containing one female and one egg mass of *M. chitwoodi* or *M. fallax* for hatching tests.
12. Put each mass on a 48 μm sieve, covered with 2 mL of the test solution and keep in small plastic bottles fitted with a perforated lid.
13. After hatching declines to <20 J2 week^{-1}, cover the remaining eggs with 10% NaOCl and homogenize; count the numbers of eggs containing unhatched J2 to determine the percentage hatch.

Manually counting nematodes under the microscope is laborious and time-consuming. Molecular methods, such as real-time PCR, can be used alone or in combination with optical counting. Accurate automated image analysis systems have been used for automatic counting of the number of J2 of *G. rostochiensis* and *G. pallida* in suspensions from hatching tests [8], and various automatic applications, using new algorithm sand machine learning techniques, have been developed, including analyzing phenotypic features like number of eggs and number of vermiform life cycle stages [9–12].

3.1.3 Fluorescence Hatching Assay

Hatching of *G. rostochiensis* involves a change in permeability of the lipoprotein membranes of the eggshell allowing leakage of trehalose out of the egg with concomitant movement of water into the unhatched J2 [1] (*see* **Note 6**). Acridine orange appears not to be capable of penetrating the lipoprotein egg membranes in the absence of a hatching stimulus, such as potato root exudates. Change in lipoprotein membrane permeability allows the passage of acridine orange through the eggshell and into the viable unhatched J2. This has been used with *G. rostochiensis* as a rapid, simple primary screen to evaluate viability and hatching potential [13, 14].

1. Cut cysts open to release the eggs.
2. Rinse eggs in GDW and added to 6 mm diameter glass cylinders, each with 30 μm mesh nylon netting fixed over the lower end to retain the eggs.
3. Place each cylinder in an excavated glass block containing acridine orange dissolved in PRD, made up to a concentration of 0.001%.
4. Keep eggs in solutions for different time periods (6 min to 24 h), long enough to initiate hatch but too short for actual eclosion.
5. Rinse eggs in GDW and transfer them into a drop of GDW on a glass slide.
6. Examine under a fluorescence microscope.
7. Where the acridine orange passes through the eggshell, the unhatched viable J2 fluoresced a bright green and could easily be distinguished from non-fluorescing J2 in eggs where there was no uptake of stain. Percentage fluorescing J2 is an indication of percentage viability and hatch.

3.1.4 Adenosine Triphosphate Hatching Assay

Measurement of adenosine triphosphate (ATP), present only in living tissue, using bioluminescent photometry has been used to determine viability of cyst nematodes [15, 16]. Huijbregts et al. [17] used HPLC to measure ATP content to estimate viability of field populations of *Heterodera schachtii*. Storey and Marks [18] used measurement of the ATP content of cysts by bioluminescent photometry as a screen for assessing the resistance of potato clones to *Globodera* spp. The multiplication rates of the nematodes were based on the ratio of ATP levels for the cyst inoculum and the progeny:

1. Macerate nematodes with 0.1 M arsenate buffer at pH 4 and 4 °C to extract ATP.
2. A photometer is used to measure the ATP content of the cysts by bioluminescence of a luciferin–luciferase extract from fireflies.

3. The bioluminescence the ATP produces is related to the amount of ATP and, therefore, the number of living organisms present.
4. Nematodes must be free of other living material, such as microorganisms and fungi, as ATP from these will interfere with the results.

3.2 Attraction

3.2.1 Agar Plate Attraction Assays

The most common attraction/repulsion assays are based on Petri dishes containing agar or nematode growth medium where nematodes and test chemicals are added to the medium. The nematodes move along the surface of the medium in response to gradients of non-volatile compounds, and the attraction/repulsion can be determined by tracking nematode movement over a set period. Several modifications of the basic technique have been used, including photographing nematode tracks and video monitoring of nematode movement.

1. A simple agar plate assay was used by Hewlett et al. [19] to study nematode responses to tannic acid solutions.
2. Use a Petri dish (10 cm diameter) containing 1.7% distilled water agar (15 mL dish), and cut circular wells (2 cm diameter) out of the agar on opposite sides.
3. Pipette the test compound on one side and control on the other, and leave for 2 days to establish gradients (*see* **Note 7**).
4. Place nematodes (20–30 in 0.05 mL water) at the center of each plate, and keep in the dark at a set temperature.
5. After 24 h the number of nematodes in the attractive and repellent zones can be counted (*see* **Note 8**).

3.2.2 Monitoring Plant-Parasitic Nematode Movement

Several developments of this basic chemotaxis assay have been made. One frequently used to monitor movement of plant-parasitic nematodes is the more complex system of Wuyts et al. [20] (*see* Fig. 1). Assessment of effects was based on nematode tracks, instead of nematodes themselves, which takes into account the movement of the nematodes over the test plates. Plates were divided in 16 segments. The technique of Wuyts et al. [20] is summarized here:

1. Fill Petri dishes (5 cm diameter) with 5 mL water agar (0.4%).
2. On opposite sides of the plates, make wells of 1 cm diameter. Fill wells with 100 μL of the compounds or concentrations to be tested. Leave for gradients to be established (30 min–3 h) before starting the test.
3. Where applicable use the compound solvent or distilled water (dH_2O) as controls. 1% acetic acid can be used as a standard repellent and 0.5 M $CaCl_2$ as an attractant.

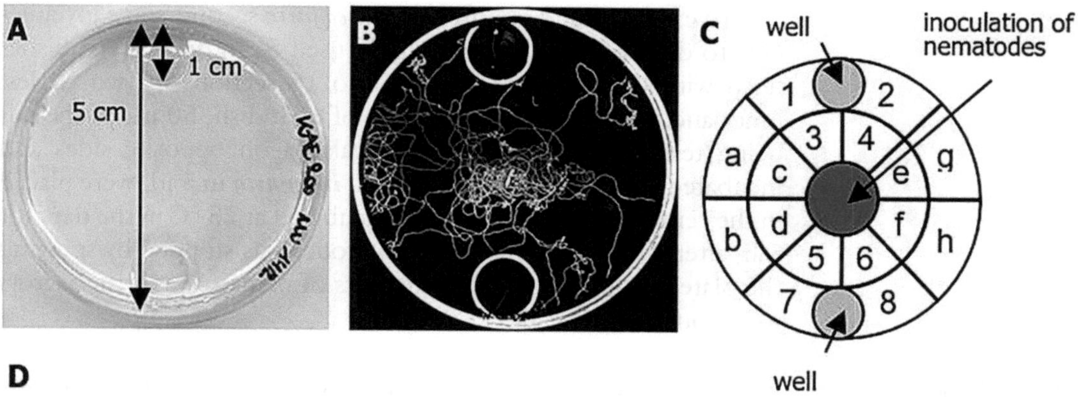

D

$$Cf = \frac{\text{attractant}}{\text{repellent}} = \frac{\text{scores for } [1]+[2]+...+[8]}{\text{scores for } [a]+[b]+...+[h]}$$

attractant	$Cf \geq 2.0$
repellent	$Cf \leq 0.5$
no effect	$0.5 \leq Cf \leq 2.0$

Fig. 1 In vitro chemotaxis assay; (**A**) Medium on which nematodes come into contact with a concentration gradient of a compound placed in two wells (see **c**); (**B**) nematode migration tracks on the medium; (**C**) grid for the analysis of nematode preferential orientation on the medium; each segment of the grid was given a score for the presence (1) or absence of nematode tracks (0); (**D**) calculation of the chemotaxis factor (Cf). (From Ref. [20])

4. Pipette test nematodes (usually 10) in the middle of the Petri dish in a minimum volume of water. Keep the plate at constant temperature in the dark for a set period; Wuyts et al. [20] used 1 h for females of *Radopholus similis* and *Pratylenchus penetrans* and 2 h for juveniles of *M. incognita*.
5. After the set period, stop nematode movement by spraying the plates with ethanol (70%). Movement tracks can then be recorded, for example, on film.
6. Tracks can be analyzed by placing a grid that divides the plates into attractive zones, close to the wells, and repellent zones (*see* Fig. 1). For each zone, eight segments are included in the grid. For each segment of the grid, a score is given for the presence (1) or absence of nematode tracks (0). The chemotaxis factor (Cf) can be calculated by dividing the sum of scores of the attractant zones by those of the repellent zones. A Cf greater than 2 means attraction of nematodes, while that lower than 0.5 means repellence.
7. As with all such assays, they should be replicated for each compound/concentration, and the experiment independently repeated in time.

To examine the response of *M. incognita* second-stage juveniles (J2) to different salts, Qi et al. [21] used 5 cm diam. Petri dishes filled with 0.8% agarose divided into 16 sections in two circles, inner and outer. In the outer circle of each dish, 50 μL of the salt being tested and dH_2O were inoculated on opposite sides and incubated for 1 h at 25 °C. Thirty *M. incognita* in 5 μL were placed in the center of the test arena and incubated at 25 °C in the dark for 5 h, after which movement of nematodes was stopped by spraying the plates with ethanol. The numbers of nematodes in the sectors were counted.

3.2.3 Pluronic Gel Attraction Assay 1

The two-dimensional nature of the Petri dish movement bioassays is a disadvantage, as is the uncertainty about how quickly a gradient is established. The concentration to which the nematode is responding is difficult to establish (*see* **Note 9**). A three-dimensional assay is better, and Pluronic gel is the most commonly used. Pluronic F-127 (PF-127) is a non-ionic surfactant, co-polymer of propylene oxide and ethylene oxide, which is transparent making it easy to monitor nematode movement to be quantified over time in response to gradients. A 23% solution of PF-127 gel is a semisolid gel at room temperature but is liquid at temperatures of 15 °C and below, and nematodes can move freely in the gel. Pluronic gel assays can be used without autoclaving for short-term experiments and autoclaved for assays lasting 3 days or longer. Pluronic gel assays are applicable to a range of nematode and plant species [22–27].

1. To make 100 mL gel for slide and Petri dish assays, 23 g PF-27 powder is added to 80 mL cold, sterile (Millipore) water at 4 °C and stirred to dissolve for 24 h.
2. Store the dissolved gel at 4 °C, and retrieve aliquots as required, either for slide assays or Petri dish assays.
3. For slide assay use 80 μL gel solution, pH 7.0, at 15 °C.
4. Different combinations of nematodes and plants can be used to determine attraction/repulsion and other behavioral aspects.
5. Wang et al. [22, 23] used the assay with J2 of *Meloidogyne hapla*; 150 J2 were pipetted onto each of 5 glass slides with a sterile 5-day-old tomato root tip (*ca* 1 cm long) in the center of the gel and covered with a glass cover slip.
6. Slides were transferred to Petri dishes lined with wet filter paper and incubated at room temperature.
7. Individual J2 can be monitored and photographed at set time intervals (Wang et al. [22] started monitoring 6 h after the start of the assay and took observations every 2 min for 20 min).

8. Roots on slides can become dehydrated, so a Petri dish (35 or 60 mm diameter) or 6- or 12-well tissue culture plate assay is easy and preferable.
9. For Petri dish assays, pour 23% Pluronic gel containing nematodes into each Petri dish (2–4 mL for 35 mm diameter and 10 mL for 60 mm diameter) at 15 °C. For tissue culture plate assays, pour 2–4 mL gel to each well of 6-well plate and 1 mL to each well of 12-well plate. The size of Petri dish or tissue culture plate chosen depends on the test root length, and the gel volume needs to be sufficient to cover the root.
10. Add sterile seedlings, and then place the dish or plate at room temperature for the gel to solidify. Photographs can be taken at set time intervals as early as at 1 h post exposure. The observation time depends on plant species, nematode species and stage, and nematode concentration. Usually, it takes 30 min to solidify, but more time is needed to stabilize the gel. To get clearer photographs, the Petri dish or plate can be inverted 6 h post exposure or later (if inverted too early, the gel is not solid enough and drops down).

3.2.4 Pluronic Gel Attraction Assay 2

Pluronic gel can be used as a medium to examine attraction to test solutions dispensed via pipette tips. The following methodology is from Jiang et al. [28]:

1. Dissolve PF-127 gel in 10 mM Tris–MES (morpholinoethanesulfonic) buffer (pH 7.1) at 23% (w/v). Using this buffer enables pH or chemical gradients to be formed around the openings of pipette tips dispensing the test solution. However, if test chemicals are not associated with pH, this buffer is not necessary.
2. Dispensing tips can be made from 200 μL pipette tips cut 10 mm from the small end and 25 mm from the large end to contain 50 μL of the test solution.
3. Petri dishes can be set up as described above, and 300 J2 (or a number as required) per mL of gel solution are added.
4. To prepare test samples, stock solution of test solutions can be mixed 1:1 with PF-127 on ice. Cooled samples can then be added to the dispensers, which are placed horizontally at the center of the Petri dish.
5. At a set time after introducing the test samples, orientation behavior of the nematodes can be photographed outside or inside the dispenser. The Petri dish can be inverted to get clearer photographs 6 h post exposure or later.
6. To determine the number of nematodes that moved inside the dispensers, remove dispensers from the gel, clean the outside with tissue paper, and place each dispenser into a Petri dish

containing 3 mL autoclaved deionized water at room temperature. After 5 h nematodes move out of the diluted gel and can be counted.

3.2.5 Movement Assays for Root Exudates

To test the effects of non-volatile compounds on nematode movement, simple sand tube assays are useful.

1. Polythene tubes (usually, height 1.0 cm) sealed at the base with nylon mesh (53 μm aperture) are filled with sand (particle size 250–600 μm) and placed upright in a small tube or sample jar containing the test solution.
2. Equilibrate the tubes with the test solution by passage of the solution (200 mL) through the tube before use.
3. Pipette an aliquot with a known number of nematodes into the top of the tube. The apparatus can be kept at a set temperature, and after a set period of time, the number of nematodes that migrated through the sand to the reservoir is counted.
4. Acrylic tubes filled with sand were used to establish linear gradients of temperature [29] and CO_2 [30] to investigate movement of species of plant-parasitic nematodes. Final distribution of the nematodes was determined by extracting nematodes from sequential slices of sand along the linear gradient.

Sand column assessments of nematode movement preclude direct observation. Microchannel devices have been used with *C. elegans* comprising poly-dimethylsiloxane (PDMS) microchannels on an agar plate, which effectively restrict nematodes to a defined space [31, 32]. Otobe et al. [33] used microstructures, consisting of networks of channels of rectangular cross sections (50 μm high, 40–200 μm wide) in 5 mm^2 areas on transparent substrates made of silicon rubber and sealed with a glass plate. This setup was used with different channel dimensions to mimic soil pore spaces and was used to examine the movement and migration of *M. incognita* J2 when the channels were filled with water.

O'Callaghan et al. [34] developed microcosm assays for testing nematode responses. They used transparent soil microcosms, made from microscope slides, a new light sheet imaging technique termed Biospeckle Selective Plane Illumination Microscopy (BSPIM) for nematode detection, and confocal laser scanning microscopy for high-resolution imaging. BSPIM enabled automatic detection of nematodes in artificial transparent "soil" [35] in the microcosms. Growing plant root systems could be scanned for nematode abundance and activity, and nematode feeding behavior and responses to various treatments could be observed in soil-like conditions.

3.2.6 Attraction Choice Assay

Several types of three-dimensional attraction assays in sand or soil have been developed to study the host finding ability of plant-parasitic nematodes. The following example, called a host finding pipette–bulb assay, was developed by Dalzell et al. [36] (*see* Fig. 2).

1. From two hydrophobically lined PVC Pasteur pipettes, cut 1 cm from the bulb terminus, and attach linearly by parafilm connection.
2. From two more Pasteur pipettes, cut 0.5 cm from the bulb terminus, and insert vertically into the initial construction through small incisions in the horizontal terminal bulbs. Fasten these in place with parafilm wrapping.
3. Cut open the top of the vertical bulbs to create "seedling barrels," and fill the bottom bulbs and central barrel with sand.
4. Fill the vertical seedling barrels with seeding compost, and plant seeds of the test plant or nothing at all (control).
5. Water both seedling bulbs equally throughout to maintain suitable conditions for the growth of plants, thus establishing a root exudate gradient, and moisten the underlying sand-filled sections in order to aid nematode migration. Usually 2 mL water is added through each barrel at least twice a day during germination and seedling/plantlet growth. The watering regime is changed to 1 mL added through each barrel four times over a 10-h period following nematode inoculation.
6. Dalzell et al. [36] used a syringe to inject *Meloidogyne incognita* J2 into the connecting central barrel through the parafilm connection. Use fresh parafilm to re-seal the area, and maintain the assays in glasshouse conditions (25 °C, 18 h light, 6 h dark).
7. Keep the pipette–bulb assays vertical in a suitable support base.
8. Partially cover the pipette–bulb assays with tinfoil so that the seedling bulbs are maintained in the light, and the lower bulbs and central barrel (which contained the nematodes) are in the dark. Maintain for 24 h in total.
9. Following the 24-h assay period, cut and discard the vertical seedling bulbs. The remaining bottom barrels are then cut into three sections: the section immediately below the host plant (the seedling bulb), the central connecting barrel, and the no-seedling bulb.
10. Transfer each individual bulb and central barrel to a separate 50 mL Falcon tube, which is subsequently filled with 40 mL of 1:1 (w/v) concentrated sucrose solution, capped, and vigorously shaken to dislodge the sand and nematodes.
11. Once the sand has been removed, the plastic bulbs and central barrel float to the top; they are then rinsed over the Falcon tube

Fig. 2 Pipette–bulb assay. The vertical bulbs were filled with compost to allow seedling growth; the bottom seedling bulbs, no seedling bulbs, and central barrel were filled with sand to allow migration of the J2 of plant-parasitic nematodes (PPN) injected into the central barrel connection. Following the 24-h assay period, the up-facing bulbs were removed and discarded, leaving the horizontal seedling bulbs, no seedling bulbs, and central barrel, which were cut from each other and washed into 50 mL Falcon tubes for sucrose centrifugation and counting of the J2 located in each bulb or central barrel. (From Ref. [36])

using the same concentrated sucrose solution and subsequently discarded. A few mL of water is added to the top of the sucrose saturated solution in the tube, which is then centrifuged at 350 *g* for 3 min. The top water layer (which contains the nematodes present in each section) is transferred to a fresh 15 mL Falcon tube, and the nematodes can be counted under a binocular microscope.

3.2.7 Attraction Assay for Volatile Components

Volatile signals have been implicated in the ability of parasitic nematodes to locate their hosts. Assays for volatile attractants are more complicated to set up than those for non-volatiles; however, there are several olfactometer assay arenas that are available. The most straightforward are the one-choice point Y olfactometers used by several authors with plant-parasitic nematodes and entomopathogenic nematodes [37]. A Y-tube olfactometer filled with sand was used to examine host-searching behavior over a 24-h period of infective juveniles of *Heterorhabditis megidis* in the presence of insect hosts and plant roots, either individually or in combination [38].

To test the response of *M. incognita* J2 to plant volatiles, Kihika et al. [39] used a modified dual choice olfactometer (*see* Fig. 3).

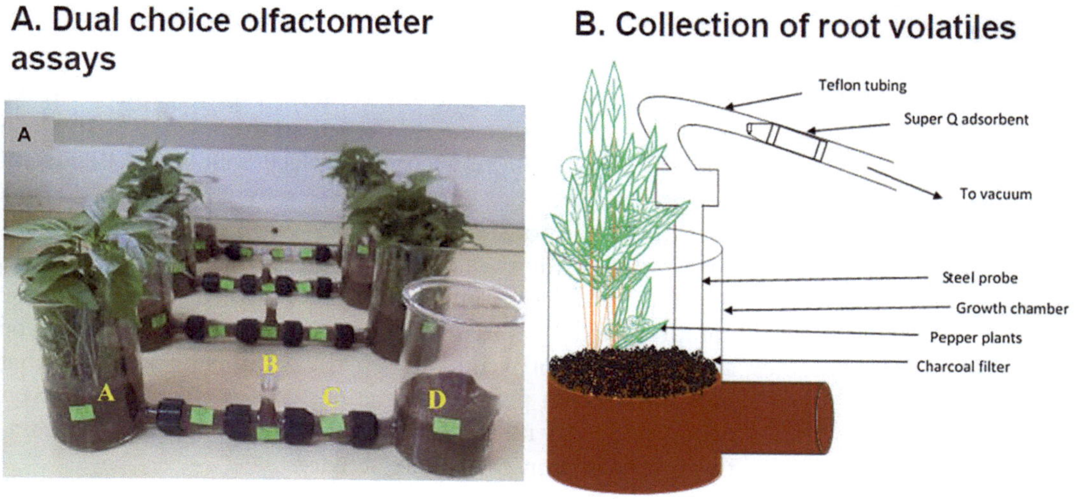

Fig. 3 Response of *Meloidogyne incognita* juveniles (J2) to root volatiles of pepper and chemical analysis of the root volatiles. (**a**) Dual choice olfactometer assays to test J2 responses to *Capsicum annum* root volatiles and synthetic blends: A stimulus chamber, (**b**) release arm, (**c**) connecting arm, (**d**) control chamber. (**b**) A schematic representation of the volatile collection setup in the laboratory. (Modified from Ref. [39])

1. The olfactometer is composed of four components: the stimulus chamber (A) and the control chamber (D) (each 85 mm diam. × 140 mm deep) with a connector (15 mm diam. × 30 mm length) fitted with an ultra-fine mesh screen filled with sterilized sand.
2. Link the central release arm (B) (20 mm diam. × 60 mm length) to detachable arms (C) (20 mm diam. × 70 mm length) that connect chambers A and D (*see* Fig. 3A).
3. For the dual choice olfactometer assays with intact plants, Kihika et al. [39] paired the growth chamber containing 30 plants with a growth chamber containing 300 g moist sand (control).
4. After 4 h (the optimal recovery time), disassemble the olfactometer, and place the sand in each detachable section on a Baermann funnel to recover and then count the nematodes.
5. Kihika et al. [39] also used the olfactometer to determine the response of J2 to collected volatiles (*see* Fig. 3B). The volatiles were collected for 24 h on a pre-cleaned Super Q (30 mg) adsorbent connected to a steel probe (17 cm long, 0.5 cm i.d.) inserted in the plant sand root zone in the glass chamber. The probe was connected to a vacuum pump that extracted volatiles from the soil.

4 Notes

1. The extent of root growth and volume of roots, rather than the hatching activity of exudate per se, are important factors. To quantify activity more accurately, place well-washed roots of pot-grown plants of a known age in a set volume of dH_2O for a set time period to provide the stock solution. Exudate is standardized, and the possible influence of soil microorganisms is negated. However, comparative data between cultivars may be compromised by variations in root growth. Activity can be adjusted for root weight giving the number of juveniles that hatch in a given period of time per g root weight. However, root weight may not be as accurate as root length for comparative quantification of exudate activity.
2. Plastic and glue are not inert, and chemicals from these materials could cause contamination. Thus, glass is preferable for constructing a hatching unit.
3. Cysts of *Globodera* spp. that have been air-dried should be soaked in water for 7 days before being used in hatching assays. Cysts of other species do not tolerate drying and should be kept moist during extraction.
4. If the population of cysts is from the field, including cysts of various ages, then variability of hatch between batches will be greater. If the populations have been treated with nematicides, variability will be even higher, and much longer times for the assay will be needed [40].
5. Species of *Meloidogyne* generally hatch well in water without the need for host root exudate stimulus, but, while hatching of *M. chitwoodi* from young plants did not require exudate stimulus, at the end of the plant-growing season, egg masses contained a percentage of unhatched J2 that required host root exudate to cause hatch [7].
6. Inside the egg, the J2 of *Globodera* and *Heterodera* spp. cyst nematodes is surrounded by perivitelline fluid, which contains trehalose [41, 42]. The presence of trehalose has been used as the basis for a viability assay for *G. rostochiensis* and *G. pallida* with a detection limit of five to ten juveniles [43, 44].

   The time to establish a gradient will vary with the solution/chemical used.
7. As with all such behavioral assays, the time period for evaluation will depend on the mobility of the tested nematodes. Some, like *Ditylenchus dipsaci*, are very active; others like *Pratylenchus* spp. are much slower moving. Often *Globodera* species need to be activated by root exudate for maximum activity.

8. Electrophysiology is a useful technique to examine the sensory responses of individual live, intact nematodes and has provided detailed analysis of responses to known concentrations of test chemicals [45]. However, the technique requires expensive equipment and is time-consuming and less amenable for routine behavioral analyses.

9. Soybean cyst nematode, *H. glycines*, responding to soybean root tip differs from root-knot nematodes. *Meloidogyne javanica and M. incognita* (2 h) respond faster to root tip than *M. hapla* (6 h) [23]. The time point for maximum count will depend on the nematode species tested and needs to be determined in pilot trials.

References

1. Perry RN (2002) Hatching. In: Lee DL (ed) The biology of nematodes. Taylor and Francis, London, pp 147–169
2. Mohan S, Kumar KK, Sutar V et al (2020) Plant root-exudates recruit hyperparasitic bacteria of phytonematodes by altered cuticle aging: implications for biological control strategies. Front Plant Sci 11:763. https://doi.org/10.3389/fpls.2020.00763
3. Twomey U, Raftery T, Devine K et al (1995) An improved procedure for assaying hatching activity of potato root exudates towards *Globodera rostochiensis*. Nematologica 41:258–268. https://doi.org/10.1163/003925995X00224
4. Kroese D, Zasada IA, Ingham RE (2011) Comparison of Meldola's Blue staining and hatching assay with potato root exudate for assessment of *Globodera* spp. egg viability. J Nematol 43:182–186
5. Hussey RS, Barker KR (1973) A comparison of methods of collecting inocula of *Meloidogyne* spp., including a new technique. Plant Dis Rep 57:1025–1028
6. McClure MA, Kruk TH, Misaghi I (1973) A method for obtaining quantities of clean *Meloidogyne* eggs. J Nematol 5:230
7. Wesemael WML, Perry RN, Moens M (2006) The influence of root exudate and host age on hatching of the root-knot nematodes, *Meloidogyne chitwoodi* and *M. fallax*. Nematology 6: 895–902. https://doi.org/10.1163/156854106779799204
8. Been TH, Meijer EMJ, Beniers AE et al (1996) Using image analysis for counting larvae of potato cyst nematodes (*Globodera* spp.). Fundam Appl Nematol 19:297–304
9. Jung SK, Aleman-Meza B, Riepe C et al (2014) QuantWorm: A comprehensive software package for *Caenorhabditis elegans* phenotypic assays. Plos One 9:e84830. https://doi.org/10.1371/journal.pone.0084830
10. Xian B, Shen J, Chen WY et al (2013) Worm-Farm: a quantitative control and measurement device toward automated *Caenorhabditis elegans* aging analysis. Aging Cell 12:398–409. https://doi.org/10.1111/acel.12063
11. Chen WY, Li WW, Dong XJ et al (2018) A review of biological image analysis. Curr Bioinform 13:337–343. https://doi.org/10.2174/1574893612666170718153316
12. Hallman J, Daub M, Wesemael W (2022) Estimating numbers. In: Perry RN, Hunt DJ, Subbotin SA (eds) Techniques for work with plant and soil nematodes. CAB International, Wallingford, pp 42–59
13. Perry RN, Feil JJ (1986) Observations on a novel hatching bioassay for *Globodera rostochiensis* using fluorescence microscopy. Rev Némat 9:280–282
14. Twomey U, Warrior P, Kerry BR et al (2000) Effects of the biological nematicide, DiTera®, on hatching of *Globodera rostochiensis* and *G. pallida*. Nematology 2:355–362. https://doi.org/10.1163/156854100509114
15. Atkinson HJ, Ballantyne AJ (1977) Changes in the adenine nucleotide content of cysts of *Globodera rostochiensis* associated with the hatching of juveniles. Ann Appl Biol 87:167–174. https://doi.org/10.1111/j.1744-7348.1977.tb01872.x
16. Storey RMJ (1982) The ATP method for rapid assessment of the efficacy of a single application of a fumigant against *Globodera* spp. in field soils. Ann Appl Biol 101:93–98. https://doi.org/10.1111/j.1744-7348.1982.tb00804.x
17. Huijbregts AWM, Gijssel PD, Munning RG et al (1996) Estimation of the field population

and the vitality of *Heterodera schachtii*, measuring ATP, ADP and AMP contents of eggs and juveniles by HPLC. Commu Soil Sci Plant Anal 27:1153–1168. https://doi.org/10.1080/00103629609369623

18. Storey RMJ, Marks RJ (1983) Screening for resistance to potato cyst nematodes using bioluminescent photometry. Ann Appl Biol 103: 321–326. https://doi.org/10.1111/j.1744-7348.1983.tb02770.x
19. Hewlett TE, Hewlett EM, Dickson DW (1997) Response of *Meloidogyne* spp., *Heterodera glycines*, and *Radopholus similis* to tannic acid. J Nematol 29:737–741
20. Wuyts N, Swennen R, De Waele D (2006) Effects of plant phenylpropanoid pathway products and selected terpenoids and alkaloids on the behaviour of the plant-parasitic nematodes *Radopholus similis*, *Pratylenchus penetrans* and *Meloidogyne incognita*. Nematology 8:89–101. https://doi.org/10.1163/156854106776179953
21. Qi YH, Meng L, Cao SF et al (2015) Chemotaxis of *Meloidogyne incognita* in response to different salts. Agric Sci 6:900–907. https://doi.org/10.4236/as.2015.69086
22. Wang C, Lower S, Williamson V (2009) Application of Pluronic gel to the study of root-knot nematode behaviour. Nematology 11:453–464. https://doi.org/10.1163/156854109X447024
23. Wang C, Bruening G, Williamson VM (2009) Determination of preferred pH for root-knot nematode aggregation using Pluronic F-127 gel. J Chem Ecol 35:1242–1251. https://doi.org/10.1007/s10886-009-9703-8
24. Dutta TK, Powers SJ, Kerry BR et al (2011) Comparison of host recognition, invasion, development and reproduction of *Meloidogyne graminicola* and *M. incognita* on rice and tomato. Nematology 13:509–520. https://doi.org/10.1163/138855410X528262
25. Sasaki-Crawley A, Curtis R, Birkett M et al (2012) The use of Pluronic F-127 to study the development of the potato cyst nematode, *Globodera pallida*. Nematology 14:869–873. https://doi.org/10.1163/156854112X631944
26. Kumari C, Dutta TK, Banakar P et al (2016) Comparing the defence-related gene expression changes upon root-knot nematode attack in susceptible *versus* resistant cultivars of rice. Sci Rep 6:22846. https://doi.org/10.1038/srep22846
27. Pokhare SS, Habash SS, Jones J et al (2019) *In vitro* life cycle of *Heterodera sacchari* on pluronic gel. Nematology 21:573–579. https://doi.org/10.1163/15685411-00003236
28. Jiang Y, Huang M, Li C (2022) Responses of infective juveniles of the soybean cyst nematode (*Heterodera glycines*) and the root-knot nematodes (*Meloidogyne hapla*, *M. incognita*) to amino acids. Nematology 24:1049–1062
29. Robinson AF (1994) Movement of five nematode species through sand subjected to natural temperature gradient fluctuations. J Nematol 26:46–58
30. Robinson AF (1995) Optimal release rates for attracting *Meloidogyne incognita*, *Rotylenchulus reniformis*, and other nematodes to carbon dioxide in sand. J Nematol 27:42–50
31. Qin J, Wheeler AR (2007) Maze exploration and learning in *C. elegans*. Lab on a Chip 2: 186–192. https://doi.org/10.1039/b613414a
32. ZhangY LH, Bargmann CI (2005) Pathogenic bacteria induce aversive olfactory learning in *Caenorhabditis elegans*. Nature 438:179–184. https://doi.org/10.1038/nature04216
33. Otobe K, Itou K, Mizukubo T (2004) Micro-moulded substrates for the analysis of structure-dependent behaviour of nematodes. Nematology 6:73–77. https://doi.org/10.1163/156854104323072946
34. O'Callaghan FE, Braga RA, Neilson R et al (2017) New live screening of plant-nematode interactions in the rhizosphere. Sci Rep 8: 1440. https://doi.org/10.1038/s41598-017-18797-7
35. Downie H, Holden OW et al (2012) Transparent soil for imaging the rhizosphere. PLoS One 7:e44276. https://doi.org/10.1371/journal.pone.0044276
36. Dalzell JJ, Kerr R, Corbett MD et al (2011) Novel bioassays to examine the host finding ability of plant-parasitic nematodes. Nematology 13:211–220. https://doi.org/10.1163/138855410X516760
37. Reynolds AM, Dutta TK, Curtis RHC et al (2011) Chemotaxis can take plant-parasitic nematodes to the source of a chemo-attractant via the shortest possible routes. J R Soc Interface 8:568–577. https://doi.org/10.1098/rsif.2010.0417
38. Boff MIC, Zoon FC, Smits PH (2003) Orientation of *Heterorhabditis megidis* to insect hosts and plant roots in a Y-tube sand olfactometer. Entomol Exp Appl 98:329–337. https://doi.org/10.1046/j.1570-7458.2001.00789.x
39. Kihika R, Murungi LK, Coyne D et al (2017) Parasitic nematode *Meloidogyne incognita* interactions with different *Capsicum annum*

cultivars reveal the chemical constituents modulating root herbivore. Sci Rep 7:2903. https://doi.org/10.1038/s41598-017-02379-8

40. Been TH, Schomaker CH (2001) Ways to improve the accuracy of hatching tests for *Globodera* spp. with special emphasis on nematicide trials. Nematology 3:773–795. https://doi.org/10.1163/156854101753625290
41. Clarke AJ, Perry RN, Hennessy J (1978) Osmotic stress and the hatching of *Globodera rostochiensis*. Nematologica 24:384–392. https://doi.org/10.1163/187529278X00506
42. Perry RN, Clarke AJ, Beane J (1980) Hatching of *Heterodera goettingiana in vitro*. Nematologica 26:493–495. https://doi.org/10.1163/187529280X00422
43. van den Elsen S, Ave M, Schoenmakers N et al (2012) A rapid, sensitive, and cost-efficient assay to estimate viability of potato cyst nematodes. Phytopathology 102:140–146. https://doi.org/10.1094/PHYTO-02-11-0051
44. Ebrahimi N, Viaene N, Moens M (2015) Optimizing trehalose-based quantification of live eggs in potato cyst nematodes (*Globodera rostochiensis* and *G. pallida*). Plant Dis 99:947–953. https://doi.org/10.1094/PDIS-09-14-0940-RE
45. Perry RN (2001) Analysis of the sensory responses of parasitic nematodes using electrophysiology. Int J Parasitol 31:909–918. https://doi.org/10.1016/S0020-7519(01)00198-9A
46. Shepherd AM (1986) Extraction and estimation of cyst nematodes. In: Laboratory methods for work with plant and soil nematodes, pp 31–49

Chapter 9

Use of Chemicals and Biological Control Agents as Activators of Plant Defense Against Endoparasitic Sedentary Nematodes

Sergio Molinari

Abstract

Full compatible interactions between crop plants and endoparasitic sedentary nematodes (ESNs) lead to severe infestation of the roots and plant growth impairing, as well as to the increase of nematode population in the soil that is a threat for the next planting crop. In the absence of activators, basic plant defense is overcome by nematode secretion of effectors that suppress defense gene expression, inhibit ROS generation and the oxidative burst used by plants to hamper nematode feeding site settlement and limit its development and reproduction. Activators can be exogenously added as a preventive measure to prime plants and strengthen their defense against ESNs. Activators can be an array of antioxidant compounds or biocontrol agents, such as mutualist microorganisms living in the rhizosphere (biocontrol fungi (BCF), arbuscular mycorrhizal fungi (AMF), plant growth-promoting bacteria (PGPB), etc.). In this chapter, methods are described for usage of both salicylic acid (SA) and its methylated form (Met-SA), and BCF/AMF as elicitors of resistance of vegetable crops against root-knot nematodes (RKNs). The rhizosphere-living BCF/AMF were recovered from commercial formulates pre-incubated in suitable growth media and provided exclusively as soil drench of potted plants. The plant hormones SA and Met-SA were provided to plants as soil drench, foliar spray, and root dip. It is indicated that activators' dosages and plant age are crucial factors in determining the success of a pre-treatment to reduce nematode infection. Therefore, dosages should be expressed as amounts of activators per g of plant weight at treatment. Thresholds exist above which dosages start to work; overdoses were found to be toxic to plants and useless as activators.

Key words Antioxidants, Biocontrol agents, Defense activators, Salicylic acid, Nematode management, Root-knot nematodes, Vegetable crops

1 Introduction

Induced resistance has for long been proposed as an effective strategy to control plant-parasitic nematodes alternative to pesticides [1]. Most of the studies have been done on plants infected by endoparasitic sedentary nematodes (ESNs), of which two large families (root-knot nematodes, RKNs, and cyst nematodes, CNs) attack and damage most of the valuable crops worldwide. Many

Sergio Molinari (ed.), *Plant-Nematode Interactions*, Methods in Molecular Biology, vol. 2756,
https://doi.org/10.1007/978-1-0716-3638-1_9,

physiological chemicals and their synthetic analogues have been found to induce plant immunity and prime host plants against ESNs [2–4]. Most investigations have focused on plant hormones (salicylic acid, SA, jasmonic acid, JA, chemicals releasing ethylene, ET) and their synthetic analogues, although derivates of amino acids have also been tested. An array of factors must be considered and evaluated to make chemical treatments effective against nematodes, such as the species and age of plants, the dose of treatment, type of application, and the way in which plants are grown [5].

Biocontrol agents (BCAs) are another class of effective defense activators against ESNs. BCAs are beneficial soil-borne microorganisms that interact with roots and improve plant health. These root-associated mutualists include biocontrol fungi (BCF) and arbuscular mycorrhizal fungi (AMF) [6, 7]. The most studied *Trichoderma* spp., a class of opportunistic fungi, belong to BCF and have been proved to lower the infection of the RKN *Meloidogyne incognita* [8]. AMF are obligate root symbionts, diffused in most of the soils that improve plant growth and can alleviate both abiotic and biotic plant stresses [9]. Commercial formulates containing AMF and BCF have been used to significantly control ESN infections to vegetable crops [10, 11]. Two factors should be taken in careful consideration to make treatments with AMF effective against ESNs: the amount of AMF provided per plant or per kg of soil, the age and the weight of the treated plants, and the time set from fungi inoculation and nematode inoculation. The rate of root colonization prior to nematode attack is crucial for an effective priming of plants [12]. In this chapter, protocols to make effective such treatments with activators are described and indicated the opportune amounts to provide plants for provoking defense responses able to significantly reduce damages and reproduction of ESNs.

2 Materials

All experiments require a glasshouse and an apparatus for temperature settings. Moreover, seedbed containers, sandy soil or topsoil for seedbeds, 100 cm^3 clay pots, or larger, wild freshly collected loamy soil to fill the pots, seeds and growing plants of vegetable crops, pH meter, balance (2–200 g), stirrers, and bars.

2.1 Soil-Drench of SA

5 mL pipettes and tips, salicylic acid (SA), KOH, glass beakers.

2.2 Spray of SA and Met-SA

Little (50–100 mL) spray bottles, SA, KOH, Met-SA, ethylic alcohol (Et-OH), Tween 20.

2.3 Use of Biocontrol Agents (BCA) as Activators

In this chapter, all the BCA activators are obtained from commercial formulations that can contain arbuscular mycorrhizal fungi (AMF), biocontrol fungi (BCF), or both. The commercial names and formulations of the used products are as follows:

1. AMF-containing formulations.
   - MICOSAT F®, named Myco in the text, consisting of 40% root fragments containing intraradical spores/vesicles and hyphae of *Glomus* spp. (*Glomus* spp. GB 67, *G. mosseae* GP11, *G. viscosum* GC 41). The commercial formulate contains also rhizobacteria and yeasts.
   - OZOR containing 500 propagules/g of *G. intraradices* CMCCRO7.
2. BCF-containing formulations.
   - TUSAL® in which 100 g of the product contain 0.5 g (1×10^8 UFC/g, *Trichoderma asperellum* T25, and 0.5 g, 1×10^8 UFC/g, *Trichoderma atroviride* T11).
   - TELLUS™ in which 100 g of the product contain 2 g each of *T. asperellum* ICC012 and *T. gamsii* ICC080 for a total of 3×10^7 UFC/g.
   - Moreover, the following materials are required: glucose, peptone, temperature-controlled orbital shaker, flasks, plastic trays.

2.3.1 Detection of Root Colonization by AMF

Temperature-controlled water bath, KOH, HCl, lactophenol, fume hood, methanol, acetic acid, distilled water, refrigerator, dissecting microscope provided with a HD photocamera.

2.4 Inoculation of Plants with Nematodes

Infected roots with root-knot nematodes (RKNs), cysts extracted from soil infected with cyst nematodes (CNs), filter apparatus (*see* Fig. 1), glass Petri Dishes (diam 10 cm), temperature-controlled growth chamber, 500 mesh sieve, stereoscope provided with a counting device (*see* Fig. 2).

2.5 Measurement of Infection Factors

Acid fuchsin, acetic acid, bleach, sodium hypochloride, glass beakers, glycerin, HCl, blendor, Eosin Yellow, refrigerator, Pectinex®, and Celluclast® (Novo Nordisk, Italy).

3 Methods

Seeds are germinated in a sterilized mixture of sand and soil in a temperature-controlled glasshouse set at 23–25 °C. Seedlings are transplanted into 100 cm^3 clay pots (100–150 g soil) filled with wild freshly collected loamy soil (*see* Fig. 3). These small pots allow experiments with a large number of different treatments and several plants as replicates per treatment.

Fig. 1 Filter apparatuses used to incubate egg masses or eggs to obtain J2s

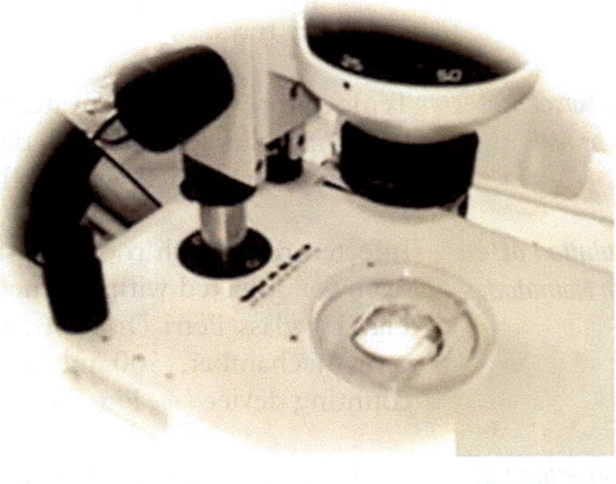

Fig. 2 Stereoscope set at 25× and slide apparatus for counting J2s or eggs contained in 1 mL suspensions

Chemicals can be provided to plants by three types of treatments:

(a) Chemical solutions soil-drenched in pots.
(b) Roots dipped in chemical solutions for a variable time interval (usually overnight).
(c) Chemical solutions sprayed onto leaves (*see* **Note 1**).

Fig. 3 Typical tomato young plant growing in 100 cm^3 clay pots (100–150 g soil) used in the experiments

Biological control agents usually consist of commercial powdered formulates containing AMF or BCF dissolved in nutrient suspensions that are soil-drenched in pots for colonization of roots.

It is very important to fix the dose of activator to be used. Doses are expressed as amounts of activator (g or mg) per g of plant fresh weight (g pfw) at treatment. Therefore, groups of plants to be treated are weighed before treatments and the average weight calculated. Usually, plants are pre-treated with activators before nematode inoculation only once; however, multiple treatments during the experimental period are possible. A fixed amount of activator per plant can be tested on different growth stages that are indicated below, associated with the corresponding ranges of plant weights:

- Seedlings (1.0–2.5 g).
- Young plants (3.5–5 g).
- Adult plants (8.5–10 g) (*see* **Note 2**).

Moreover, different amounts of activators can be tested on one growth stage (*see* **Note 3**). Doses suitable to trigger plant defense against endoparasitic sedentary nematodes (ESNs) may vary according to the examined plant species.

3.1 Use of Chemicals as Activators

A wide array of chemicals can be used as activators of plant defense against ESNs. The most effective compounds at the appropriate concentrations show pro-inflammatory properties, favor reactive oxygen species (ROS) generation, prime plants, and activate their immune system. In this chapter, protocols will be proposed for the use of salicylates, such as salicylic acid (SA) and methyl-salicylic acid (Met-SA) as activators of defense against ESNs and root-knot nematodes (RKNs) in particular. Such protocols can be applied to other chemical treatments in their general procedures.

SA solutions may be sprayed onto leaves of plants, soil-drenched in the pots, or used to dip roots overnight, while Met-SA is usually sprayed to plants.

3.1.1 Soil-Drench of SA

1. Concentrated aqueous solutions of SA (4–5 mg/mL) should be prepared, and pH brought to approx. 6.0 by adding KOH (not NaOH as Na^+ is toxic to plants).
2. Usually control plants and plants to be treated are located in lines of 6 in metal containers inserted in temperature-controlled benches settled in a glasshouse, according to Fig. 4.
3. Average weight of plants of similar size is previously measured.
4. Soil-drenched SA is pipetted (not more than 5 mL) on the surface of pot soil.
5. Soil-drenched SA has been tested on tomato, eggplant, and pepper: doses effective in priming plants against RKNs were 4.0–10.0 mg/g pfw for tomato, 3.0–6.5 for eggplant, and 1.5–1.8 for pepper.
6. Doses exceeding the abovementioned ranges may be toxic to plants; conversely, lower doses may be not effective to induce resistance against RKNs.

3.1.2 Spray of SA and Met-SA

1. Concentrated aqueous solutions of SA (10–20 mg/mL) should be prepared, and pH brought to approx. 6.0 by adding KOH (not NaOH as Na^+ is toxic to plants).
2. Solutions (25–50 mL) containing suitable amounts of SA/MetSA are usually sprayed to groups of six potted plants, used as replicates for one treatment.
3. Average weight of six plants of similar size is previously measured.

Fig. 4 Potted plants located in metal containers inserted in temperature-controlled benches

4. Before spraying the plants, surface of the pots is covered by aluminum foils to prevent drops of spray from falling into the soil.
5. SA sprays have been tested on tomato: effective doses should be in the range 3.0–4.0 mg SA/g pfw.
6. Met-SA has been tested on tomato, eggplant, and pepper: effective doses were 2.0–5.0 mg/g pfw for tomato, 3.0–5.0 for eggplant, and 0.5–1.5 for pepper.
7. Doses exceeding the abovementioned ranges may be toxic to plants; conversely, lower doses may be not effective to induce resistance against RKNs.

3.1.3 Roots Dipped in SA Solutions

1. 200 mL of SA solution (0.45–0.55 mg/mL, pH 6.0–6.5) are poured in plastic beakers in which root systems of six young plants are immersed overnight—groups of six control plants are immersed in distilled water.
2. The groups of six plants are previously weighed.

3. After incubation, root systems are thoroughly rinsed with tap water; plants are transferred into clay pots that are filled with wild freshly collected loamy soil.
4. Doses are applied as amounts of SA in mg per g of plant fresh weight (pfw) at treatment.
5. Doses effective against RKNs in tomato were 3.5–4.5 mg SA/g pfw.
6. Doses exceeding the abovementioned ranges may be toxic to plants; conversely, lower doses may be not effective to induce resistance against RKNs.

3.2 Use of Biocontrol Agents (BCAs) as Activators

Protocols are described for use of BCF and AMF as activators of resistance to sedentary endoparasitic nematodes. The BCF used belong to *Trichoderma* spp., and the AMF used belong to *Glomus* spp.

1. Opportune amounts of commercial powder formulations are dissolved in distilled water containing 0.7 g/L glucose and peptone.
2. Formulation suspensions are put in flasks and incubated for 3 days in a temperature-controlled orbital shaker in dark at 25 ° C.
3. Stirring suspensions are soil-drenched by pipetting no more than 5 mL into pots containing growing plants; controls are provided with 0.7 g/L solutions of glucose and peptone incubated under the same conditions of formulation suspensions; opportune amounts of formulations can be added to 10–12 kg soil in plastic trays in which 9–12 plants are planted.
4. Doses are applied as amounts of formulation in g per g of plant fresh weight (pfw) at treatment or formulate in g per kg of soil.
5. Ranges of doses effective against ESN infection are shown in Table 1 when young plants (3–5 g) are used.

3.2.1 Detection of Root Colonization by AMF

1. Freshly collected roots untreated and treated with AMF-containing commercial formulates are gently washed with water to remove attached soil particles and soil debris.
2. Roots are chopped in short pieces and immersed in 10% KOH for 45 min at 90 °C in a water bath located under a fume hood.
3. Roots are thoroughly rinsed with water and acidified in 1% HCl.
4. Staining is performed by soaking the roots in 0.05% lactophenol blue in a 90 °C water bath for 30 min, and roots are destained in 30% methanol/10% acetic acid distilled water solution.

Table 1
Ranges of doses of BCA-containing commercial formulates effective in reducing infection by endoparasitic sedentary nematodes in young plants

	Myco		Ozor		Tellus		Tusal	
	g/g pfw	g/kg soil	g/g pfw	g/kg soil	g/g pfw	g/kg soil	g/g pfw	g/kg soil
Tomato	0.20–0.25	4.0–6.0	0.5–1.0	2.0–5.0	0.3–0.4	4.0–6.0	0.02–0.06	0.5–1.0
Eggplant	0.20–0.25	4.0–6.0	0.1–0.5	1.0–2.5	0.1–0.2	0.5–1.5	0.05–0.10	0.5–1.0
Pepper	0.18–0.20	4.0–6.0	0.08–0.10	0.25–0.50	0.05–0.20	0.5–1.5	0.05–0.15	0.5–2.0

Doses are expressed either as g formulate per g of plant fresh weight at treatment (pfw) or as g formulate per kg of soil

5. Destained roots are placed in a refrigerator before being observed under a dissecting microscope provided with a HD photocamera.
6. Discrete blue-stained areas containing typical fungal structure (i.e., hyphae, vesicles/spores and arbuscules) can be observed.
7. Typical blue-stained area can be counted per gram fresh weight of root to express the intensity of root colonization.

3.3 Inoculation of Plants with Nematodes

To ensure uniformity of infection for all plants, inoculation must be carried out by living active invading second-stage juveniles (J2s) of both root-knot and cyst nematodes (*see* **Note 4**).

1. Freshly hatched RKN J2s are obtained by incubating manually collected egg masses or extracted eggs in the filter apparatuses shown in Fig. 1, depending on the amounts required. CN J2s are obtained from cysts extracted from soils.
2. Tap water is used to fill glass Petri Dishes (diam 10 cm) up to the lower surface of the filters.
3. Filters are placed in a growth chamber at 25 °C in the dark.
4. Living J2s are collected after 2–3 days, concentrated if necessary by filtering the suspension with a 500 mesh sieve.
5. J2 suspensions are stirred, and the space of an apparatus made by 2 overlapped slides filled; squaring of the superior slide defines a volume of 1 mL.
6. J2s are counted per mL at 25× magnification under a stereoscope (*see* Fig. 2).
7. Aliquots of stirring J2 suspensions are pipetted into two holes done at the base of plants in pots or in trays; usually 300 J2s are inoculated per plant.
8. Inoculation is carried out 1–5 days after chemical or BCA treatments; to allow a full root colonization by AMF,

inoculation can be postponed up to 15 days after treatments with AMF-containing formulates, such as Ozor.

9. Plants inoculated with RKNs are located in the temperature-controlled benches shown in Fig. 4 at 24–26 °C soil temperature, while plants inoculated with CN at 22–24 °C.

3.4 Measurement of Plant Growth Parameters

Effects of chemical and BCA treatments on plant growth are detected by measuring the fresh weights of shoots (SW) and roots (RW) in grams and the length of plants (PL) in cm.

1. Uproot plants and get rid of the soil by immersing root systems in water.
2. Gently wipe roots and measure plant length.
3. Excise roots from shoots by a scissor.
4. Weigh roots and shoots (*see* **Note 5**).
5. Decrease in plant weight in plants treated with defense activators may be due to the fitness costs associated with the elicitation of highly effective defenses (priming) against nematodes. When priming fitness costs caused ≥20% decrease in plant weights with respect to not primed plants, treatments should be considered as not effective, regardless of their effect on nematode infection. Direct toxicity of high doses of activators to plants is possible and is indicated by plant high mortality or impaired growth, also in the absence of nematodes.

3.5 Measurement of Infection Factors

Nematode infection is characterized by either the reproduction rate of the inoculated nematodes or the level of damage caused to plants by the degree of root galling. Reproduction rate indicates the potential (reproduction potential, RP) of a population to multiply its initial population density (P_i) that in the present protocol is 300, the number of inoculated J2s. Therefore, RP was calculated as:

$$\mathrm{RP} = P_f / P_i$$

where P_f is the final population density; moreover,

$$P_f = \mathrm{EMs} \times \mathrm{FF}$$

P_f, in small pots as those used in this protocol, may be calculated as the number of egg masses (EMs) by female fecundity (FF), that is, the average number of eggs contained in one EM, that is, the numbers of total eggs detected at harvest [1].

The degree of root galling depends on the number of galls produced on a root system. Counting of root galls is not suitable especially for heavy infested roots. However, numbers of root galls are proportional to the number of sedentary developing individuals inside the roots. It is possible to extract such sedentary forms (SFs,

J3s, J4s, and swollen females) from roots and count them per root system or per g root fresh weight.

Inoculated plants are usually harvested 40 days after inoculation. At this time of nematode life cycle, under the plant growth conditions described in this protocol, only the inoculated J2s are able to develop in egg-laying gravid females and reproduce forming the typical egg masses visible on the roots; a second generation of nematodes is produced in about 1 month by the eggs laid in the pot or tray soils; these eggs hatch in the soil, enter the roots, and develop; however, they are not able to reproduce before harvest of plants. SFs counts include these newly formed sedentary individuals, and that is why total SF numbers normally exceed the number of inoculated J2s, while EM numbers do not do it.

It is possible to detect the motile J2s that entered into the roots within the first 3–5 days after inoculation or the sedentary forms that developed at 7-14-21 days after inoculation, under a stereoscope after coloration by the sodium hypochlorite–acid fuchsin method [13]:

1. Prepare stock solutions with acid fuchsin 3.5 g in 250 mL acetic acid and 750 mL distilled water.
2. Put weighed root samples in a beaker.
3. Put 50 mL of tap water +10 mL of bleach.
4. Soak roots for 4 min.
5. Rinse roots for 45 s in running water, and soak in tap water for 15 min.
6. Drain and transfer roots to a glass beaker with 50 mL water.
7. Add 1 mL acid fuchsin stock solution and boil for 30 s.
8. Cool, drain, and rinse with running water.
9. Place roots in 30 mL glycerin acidified with a few drops of 5 N HCl, and heat to boiling for destaining.
10. Macerate roots in a blender, sieve to separate nematode, and count nematodes (*see* **Note 6**).

Procedures to measure infection factors at 40 days after inoculation are as follows:

1. From plants coming from each treatment group, a suitable amount of root systems are chopped together to have one sample for detection of infection factors.
2. One sample is divided into three sub-samples that are weighed and used for (i) extraction and counting of EMs; (ii) extraction and counting of SFs; and (iii) extraction and counting of eggs.
3. For EM detection, root tissue is immersed in a solution (0.1 g L^{-1}) of the colorant Eosin Yellow and put in a refrigerator for at least 1 h. EMs are red-colored and easily visible under

Table 2
Plant growth and infection factors detected 40 days after inoculation with RKN J2, in untreated (cntr) and chemical-treated (salicylic acid, SA; methyl-salicylic acid, MetSA) tomato young plants

	cntr	SA-spray	cntr	MetSA-spray	cntr	SA-SD
SW (g)	8.2 ± 2.8	10.9 ± 2.7**(33)	10.3 ± 4.8	10.7 ± 4.7	11.0 ± 3.9	12.4 ± 5.6**(13)
RW (g)	2.2 ± 0.5	1.8 ± 0.7**(−18)	2.3 ± 0.4	2.2 ± 1.0	2.8 ± 1.6	2.6 ± 1.6
EM/rs	64 ± 24	50 ± 28**(−21)	112 ± 56	64 ± 36**(−43)	143 ± 75	78 ± 46**(−45)
SF/rs	237 ± 58	132 ± 67**(−44)	495 ± 219	237 ± 126** (−52)	514 ± 281	272 ± 200** (−47)
FF	337 ± 148	259 ± 147**(−23)	302 ± 86	297 ± 106	360 ± 134	406 ± 200
RP	167 ± 86	102 ± 79**(−39)	133 ± 34	69 ± 27**(−48)	219 ± 106	137 ± 94**(−37)

Effective doses of SA and MetSA were applied to plants by spraying to leaves; effective doses of SA were also soil-drenched in pots (SD). Plant growth was detected by shoot weight (SW) and root weight (RW); nematode infection by sedentary forms per root system (SF/rs), nematode reproduction by reproduction potential (RP); RP is the result of EM/rs (egg mass per root system) by FF (female fecundity) divided by the number of inoculated J2s (*Pf/Pi*). Mean values ± standard deviations obtained from untreated and BCA-treated tomato plants are differentiated by a paired *t*-test (*$P < 0.05$; **$P < 0.01$). In parentheses significant difference is expressed in percentages

a stereoscope (×6 magnification); then, they are manually separated from the roots by forceps and counted.

4. SF extraction is carried out by incubating the roots in a diluted mixture of pectinase and cellulase enzymes at 37 ° C (Pectinex® and Celluclast®) in an orbital shaker; after a brief homogenization in physiological solution, SFs are collected on a 90 μm sieve.
5. Aliquots (2 mL) of stirring SF suspensions of known volumes are pipetted into small square Petri dishes and SFs counted under a stereoscope (×12 magnification); averages of a minimum of five counts are calculated.
6. Eggs were extracted by sodium hypochloride and counted (1 mL samples) under a stereoscope (×25 magnification) with the same apparatus shown in Fig. 2 [14].

Effective doses of salicylates (Table 2) and BCA (Table 3) can actually reduce to the half the severity of nematode infection, and this result is obtained by a single pre-treatment of effective doses. Additional treatments of salicylates during nematode parasitism did not prove to further reduce infection, also because doses higher than the effective ones can be toxic to plants. Normally, heavy-parasitized roots weigh more than low-parasitized ones, although plants with lower symptoms grow better showing higher shoot weights. This may not happen if infection is very severe and the reduction caused by the activators is not sufficient to make plants recovered from infection. A normally occurring reaction to a

Table 3
Plant growth and infection factors detected 40 days after inoculation with RKN J2, in untreated (cntr) and BCA-treated (Tellus, Tusal, Ozor) tomato young plants with effective doses

	cntr	Tellus	cntr	Tusal	cntr	Ozor
SW (g)	11.5 ± 2.4	10.5 ± 2.6	10.1 ± 3.0	10.4 ± 2.6	9.1 ± 1.7	8.4 ± 2.1
RW (g)	2.1 ± 1.0	2.6 ± 0.5	2.2 ± 1.7	2.1 ± 1.2	1.5 ± 0.9	0.8±0.3** (−48)
SF/rs	278 ± 66	147 ± 39**(−47)	269±86	113±41**(−58)	237±115	65±53** (−73)
RP	54 ± 26	88±26*(51)	79±27	68±27	57 ± 29	78± 28* (36)
EM/rs	164±50	182±89	112±67	49±34** (−57)	144±67	61±52** (−58)
FF	236±127	420±242* (78)	385 ± 86	866±339** (137)	270±107	344±218* (27)

Plant growth was detected by shoot weight (SW) and root weight (RW); nematode infection by sedentary forms per root system (SF/rs), nematode reproduction by reproduction potential (RP); RP is the result of EM/rs (egg mass per root system) by FF (female fecundity) divided by the number of inoculated J2s (*Pf/Pi*). Mean values ± standard deviations obtained from untreated and BCA-treated tomato plants are differentiated by a paired *t*-test (*$P < 0.05$; **$P < 0.01$). In parentheses significant difference is expressed in percentages

decreased density of growing nematodes in the root is an augmented fecundity of egg-laying females because of a major availability of nutrients. In some few cases of a modest reduction of nematode settlement in the roots by activator treatments, higher female fecundity may lead to an unchanged or even higher reproduction rate in the treated compared with untreated plants. The suppressive action of activators on nematode parasitism is not exerted through impediment of penetration into the roots but through restriction of feeding site numbers and consequent nematode development up to reproductive females [10].

Once a feeding site is established, development and reproduction of successful juveniles are apparently not impeded. The age of the plants at treatment can influence the sensitivity to both chemical and BCA activators [3, 11]. Effects of different doses of the commercial BCA formulate Myco on plant growth and RKN infection are reported in Molinari et al. [12], while effect of Myco and Tusal on the production of cysts by the potato cyst nematode *Globodera rostochiensis* is reported in Molinari [11].

4 Notes

1. The level of chemical provided to plants may be very different if chemical is absorbed by roots from a water solution or from soil or uptaken by leaves and transferred downward to roots. The best type of application must be evaluated in preliminary tests.
2. Such weights are referred to vegetable plants such as tomato, eggplant, and pepper.

3. It is recommendable to use plants at the same growth stages when the best doses of a chemical are being evaluated as activators of defense against nematodes. Young plants are the most suitable because they are more sensitive to nematode infection than adult plants and show a more active immune system than seedlings.
4. Inoculations with extracted eggs are not suitable when the degree of infection must be compared among different groups of plants; percentages and time of hatch in each group cannot be predicted.
5. Other plant growth factors to be measured are the number of leaves, leaf groups, content of chlorophyll, etc.
6. Destained intact roots can be monitored for red-colored nematode life stages under a dissecting microscope.

References

1. Molinari S (2011) Natural genetic and induced plant resistance, as a control strategy to plant-parasitic nematodes alternative to pesticides. Plant Cell Rep 30:311–323
2. Molinari S, Fanelli E, Leonetti P (2014) Expression of tomato salicylic acid (SA)-responsive pathogenesis-related genes in *Mi-1*-mediated and SA-induced resistance to root-knot nematodes. Mol Plant Pathol 15: 255–264. https://doi.org/10.1111/mpp.12085
3. Molinari S (2016) Systemic acquired resistance activation in Solanaceous crops as a management strategy against root-knot nematodes. Pest Manag Sci 72:888–896. https://doi.org/10.1002/ps.4063
4. Vieira dos Santos MC, Curtis RHC, Abrantes I (2013) Effect of plant elicitors on the reproduction of the root-knot nematode *Meloidogyne chitwoodi* in susceptible hosts. Eur J Plant Pathol 136:193–202
5. Molinari S (2023) Dose-dependent effects of antioxidants on root-knot nematode infection in vegetable crops and dosage standardization for preventive measures. Agronomy 13:746. https://doi.org/10.3390/agronomy13030746
6. Shoresh M, Harman GE, Mastouri F (2010) Induced systemic resistance and plant responses to fungal biocontrol agents. Annu Rev Phytopathol 48:21–433
7. Shouteden N, De Waele D, Panis B, Vos CM (2015) Arbuscular Mycorrhizal Fungi for the biocontrol of plant-parasitic nematodes: a review of the mechanisms involved. Front Microbiol 6:1280. https://doi.org/10.3389/fmicb.2015.01280
8. Leonetti P, Zonno MC, Molinari S, Altomare C (2017) Induction of SA-signaling pathway and ethylene biosynthesis in Trichoderma harzianum-treated tomato plants after infection of the root-knot nematode *Meloidogyne incognita*. Plant Cell Rep 36:621–631
9. Cameron DD, Neal AL, van Wees SCM, Ton J (2013) Mycorrhiza-induced resistance: more than the sum of its parts? Trends Plant Sci 18: 539–545
10. Molinari S, Leonetti P (2019) Bio-control agents activate plant immune response and prime susceptible tomato against root-knot nematodes. PLoS One 14:e0213230. https://doi.org/10.1371/journal.pone.0213230
11. Molinari S (2021) Factors determining the variability of performance of bio-control agents against root-knot nematodes in vegetable plants. Agronomy 11:1602. https://doi.org/10.3390/agronomy11081602
12. Molinari S, Akbarimotlagh M, Leonetti P (2022) Tomato root colonization by exogenously inoculated arbuscular mycorrhizal fungi induces resistance against root-knot nematodes in a dose-dependent manner. Int J Mol Sci 23:8920. https://doi.org/10.3390/ijms23168920
13. Byrd DW Jr, Kirkpatrick T, Barker KR (1983) An improved technique for clearing and staining plant tissue for detection of nematodes. J Nematol 15:142–143
14. Molinari S (2009) Bioassays on plant-nematode interactions. In: Narwal SS, Sampietro DA, Catalàn CAN, Vattuone MA, Politycka B (eds) Plant Bioassays. Enfield Science Publisher, pp 293–326

Chapter 10

ROS Detection and Quantification in Plant–Nematode Interactions

Karen De Kock, Jasper Matthys, and Tina Kyndt

Abstract

Reactive oxygen species (ROS) accumulation is one of the earliest hallmarks upon successful pathogen recognition in plants. H_2O_2 is considered the most important ROS in plant defense considering its relatively high stability and capacity to cross long distances in the plant. However, ROS also play roles in cell development and could hence facilitate nematode feeding site development. Several methods to analyze the cellular redox state exist, among which ROS detection and quantification and the evaluation of ROS scavenging enzyme activity (peroxidase activity, catalase activity, etc.). Here, we describe DAB staining, which is used to detect and localize ROS *in planta* upon an external trigger. Furthermore, ROS quantification using the FOX assay is described. Both methods have been used extensively in research and yield repeatable results in various plants.

Key words ROS, Oxidative stress, H_2O_2, DAB staining, FOX assay, Xylenol orange

1 Introduction

One of the earliest plant cellular responses following successful pathogen recognition is the oxidative burst [1]. This phenomenon, characterized by a strong and rapid increase in reactive oxygen species (ROS) in the apoplast, can lead to severe damage to lipids, proteins, carbohydrates, and nucleic acids [2]. Notwithstanding, the essential role of ROS in defense, by acting as toxins for invading pathogens and defense signaling in plants, has been demonstrated repeatedly [3–6]. To finely regulate and control ROS production, several enzymatic and non-enzymatic ROS scavenging tools are incorporated in the plant cellular redox machinery, among which catalases (CAT), superoxide dismutases (SOD), anthocyanins, carotenoids, tocopherols, the ascorbate-glutathione cycle (AsA-GSH), and ascorbate peroxidases (APX) [7].

The main biochemical sources of pathogen-induced ROS are respiratory burst oxidase homologs (RBOH), which are plasma

Sergio Molinari (ed.), *Plant-Nematode Interactions*, Methods in Molecular Biology, vol. 2756,
https://doi.org/10.1007/978-1-0716-3638-1_10,

membrane-bound NADPH oxidases that promote electron transfer to O_2 to form the unstable superoxide anion (O_2^-) which subsequently dismutates to hydrogen peroxide (H_2O_2) and oxygen (O_2) [8]. Several reports have documented a fundamental role for the *Rboh* gene family in ROS generation upon pathogen-associated molecular pattern (PAMP)-triggered immunity (PTI) stimulation with nematode-associated molecular patterns (NAMPs) and nematode infection [6, 9–11]. Due to its stability and capability to cross cell membranes through aquaporins, H_2O_2 is considered the most valuable ROS functioning as a messenger [12]. The role of NADPH-induced H_2O_2 production in plant resistance or susceptibility for a pathogen, however, is dependent on its intensity and the specific plant–pathogen interaction. Indeed, H_2O_2 accumulation, potentially leading to programmed cell death, is generally considered beneficial for pathogens with a necrotrophic lifestyle, whereas this hampers the *in planta* growth of biotrophic pathogens. Multiple exceptions to this rule have been reported, among which the potentially dual role of H_2O_2 in plant interactions with parasitic nematodes.

Chopra et al. [9] report a promoting role for NADPH-generated ROS in the infection of *Arabidopsis* with the beet cyst nematode *Heterodera schachtii*, which uses ROS to reprogram the plant's indole metabolism necessary for feeding site induction and syncytium development. However, other researchers observed reduced susceptibility in GSH-deficient soybean plants in which ROS accumulates at high levels upon infection, leading to developmental arrestation of the soybean cyst nematode *H. glycines* [16]. This suggests that ROS could play a dual role in plant–cyst nematode interactions.

On the other hand, NADPH-produced ROS hampered development of the root-knot nematode *Meloidogyne incognita* [9, 11, 13], which is in accordance with the observed inhibition of ROS scavenging enzymes, among which catalase and ascorbate peroxidase, in resistant tomato plants upon *M. incognita* infection [5, 14, 15]. A clear time-course of ROS production in tomato upon root-knot nematode inoculation has been provided by Melillo et al. [10] in which a first H_2O_2 peak occurs early upon nematode detection (12 h after inoculation), which only intensifies at 24 h and persists until at least 48 h after inoculation in incompatible plant–nematode interactions, leading to cell death and inhibition of giant cell development. In root-knot nematode-induced galls in rice, increased H_2O_2 levels were detected at 3 days post inoculation, followed by an induced accumulation of ascorbate and its oxidized form, dehydroascorbic acid (DHA), at 7 days post inoculation, suggesting the role for ascorbate as an important ROS scavenger during the plant–nematode interaction [17]. Indeed, follow-up research indicated evidence for the oxidized form of AsA to enhance resistance against the root-knot nematode *M. graminicola* and the cyst nematode

H. schachtii in rice and sugar beet, respectively. External application of DHA, for example, activates the plant's defense mechanisms and strengthens the production of H_2O_2 [17–19].

The above-described examples exemplify the important role for H_2O_2 generation and accumulation in plant–nematode interactions and demands for a straightforward protocol to detect and quantify H_2O_2 in plants. Therefore, we provide two easy-to-follow protocols to detect and quantify ROS that only require standard lab material and guarantee repeatable results in various plant species, among which crop plants such as sugar beet, rice, and tomato.

The protocol used for ROS detection is based on the polymerization of 3,3′-diaminobenzidine (DAB) in the presence of ROS. The resulting polymer is dark brown, immobile, and extremely stable, which makes it readily visible. The protocol used for H_2O_2 quantification is based on the FOX assay. The FOX reagent is an acidic solution containing Fe^{2+} ions and xylenol orange. In an acidic environment, H_2O_2 oxidizes Fe^{2+} to Fe^{3+} forming a purple complex with xylenol orange. This can be quantified by measuring the absorbance in a spectrophotometer preferentially at 560 nm (or alternatively at 595 nm). Both assays have been successfully applied, and the results have been published in several research articles [17–20].

2 Materials

2.1 ROS Detection (DAB Staining)

For efficient handling of the protocol, we recommend the following standard lab equipment:

2.1.1 General Material

1. Tweezers and scissors.
2. 6-well plates.
3. Desiccator attached to a vacuum pump.
4. Orbital shaker.
5. Hot water bath.
6. Petri plates or microscopic glass slides.
7. Light microscope with camera.
8. Distilled H_2O (dH_2O).

2.1.2 Chemicals

1. 3,3′-diaminobenzidine tetrahydrochloride salt ($C_{12}H_{14}N_4.4HCl$) (*see* **Notes 1** and **2**).
2. Dibasic potassium phosphate (K_2HPO_4).
3. Tween 20.
4. Ethanol (C_2H_5OH).
5. Acetic acid (CH_3COOH).
6. Glycerol.

2.2 H_2O_2 Quantification (FOX Assay)

2.2.1 General Material

For efficient handling of the protocol, we recommend the following standard lab equipment:

1. Calibrated set of micropipettes.
2. Vortex.
3. Plastics: 5 mL Eppendorf tubes, 2 mL Eppendorf tubes.
4. Polystyrene 96-well plate.
5. 96-well plate spectrophotometer.
6. Distilled H_2O (dH_2O).
7. Centrifuge with cooling option.

2.2.2 Sampling and Grinding

1. Mortars and pestles.
2. Small spoons or spatulas.
3. Liquid nitrogen.

2.2.3 Reagents

1. Trichloroacetic acid (CCl_3COOH).
2. Sorbitol ($C_6H_{14}O_6$).
3. Xylenol orange tetrasodium salt ($C_{31}H_{28}N_2Na_4O_{13}S$).
4. Ethanol (C_2H_5OH).
5. Ferrous ammonium sulfate ($(NH_4)_2Fe(SO_4)_2.6H_2O$) (Mohr's salt).
6. Sulfuric acid (H_2SO_4).
7. Hydrogen peroxide (H_2O_2) (*see* **Note 3**).

3 Methods

These protocols can be used for both root and shoot material of various plants under various conditions (untreated, treated with chemical/biological compounds, uninoculated, inoculated). Sample time points depend on the aim of the experiment. However, it should be noted that the ROS burst is a rapid response of the plant immune system upon external triggers. Hence, one should consider an appropriate (range of) sampling time point(s) prior to executing the experiment. H_2O_2 accumulation upon PTI triggering or nematode infection has been reported to occur already several minutes after treatment [6] and can last until several days after nematode inoculation [17–19, 21].

3.1 ROS Detection (DAB Staining)

This protocol describes the detection of ROS in shoot material but can be applied in root material as well (*see* **Note 4**) [11]. When working with leaves, the second youngest leaf is ideal for treatment and DAB staining. ROS can be detected upon treatment of the plants with certain compounds and/or after inoculation. Best

Fig. 1 Rice leaf taped onto a horizontal surface covered with aluminum foil. The location of spot treatment is marked (see arrows), and two droplets of the treatment have been pipetted onto the ad axial surface of the leaf

results are obtained when staining the plant part close to the site of treatment/inoculation. In case of investigating the ROS burst upon a (chemical) treatment, we recommend placing the plants (still in their growth substrate) horizontally (e.g., on a table) and taping the leaves onto the horizontal surface (*see* Fig. 1) at least 12–24 h prior to treatment. Treatment can be done by spraying the leaf (example of result shown in Fig. 2a, b) or pipetting a droplet of 10 μL onto the adaxial side of the leaves (example of result shown in Fig. 2c). Visualization of the ROS burst is done on the tissue around the treatment spot. Wounding the leaflet with a scalpel might serve as a positive control, where the edges of the wound should show intense staining (*see* Fig. 2d). We recommend using multiple biological replicates (± 6) per treatment as ROS bursts can vary greatly under slightly different environmental conditions.

3.1.1 Preparation of the Reagents

1. DAB staining solution, for 100 mL (*see* **Note 5**): Dissolve 100 mg DAB tetrahydrochloride in dH_2O, add 1 mL 1 M K_2HPO_4 to the DAB solution, add 50 μL Tween 20, and mix well.
2. Bleaching solution: Mix 60 mL (100%) EtOH, 20 mL acetic acid, and 20 mL glycerol.

3.1.2 DAB Staining

1. When working with elongated leaves (e.g., from monocots like rice), mark the location at which you will pipette a droplet of treatment (*see* **Note 6**).

Fig. 2 (**a**) Microscopic image of *Arabidopsis* leaf 1 hpt with 0.5 mM riboflavin (spray treatment) (picture by Jasper Matthys); (**b**) microscopic image of *Arabidopsis* leaf 1 hpt with 500 mg/L COS (chitosan oligosaccharide, spray treatment) showing intense staining of trichomes (picture by Jasper Matthys); (**c**) microscopic image of rice leaf showing intense DAB staining 24 hpt with 20 mM DHA, a recently discovered induced resistance (IR) compound [18] (picture by Jasper Matthys); (**d**) tomato leaves wounded with a scalpel showing intense DAB staining, indicating ROS accumulation at the edges of the wound. (Picture by Willem Desmedt)

2. Treat the root or leaf tissue by spraying or pipetting the chosen chemical or biological agent on the marked location.
3. Fill a sufficient number of Petri plates (for elongated leaves) or 6-well plates (for smaller plant leaves) with freshly made DAB staining solution.
4. Harvest treated plants at the chosen (set of) time point(s) after treatment according to the aim of the experiment (*see* **Note 7**): carefully cut the (marked piece of) leaf that was treated with the help of tweezers and scissors, and put directly into the wells/ plates containing DAB solution. Ensure that the tissue is submerged.
5. Quickly place the plate(s) without lid in a desiccator attached to a vacuum pump, and apply a gentle vacuum of 20 mbar for 5 min to vacuum-infiltrate the leaves with DAB (*see* **Note 8**).

6. Cover the plate(s) with a lid, wrap in aluminum foil, and incubate on an orbital shaker for 4 h at room temperature (*see* **Note 9**).
7. Remove the staining solution with a pipette, and replace with the bleaching solution.
8. Place the plate(s) in a hot water bath at 95 °C for 15–20 min until the leaves are fully bleached.
9. Replace the bleaching solution with fresh bleaching solution, and again place the plate(s) in a hot water bath at 95 °C for 15–20 min.
10. Discard the bleaching solution. DAB-stained leaves can now be stored (*see* **Note 10**) or visualized (under a microscope).
11. Results can be analyzed by calculating the percentage of treated spots that resulted in DAB staining.

3.2 H_2O_2 Quantification with the FOX Assay

To minimize inherent variability and increase statistical power, pool plant tissues of several plants (preferably 4–6 plants) per biological replicate, and analyze 5–6 independent biological replicates per treatment (*see* **Note 11**).

3.2.1 Sampling and Grinding

1. Sample plant material at a chosen (set of) time point(s), and place in 5 mL Eppendorf tubes in a container filled with liquid nitrogen.
2. Precool mortar and pestle with liquid nitrogen, and grind plant material to a fine powder. Use a different precooled mortar and pestle for each sample, or wash with water and 70% EtOH sequentially between each sample. Store samples at −80 °C or in liquid nitrogen for subsequent handling. Perform the FOX assay as soon as possible after sampling (*see* **Note 12**).

3.2.2 Preparation of the Reagents

1. Prepare 5% (w/v) of trichloroacetic acid (TCA) in distilled H_2O. Store on ice.

   Prepare the FOX reagent:

   Solution A, for 100 mL: Add 1.82 g sorbitol, 9.5 mg xylenol orange tetrasodium salt, and 1 mL EtOH to 99 mL H_2O. Store on ice.

   Solution B, for 2 mL: Add 19.6 mg Mohr's salt and 280 μL H_2SO_4 to 1720 μL dH_2O (*see* **Note 13**). Store on ice.

   Mix 100 volumes of A with 1 volume of B prior to use (*see* **Notes 14** and **15**). Store on ice.
2. Prepare the blank control reagent (BLK FOX) (*see* **Note 16**): Prepare the same as the FOX reagent (repeat step 2), but without adding xylenol orange. Store on ice.
3. Prepare a 1:5 H_2O_2 serial dilution for the calibration curve:

   Start by diluting 30% H_2O_2 to a stock solution of 1 M, and add 102 μL H_2O_2 to 5% TCA. Store in dark and on ice, dilute to 10 mM by adding 10 μL of 1 M stock to 990 μL TCA solution, dilute to 100 μM by adding 10 μL of 10 mM stock to 990 μL TCA solution, and prepare a 1:5 serial dilution in 5% TCA from 100 μM to 0 μM H_2O_2 (100, 20, 4, 0.8, 0.16, 0.032, 0.0064, 0) (*see* **Note 17**).

Prepare three replicates of the serial dilution. Store in the dark and on ice.

3.2.3 The FOX Assay

1. Precool labeled 2 mL Eppendorf tubes in liquid nitrogen.
2. In a precooled Eppendorf tube, weigh 70–100 mg of grinded plant material, write down the weight of each sample, and add 10 μL 5% TCA solution per mg plant material (e.g., 1 mL to 100 mg material). Store samples on ice.
3. Vortex samples briefly. When working with plant root tissue, a sonication step is recommended (*see* **Note 18**).
4. Centrifuge the samples for 10 min at 4 °C and 20,800 *g*.
5. In a 96-well plate, pipette 100 μL FOX reagent with xylenol orange in 3 wells for each sample (three technical replicates per sample). In the next three wells, add 100 μL BLK FOX reagent. Repeat this for all samples. This results in 6 wells per sample (3 with xylenol orange and 3 blanks) (*see* **Note 19**).
6. Add 20 μL of the TCA extract (upper layer after centrifugation in step 4) to each of the 6 wells per sample (3 with FOX and 3 with BLK FOX) (*see* **Note 20**).
7. Prepare the standard curve in the same manner as your samples. Instead of adding TCA extract, add 20 μl of each dilution of the 1:5 dilution series to 100 μL FOX or 100 μL BLK FOX.
8. Incubate the plate for at least 30 min in the dark and at room temperature.
9. Measure the absorbance at 560 nm in a spectrophotometer (alternatively, 595 nm can also be used).
10. Calculate the H_2O_2 concentration per sample:

    Create a calibration curve by using the known H_2O_2 concentrations and measured absorbances of the 1:5 dilution series samples. Take the median or the mean of the three calibration curves. Calculate the slope, calculate the median or mean absorbance of the three technical replicates for each biological replicate (*see* **Note 21**), and subtract their respective blank absorbance values. Calculate the H_2O_2 concentration for each biological replicate by using the slope of the linear part of the calibration curve, and recalculate the H_2O_2 concentration to the concentration per mg sample (*see* **Note 22**).

4 Notes

1. Regular DAB instead of DAB tetrahydrochloride salt can be used. In this case, add DAB to dH_2O, and add HCl until the pH drops to 3.0. When DAB is dissolved, bring the pH back to normal by adding K_2HPO_4.
2. DAB is toxic and irritating. Wear protective equipment when preparing and handling the solution.
3. H_2O_2 is highly light sensitive. For long-term storage (closed bottle, maximum 2 years; opened bottle, maximum 2 months), wrap the H_2O_2 solution in aluminum foil, and keep at 4 °C.
4. DAB staining can be done in root tissue. However, uptake of DAB in roots appears to be limited to soft, thin areas (e.g., root tips), and the low baseline ROS in roots may require a longer incubation time, depending on the plant species.
5. DAB solution should always be made fresh since it rapidly oxidizes. Ideally, prepare the DAB solution maximum 1 h beforehand. Store DAB solution in a glass bottle wrapped in aluminum foil since DAB easily oxidizes under the influence of light in the presence of chemical contaminants.
6. Be careful when handling plant material for DAB staining. Any wounding caused by pipettes, tweezers, scissors, markers, etc. will induce ROS and hence result in staining. To prevent false-positive results due to these artifacts, place your plants horizontally, and mark the location of treatment 12–24 h beforehand.
7. H_2O_2 accumulation can be detected already several minutes or hours after treatment but can last until several days after treatment. One should consider an appropriate (range of) time point(s) before executing the experiment.
8. Only start timing when vacuum-infiltration of the leaves has commenced. This can be observed by small bubble formation at the edge of the leaves.
9. When DAB staining remains undetectable after 4 h, extend the incubation time to 6, 8, or 12 h. This can be required for treatments that do not induce an intense ROS burst. However, be careful for artifacts when applying longer incubation times since noise will appear due to the low baseline levels of ROS that are consistently present in plant cells.
10. DAB-stained plant material can be stored in fresh bleaching solution for up to 1 week at 4 °C.
11. H_2O_2 quantification is a very sensitive method for analyzing changes in the plant's environment. Hence, slight environmental differences result in significant quantifiable H_2O_2 changes. Keep the environmental conditions of the plants as stable and

as equal as possible, and analyze several plants per biological treatment (pool plants) to minimize inherent variability. More biological replicates can be included to increase statistical power.

12. H_2O_2 is extracted using a 5% TCA solution. Even though this acid stabilizes H_2O_2 to some extent, the molecule remains highly reactive and disappears quickly. An estimation of 2% of the initial H_2O_2 disappears per minute, even in stabilized conditions [22]. Always keep samples, including those for the 1:5 dilution series, on ice, and work quickly.
13. Do not add water to sulfuric acid. First slowly add sulfuric acid to 1 mL water, and then top up to 2 mL. Add Mohr's salt last.
14. Only mix solution A and B just before use. Do not mix the two solutions at the beginning of the assay.
15. The FOX reagent is light sensitive. Store the solution wrapped in aluminum foil on ice.
16. A blank control is used to correct the absorbance at 560 nm (or 595 nm) inherently present in the samples.
17. It is possible that the calibration curve reaches a plateau phase at high H_2O_2 concentrations (at around 100 μM or higher). In that case, remove these data points from the analysis. Only the linear part of the calibration curve is to be used in further data analysis.
18. For root samples, a sonication step of 2×1 min (40 kHz) is recommended to facilitate disruption of the cell walls.
19. Depending on the number of samples and considering the number of technical replicates, pipetting the FOX and BLK FOX reagents can be somewhat time-consuming. Using a multichannel pipette enables to work faster while still being precise.
20. When working with samples from different conditions, avoid biased results of H_2O_2 accumulation due to its fast degradation over time. Start by pipetting the TCA extract (with FOX and with BLK FOX) of each treatment of biological replicate 1, then continue with biological replicate 2 of each treatment, and so on.
21. When observing a large variability between the technical replicates of one biological replicate, calculate the median instead of the mean.
22. Recalculating the H_2O_2 concentration to the concentration per mg sample can easily be done by dividing the H_2O_2 concentration (in μM) by 10^5, since 10 μL TCA/mg sample was added consistently in step 2 of the protocol. Hence, no

additional corrections for sample weight should be made during data analysis.

References

1. Wojtaszek P (1997) Oxidative burst: an early plant response to pathogen infection. Biochem J 322:681–692. https://doi.org/10.1042/bj3220681
2. Paciolla C, Paradiso A, de Pinto MC (2016) Cellular redox homeostasis as central modulator in plant stress response. In: Gupta D, Palma J, Corpas F (eds) Redox state as a central regulator of plant-cell stress responses. Springer. https://doi.org/10.1007/978-3-319-44081-1_1
3. Ali M, Cheng Z, Ahmad H, Hayat S (2018) Reactive oxygen species (ROS) as defenses against a broad range of plant fungal infections and case study on ROS employed by crops against Verticillium dahliae wilts. J Plant Interact 13:353–363. https://doi.org/10.1080/17429145.2018.1484188
4. Chamnongpol S, Willekens H, Moeder W, Langebartels C, Sandermann H, Van Montagu M, Inzé D, Van Camp W (1998) Defense activation and enhanced pathogen tolerance induced by H_2O_2 in transgenic tobacco. PNAS 95:5818–5823. https://doi.org/10.1073/pnas.95.10.5818
5. Leonetti P, Molinari S (2020) Epigenetic and metabolic changes in root-knot nematode-plant interactions. Int J Mol Sci 21:1–17. https://doi.org/10.3390/ijms21207759
6. Mendy B, Wang'ombe MW, Radakovic ZS, Holbein J, Ilyas M, Chopra D, Holton N, Zipfel C, Grundler FMW, Siddique S (2017) Arabidopsis leucine-rich repeat receptor–like kinase NILR1 is required for induction of innate immunity to parasitic nematodes. PLoS Pathog 13:1–22. https://doi.org/10.1371/journal.ppat.1006284
7. Gupta DK, Palma JM, Corpas FJ (2016) Redox state as a central regulator of plant-cell stress responses. In: Gupta D, Palma J, Corpas F (eds) Redox state as a central regulator of plant-cell stress responses. Springer. https://doi.org/10.1007/978-3-319-44081-1
8. Torres MA, Dangl JL, Jones JDG (2002) Arabidopsis gp91phox homologues Atrbohd and Atrbohf are required for accumulation of reactive oxygen intermediates in the plant defense response. PNAS 99:517–522. https://doi.org/10.1073/pnas.012452499
9. Chopra D, Hasan MS, Matera C, Chitambo O, Mendy B, Mahlitz SV, Naz AA, Szumski S, Janakowski S, Sobczak M, Mithöfer A, Kyndt T, Grundler FMW, Siddique S (2021) Plant parasitic cyst nematodes redirect host indole metabolism via NADPH oxidase-mediated ROS to promote infection. New Phytol 232:318–331. https://doi.org/10.1111/nph.17559
10. Melillo MT, Leonetti P, Bongiovanni M, Castagnone-Sereno P, Bleve-Zacheo T (2006) Modulation of reactive oxygen species activities and H_2O_2 accumulation during compatible and incompatible tomato-root-knot nematode interactions. New Phytol 170:501–512. https://doi.org/10.1111/j.1469-8137.2006.01724.x
11. Siddique S, Matera C, Radakovic ZS, Hasan MS, Gutbrod P, Rozanska E, Sobczak M, Torres MA, Grundler FMW (2014) Host-pathogen interactions: parasitic worms stimulate host NADPH oxidases to produce reactive oxygen species that limit plant cell death and promote infection. Sci Signal 7:1–10. https://doi.org/10.1126/scisignal.2004777
12. Petrov VD, Van Breusegem F (2012) Hydrogen peroxide-a central hub for information flow in plant cells. AoBP 12:1–13. https://doi.org/10.1093/aobpla/pls014
13. Feng B, Shan L (2014) ROS open roads to roundworm infection. Sci Signal 7:8–10. https://doi.org/10.1126/scisignal.2005273
14. Abd-Elgawad M, Kabeil S, Fanelli E, Molinari S (2012) Different levels of anti-oxidant enzyme activities in tomato genotypes susceptible and resistant to root-knot nematodes. Nematropica 42:330–336
15. Molinari S (2001) Inhibition of H_2O_2-degrading enzymes in the response of *Mi*-Bearing tomato to root-knot nematodes and salicylic acid treatment. Nematol Medit 29:235–239. http://journals.fcla.edu/nemamedi/article/viewFile/86672/83588
16. Chen X, Li S, Zhao X, Zhu X, Wang Y (2020) Modulation of (homo) glutathione metabolism and H_2O_2 accumulation during soybean cyst nematode infections in susceptible and resistant soybean cultivars. Int J Mol Sci 21. https://doi.org/10.3390/ijms21020388
17. Singh RR, Verstraeten B, Siddique S, Tegene AM, Tenhaken R, Frei M, Haeck A, Demeestere K, Pokhare S, Gheysen G, Kyndt T (2020) Ascorbate oxidation activates

systemic defence against root-knot nematode Meloidogyne graminicola in rice. J Exp Bot 71: 4271–4284. https://doi.org/10.1093/jxb/eraa171

18. Chavan SN, De Kesel J, Desmedt W, Degroote E, Singh RR, Nguyen GT, Demeestere K, De Meyer T, Kyndt T (2022) Dehydroascorbate induces plant resistance in rice against root-knot nematode Meloidogyne graminicola. Mol Plant Pathol 23:1303–1319. https://doi.org/10.1111/mpp.13230
19. Singh RR, Nobleza N, Demeestere K, Kyndt T (2020) Ascorbate oxidase induces systemic resistance in sugar beet against cyst nematode Heterodera schachtii. Front Plant Sci 11:1–15. https://doi.org/10.3389/fpls.2020.591715
20. Desmedt W, Jonckheere W, Nguyen VH, Ameye M, De Zutter N, De Kock K, Debode J, Van Leeuwen T, Audenaert K, Vanholme B, Kyndt T (2021) The phenylpropanoid pathway inhibitor piperonylic acid induces broad-spectrum pest and disease resistance in plants. Plant Cell Environ 44:3122–3139. https://doi.org/10.1111/pce.14119
21. Kyndt T, Zemene HY, Haeck A, Singh R, De Vleesschauwer D, Denil S, De Meyer T, Höfte M, Demeestere K, Gheysen G (2017) Below-ground attack by the root knot nematode Meloidogyne graminicola predisposes rice to blast disease. Mol Plant-Microbe Interact 30:255–266. https://doi.org/10.1094/MPMI-11-16-0225-R
22. Cheeseman JM (2006) Hydrogen peroxide concentrations in leaves under natural conditions. J Exp Bot 57:2435–2444. https://doi.org/10.1093/jxb/erl004

Chapter 11

Detection of Pathogenesis-Related (PR) Genes in Susceptible Tomato Plants Infected by *Meloidogyne* spp.

Paola Leonetti

Abstract

Meloidogyne species, as infective second-stage juveniles (J2s) larvae, are parasites able to attack host of relevant agronomic interest such as tomato plants. The identification of gene expression markers, useful to investigate the levels of root-knot nematode infection in the roots, is a fundamental tool in plant–pathogen interaction. The laboratory methods for analyzing the differential expression of pathogenesis-related (PR) genes constitute powerful tools for detecting the induced systemic acquired resistance defense response to *M. incognita* in infected plants and can be extended to all pathogen infection markers to obtain an early and sustainable control.

Key words Plant, Pathogen interaction, Tomato, Root-knot nematodes, PR gene expression, qRT-PCR

1 Introduction

Phytoparasitic nematodes are among the main parasites affecting temperate and tropical regions; in particular, root-knot nematodes (RKNs) are characterized by their capability of infecting many host plants of great agronomic interest. Infection of *Meloidogyne* spp., the largest group of RKNs, follows several steps, including penetration and intracellular migration through the elongation zone of the host root tip and induction of the feeding site with the development of the characteristic galls on tomato roots (*Solanum lycopersicum L*), which justifies their name [1].

The procedures for the study of plant–nematode interaction could start from the observation and collection, under the stereomicroscope, of motile and viable infectious juvenile larvae (J2s) of *Meloidogyne* spp. (*see* Fig. 1a). Nematode population is maintained on susceptible tomato plants or on other compatible hosts until the nematode life cycle is completed (*see* Fig. 1b); through the production by gravid females of gelatinous masses of eggs, J2s are released,

Sergio Molinari (ed.), *Plant-Nematode Interactions*, Methods in Molecular Biology, vol. 2756,
https://doi.org/10.1007/978-1-0716-3638-1_11,

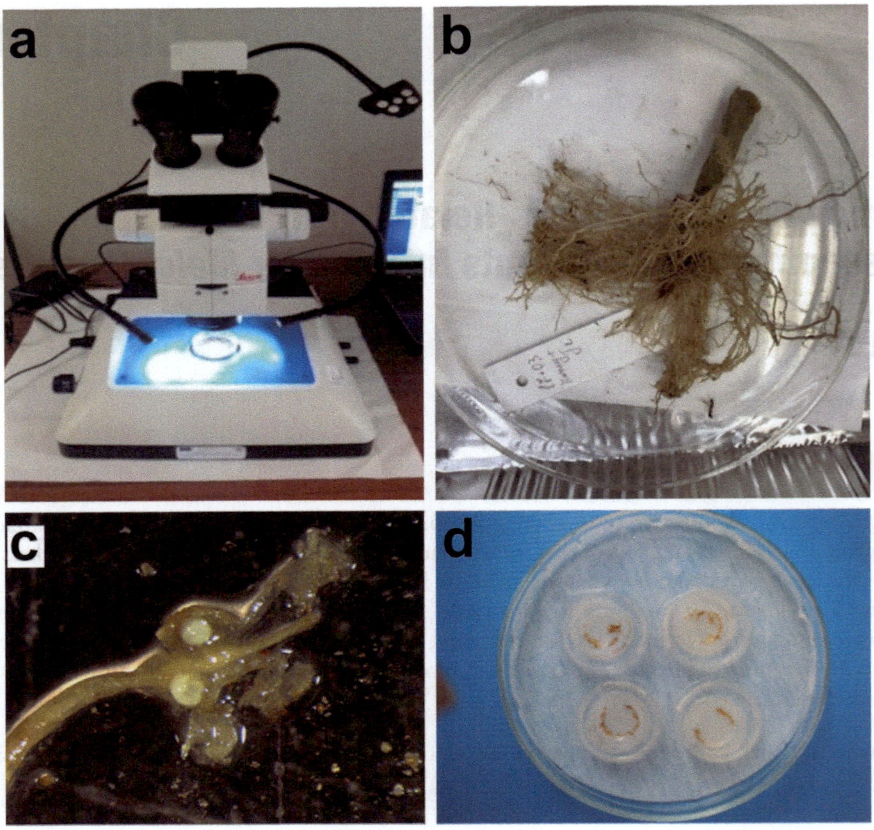

Fig. 1 The viability and motility of the infectious J2s are assessed by Leica M 125 stereomicroscope observation (**a**). J2s are used to maintain the population by infecting preferential host plants. Tomato roots with *M. incognita* galls (**b**). Enlarged detail of root apex with two pyriform females with gelatinous masses containing eggs (**c**). The masses collected and placed in sieves (**d**) are placed to hatch at 27 °C. The infection in time-course experiments will be checked with the stereomicroscope

and the inoculum regenerated (*see* Fig. 1c) and collected; J2s pass through the meshes of the sieves (*see* Fig. 1d) and are counted and used to inoculate susceptible and/or resistant plants pre-treated or not with hormone, hormone-like compound (salicylic acid, SA; benzothiadiazole, BTH), microorganism (bacteria or fungi), and factors promoting defense response, for interaction studies [2–5].

In this chapter, the procedures for growing tomato plants in growth chambers are illustrated (*see* Fig. 2a). Tomato plants are inoculated with about 300 living and motile J2s (*see* Fig. 2b); few mLs of J2 suspensions are poured in each pot, and the plants are allowed to grow (*see* Fig. 2c) for the scheduled number of days (days post infection, dpi), and then plant tissues are collected to be subjected to molecular analysis. Parameters for fitness of radical apparatus, such as length and weight of roots (RL, RW), and those indicating infection levels, such as egg mass (EM) and sedentary form (SF) numbers, which indicates the numbers of individuals

Fig. 2 Tomato seedlings grown in 2.5 cm diameter clay pots containing 60 gr of sterilized quartz sand, watered with Hoagland's solution. Experiments were conducted in a growth chamber at 24 ± 1 °C, with a photoperiod of 16 and 8 h at 200 lmol/m2/s (**a**). After 2 weeks plants were transferred into 100 cm diameter pots, filled with a mixture of sterilized loamy soil and river sand (1:1, v: v), and inoculated with 300 J2s (**b**, **c**). Control plants were watered only with Hoagland's solution. Uninfected (**d**) and *M. incognita*-infected (**e**) roots were collected, frozen in N_2, and prepared for acid nucleic extraction

developed into the roots, as an index of the heavier secondary infection, are measured to ascertain the state of the seedlings and the success of the infection [2, 4, 5]. Root samples from either control plants or infected plants are examined under a stereomicroscope and frozen and subjected to nucleic acid extraction for gene expression studies (*see* Fig. 2d-e) [2–5]. To define the mechanisms of reprogramming of many host defense genes and the associated signalling pathways, the technologies for measuring the expression of pathogenesis-related (PR) genes, which is a very important marker for the activation of SA-dependent systemic acquired resistance (SAR) of the plants, are described here, within the interaction model reported in [2, 4, 5]. Total RNAs are extracted from control and infected roots of plants pre-treated or not with SA and collected at different days after inoculation (*see* Fig. 3a). cDNAs are retro-transcribed to be subjected to real-time PCR (qRT-PCR) analysis for gene expression studies (*see* Figs. 3b and 4a,b). The changes in expression levels of PR genes are one of the earliest defense responses of infection of *M. incognita* avirulent (avr) populations versus susceptible tomato plants pre-treated with SAR inductors [2–4]. PR1b1 (P6) (unknown function of protein activity), PR2 (β-1,3-glucanase), PR3 (chitinase), the hypersensitive cell death inducer PR4b (P2), as well as PR5 (thaumatin-like) gene expression patterns can be analyzed to investigate nematode–plant interactions (*see* Table 1) [2–5].

The study of the variations in the level of PR gene expression, as a marker of the immunity response induced in plants by *M. incognita* J2s attack, provides a valid tool for the characterization of the effect of promoters/inducers on the systemic defense responses of plants. Enhanced expression of PR genes gives an effective support in controlling these pathogens which are capable of overcoming the defense mechanisms of the hosts. The validation of this protocol, which involves many steps, from the use of the latest generation stereoscopic systems, such as Leica M 125 equipped with a LEICA IC80 HD photo camera (Leica Geosystem, Germany), to the use of qRT-PCR, can be useful for the study of any key gene of other crucial mechanisms acting in plant–pathogen interactions [1–6].

2 Materials

2.1 Biological Material

1. Tomato seedlings (*Solanum lycopersicum* L.) susceptible (cv. Roma VF) to RKNs.
2. Avirulent (avr) *M. incognita* field population J2s (*MiVfield*).

2.2 Tissue Sampling and Grinding of Uninfected and J2 Infected Roots

1. Leica M 125 stereoscope equipped with a LEICA IC80 HD photo camera or another equivalent stereomicroscope, sterile needles, and forceps.
2. Autoclaved mortar and pestle.
3. Liquid nitrogen.

2.3 Extraction and Quantification of RNAs from Roots

1. Molecular grade water.
2. 1.5 ml Eppendorf tubes.
3. Thermoblock.
4. Bench-top microcentrifuge.

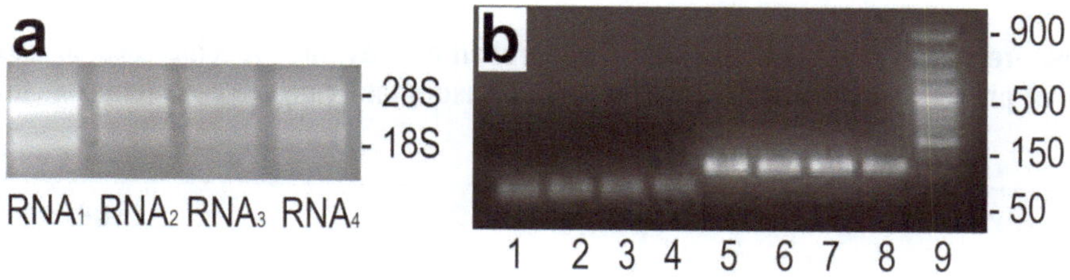

Fig. 3 Gel analysis of total RNA extracted (**a**) from pools of two biological replicas of uninfected (RNA1, 2) and two pools of *M. incognita*-infected tomato roots (RNA3, 4). (**b**) Gel analysis of cDNAs retro-transcribed from RNA1, 2, 3, 4, respectively, cDNA1, 2, 3, 4 amplified with the housekeeping gene primers (actin 7) (lines 1–4) and cDNA1, 2, 3, 4 amplified with PR4b (P2) gene primers (lines 5–8) with 50 bp commercial ladder in line 9

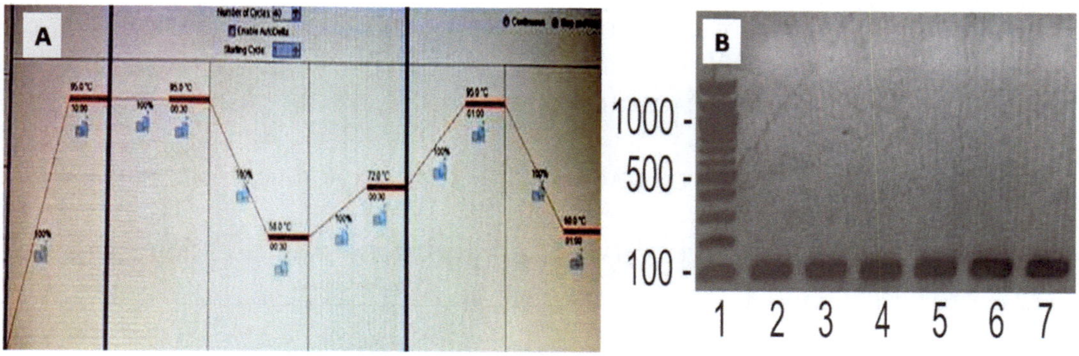

Fig. 4 Gel analysis of qRT-PCR amplification product from two cDNAs generated from *M. incognita*-infected tomato roots, cDNA 3 (lines 2–4) and cDNA4 (lines 5–7) amplified with PR4b (P2) primers, in three technical replicates. 100 pb ladder line 1

5. Spectrophotometer or nanodrop for DNA/RNA quantification.
6. RNA-easy Plant Mini Kit (Qiagen).

2.4 Check of RNA Quality

1. Molecular grade water.
2. Agarose.
3. 1X TBE buffer.
4. Ethidium bromide or equivalent green DNA–RNA dye.

2.5 cDNA Synthesis

1. Molecular grade water.
2. Sterile 0.2 mL PCR tubes.
3. Sterile filter pipette tips.
4. QuantiTec reverse transcription kit (Qiagen).

2.6 Real-Time PCR

1. Molecular grade water.
2. Primers for housekeeping gene and PR genes (*see* Table 1).

Table 1
Pathogenesis-related (PR) gene name, associate protein activity, and specific reverse and forward sequences of primers used in quantitative reverse transcriptase-polymerase chain reaction (qRT-PCR)

Gene acronym and protein activity	Primer sequence for	Primer sequence rev	NCBI reference sequence
PR1b1 (P6) - unknown	GATCGGACAACGTCCTTAC	GCAACA TCAAAAGGGAAA TAAT	NM_001247385.2
PR2 - B-1,3-glucanase	AAGTATATAGCTGTTGGTAA TGAA	TTTCCCAAACCAGC TTCACCG	NM_001312890.1
PR3 -chitinase	AACTATGGGCCATG TGGAAGA	GGCTTTGGGGA TTGAGGAG	NM_001247474.2
PR4b (P2) -ribonuclease	TGACCAACACAGGAACAGGA	GCCCAATCCATTAG TGTCCA	NM_001247154.1
PR5 - thaumatin- like	ACAACTGTCCATACACCGT	AGACTCCACCACAA TCACC	NM_001247422.3
Actin 7	CAGCAGATGTGGATCTCAAA	CTGTGGACAA TGGAAGGAC	NM_001308447.1

3. SYBR® Select Master Mix (Applied Biosystems®) or compatible reaction mix.
4. 96-well plate or strip tubes.
5. Centrifuge for plate or strip tubes.
6. Applied Biosystems® StepOne™ instrument or compatible instrument.

3 Methods

3.1 Sterilization of Materials, Tomato Seedlings, and Nematode Inoculum Preparation

1. Sterilize pots, quartz and river sand, tongs, and loops, by autoclaving for 30 minutes at 121 °C and 2 bar pressure, and mortar and pestle for RNA, for 3 h at 150 °C in laboratory stove.
2. The used *M. incognita* population was species identified by isozyme electrophoretic patterns of esterase and malate dehydrogenase [7]. Prepare *M. incognita* J2s by incubation of egg masses in tap water at 27 °C (*see* Fig. 1a–d), and store at 4 °C for using them as inoculum source for infection experiments.
3. Prepare Hoagland's solution and autoclaved water (*see* **Note 1**).
4. Sterilize seed surface and quartz sand. Transplant the seedlings to 25 cm^3 clay pots or 100 cm^3 clay pots, filled with 60 g or

150 g of sterilized quartz/river sand. Grow up seedlings (*see* Fig. 2a, c) of tomato cv. Roma VF (fully susceptible to RKNs) to the fifth/sixth compound leaf stage in a controlled temperature chamber (24 ± 1 °C), under a regular photoperiod of 16 h/day:8 h/night and watering with Hoagland's plant nutrient solution.

3.2 Penetration Experiments: Tissue Collection of Control and Infected Plants

1. Conduct the penetration experiments in a growth chamber at 24 ± 1 °C. Check under the stereomicroscope the vitality and motility of the J2, left at room temperature before inoculation at least 2 h (*see* Fig. 2b).
2. Inoculate 5-day-old susceptible tomato plantlets with a water suspension of 300 J2s (*see* Fig. 2c). Detection of nematode infection could be performed 3, 5, 7, 21, and 40 dpi in time-course experiments.
3. Grow plants in pots filled with sterilized quartz sand in the experiments in which harvest was predicted to occur 3, 5, and 7 days after nematode inoculation; transfer plants in larger pots filled with a mixture of sterilized loamy soil and river sand (1,1, v,v) when harvest is predicted at 21 and 40 dpi. Alternately, water with sterile water or sterile Hoagland's solution when needed. Examine the tomato roots by stereomicroscope, at different days (dpi) post inoculation (*see* Fig. 2d, e) at the time point predicted by the protocol.
4. Under the stereoscope collect the samples cutting the infected root tips in a sterile plate, and place them in a sterile 1.5 mL Eppendorf on ice; weigh quantities of excised plant material, and quickly place them in liquid nitrogen. Do the same for the control plant roots. Repeat the procedure within the infection times provided for any time course.
5. Grind the tissues until obtaining a fine powder; make sure your samples won't defrost, by cooling down mortar and pestle. Aliquot the samples into sterile 1.5 mL Eppendorf tubes (about 50 mg each), put all of them in liquid nitrogen, and proceed with RNA extraction or preserves at −80 °C.

3.3 RNA Extraction, Quantification, and Quality Check: cDNA Synthesis

When RNA is extracted in time-course experiments from control and SAR promoters and inductors pre-treated infected roots (*see* **Note 2**), all the material must be sterile, and the workspace must be carefully cleaned. Use only sterile pipette tips and Eppendorf tubes.

1. Adopt RNA extraction protocol for total RNA from leaves, leaves and shoots, or root extraction using RNeasy Plant Mini Kit (Qiagen, Germany) following the manufacturer's instructions (*see* Fig. 3a); at least two biological replicates should be planned, containing pooled tissue for each treatment and/or inoculation time point.

2. Quantify concentration testing 1 μL RNA suspension in NanoDrop spectrophotometer (Thermo Scientific, Wilmington, DE, USA), and proceed with measurement. To accurately assess sample quality, 260/280 or 260/230 ratios must be analyzed in combination with the overall spectral quality. Pure nucleic acids typically produce absorbance at 260/280 of about 2.0 for RNA.
3. Check the RNA quality through electrophoretic separation of the extracted samples on 1.2% agarose gel containing ethidium bromide or equivalent green DNA–RNA dye.
4. Perform cDNA synthesis using the QuantiTec Reverse Transcription kit. Synthesize first-strand cDNA from 1 μg of total RNA with random hexamers, following the manufacturer's instructions. DNAasi treatment is included in the protocol (Qiagen, Germany). In order not to waste material, test cDNAs in a first qRT-PCR in order to settle primers, suitable temperatures, and number of cycles and to verify the presence of a single amplification band. The actin gene (BT013524) can be used as the reference gene (*see* Fig. 3b).

3.4 qRT-PCR Analysis in StepOnePlus™ Real-Time PCR and PR Expression Gene Analysis

1. Using molecular grade water, prepare the primers in 100 μM master stocks, and from them, the working solution of primers is diluted to 20 μM of premixed forward and reverse primers. The oligonucleotides to be used for quantitative PCR (qRT-PCR) are listed in Table 1 and selected from literature [8, 9] or designed using a bioinformatics site (https://www.bioinformatics.nl/cgi-bin/primer3plus/primer3plus.cgi).
2. Carry out qRT-PCR by means of a StepOnePlus™ Real-Time PCR system.
3. The Applied biosystems SYBR® Select Master Mix is formulated to provide superior specificity and sensitivity. Carefully follow the handling instructions wearing appropriate protective eyewear, clothing, and gloves. Mix the SYBR® Select Master by swirling the bottle. Perform each qRT-PCR reaction from different biological starting material and three technical replicates. Place frozen cDNA samples and primers on ice to thaw. After the samples are thawed, vortex them, and then centrifuge the tubes briefly.
4. Prepare the appropriate number of reactions. A single 20 μL (final volume) qRT-PCR includes 10 μL SYBR® Select Master Mix, 1.5 μL of cDNA template, and 0.2–0.5 μL each of forward and reverse primers. The suggested PCR cycling parameters are pre-incubation at 95 °C for 10 min, 40 cycles at 95 °C for 30 s, 58 °C for 30 s, and 72 °C for 30 s, with the final step at 95 °C, 60 °C (1 min), and 95 °C for 30 s (*see* Fig. 4a). The negative controls should include water and qRT-PCR Master

mix. The specificity of the PCR products can be verified by dissociation melting curve analysis after 40 cycles and agarose gel electrophoresis.

5. The amplification products (12–15 μL) could be loaded on ethidium bromide-stained or equivalent green DNA–RNA dye in agarose gel eluted and sequenced if necessary. An example of qRT-PCR results (two extracted RNAs pooled from roots of infected plants, as biological and technical replicas were loaded on 1.2% agarose gel obtained by amplification with PR2 primers (*see* Fig. 4b).
6. Collect data, and analyze the results using a relative quantitation comparing a target against an internal standard. It is possible to perform either relative quantization using a standard curve method or the relative fold change according to the $2^{-\Delta\Delta CT}$ comparative CT method [10]. Validate data with statistical studies.

4 Notes

1. Preparation of Hoagland's solution: Working solutions of 1 M KH_2PO_4 (34 gr/250 mL H_2O), 1 M KNO_3 (50.55 gr/550 mL H2O), 1 M $Ca(NO_3)_2$ (118.07 gr/550 mL H_2O), 1 M $MgSO_4$ (61.625 gr/250 mL H_2O). Sol A: H_3BO_{23} (dissolve 34 gr/250 mL H_2O), $MnCl_2$ x $4H_2O$ (dissolve 0.181 gr/100 mL H_2O), $ZnSO_4$ x $7H_2O$ (dissolve 0.022 gr/100 mL H_2O), $CuSO_4$ x $5H_2O$ (dissolve 0.008 gr/100 mL H_2O), $Na_2MoOS_2 \times 2H_2O$ (dissolve 0.002 gr/100 mL H_2O).

   For 1 L solution: Mix 1 mL KH_2PO_4 (1 M), 5 mL KNO_3 (1 M), 5 mL $Ca(NO_3)_2$ (1 M), 2 mL $MgSO_4$(1 M), 1 mL Sol A (1 M), 1 mL $FeSO_4$* $7H_2O$ (89.5 mM) (or$FeCl_3$* $6H_2O$); pH 6 0.5 will be obtained by adding a suitable quantity of distilled water [11].
2. Pre-treatment with SAR inductors or plants soil-drench with a commercial formulate are described in details in literature.

References

1. Williamson VM, Gleason CA (2003) Plant-nematode interactions. Curr Opin Plant Biol 6:327–333
2. Molinari S, Fanelli E, Leonetti P (2014) Expression of tomato salicylic acid (SA)-responsive pathogenesis-related genes in Mi-1-mediated and SA-induced resistance to root-knot nematodes. Mol Plant Pathol 15: 255–264
3. Melillo MT, Leonetti P, Veronico P (2014) Benzothiadiazole effect in the compatible tomato-Meloidogyne incognita interaction: changes in giant cell development and priming of two root anionic peroxidases. Planta 240: 841–854
4. Leonetti P, Zonno MC, Molinari S, Altomare C (2017) Induction of SA-signaling pathway and ethylene biosynthesis in Trichoderma

harzianum-treated tomato plants after infection of the root-knot nematode *Meloidogyne incognita*. Plant Cell Rep 36:621–631

5. Molinari S, Akbarimotlagh M, Leonetti P (2022) Arbuscular Mycorrhizal fungi induces resistance against root-knot nematodes in a dose-dependent manner. Int J Mol 2:8920
6. Leonetti P, Molinari S (2020) Epigenetic and metabolic changes in root-knot nematode-plant interactions. Int J Mol Sci 21:7759
7. Molinari S, Lamberti F, Crozzoli R, Sharma SB, Sanchez M, Portales L (2005) Isozyme patterns of exotic *Meloidogyne* spp. populations. Nematol Medit 33:61–65
8. Uehara T, Sugiyama S, Matsura H, Arie T, Masuta C (2010) Resistant and susceptible responses in tomato to cyst nematode are differentially regulated by salicylic acid. Plant Cell Physiol 51:1524–1536
9. Song Y, Chen D, Lu K, Sun Z, Zeng R (2015) Enhanced tomato disease resistance primed by arbuscular mycorrhizal fungus. Front Plant Sci 6:786
10. Livak KJ, Schmittgen TD (2001) Analysis of relative gene expression data using real time quantitative PCR and the 2–ΔΔCT method. Methods 25:402–408
11. Hoagland DR, Arnon DI (1950) The water-culture method for growing plants without soil. California Agricultural Experiment Station Circular 347:1–32

Chapter 12

Conceptual Framework of Epigenetic Analyses of Plant Responses to Sedentary Endoparasitic Nematodes

Tracy Hawk, Sobhan Bahrami Zadegan, Selin Ozdemir, Peitong Li, Vince Pantalone, Meg Staton, and Tarek Hewezi

Abstract

Epigenetic modifications including miRNA regulation, DNA methylation, and histone modifications play fundamental roles in establishing the interactions between host plants and parasitic nematodes. Over the past decade, an increasing number of studies revealed the key functions of various components of the plant epigenome in the regulation of gene expression and shaping plant responses to nematode infection. In this chapter, we provide a conceptual framework for methods used to investigate epigenetic regulation during plant–nematode interactions. We focus specifically on current and emerging methods used to study miRNA regulation and function. We also highlight various methods and analytical tools used to profile DNA methylation patterns and histone modification marks at the genome level. Our intention is simply to explain the advantages of various methods and how to overcome some limitations. With rapid development of single-cell sequencing technology and genome editing, advanced and new methodologies are expected to emerge in the near future to further improve our understanding of epigenetic regulation and function during plant–nematode interactions.

Key words miRNA, DNA methylation, Histone modifications, Plant-parasitic nematodes

1 Introduction

Plant-parasitic cyst and root-knot nematodes are sedentary endoparasites of the root system in a wide spectrum of agronomically important plant species. These obligate pathogens are involved in complex and extended interactions with host plants where the nematode induces the formation of a specialized feeding structure required for nematode development and completion of the life cycle. One main characteristic of successful interaction between cyst and root-knot nematodes with host plants is the formation of functional feeding cells, so called syncytium and giant cells, respectively. Direct experimental evidence indicates that nematode infection and differentiation of root cells into syncytium or giant cells are

Sergio Molinari (ed.), *Plant-Nematode Interactions*, Methods in Molecular Biology, vol. 2756,
https://doi.org/10.1007/978-1-0716-3638-1_12,

mediated through massive changes in host transcriptomes and signaling networks in which epigenetic mechanisms play key roles [1–3].

Epigenetic mechanisms of gene regulation include regulation by non-coding microRNA genes (miRNAs), DNA methylation, and histone modifications. miRNAs are 21-nucleotide (nt) RNA molecules processed from primary transcripts that form characteristic stem loop-like structures. miRNAs post-transcriptionally silence target genes, which contain complementary sequences for miRNA binding and subsequent mRNA degradation or translational repression. DNA methylation is the addition of a methyl group (CH_3) to DNA cytosine bases in symmetric (CG and CHG) and asymmetric (CHH) sequence contexts where H represents any nucleotide but G. These three methylation contexts are established by sequence-specific methyl-transferases [4]. Histone modifications are also important epigenetic marks that involve a variety of biochemical modifications to histone proteins such as methylation, acetylation, phosphorylation, and ubiquitylation, which commonly occur at lysine (K) residues, leading to substantial changes in chromosome structure and gene expression [5].

During the last decade, great progress has been made in characterizing various components of epigenetic mechanisms. In this chapter, we discuss current and emerging methods used to investigate the function of miRNA genes, DNA methylation, and histone modifications during plant–nematode interactions.

2 Materials

2.1 miRNA

1. Total and small RNA isolation kits. TRIzol-based RNA extraction methods can be used.
2. Poly(A) polymerase.
3. Oligo(dT) adapter (5′-GCGAGCACAGAATTAATACGACTCACTATAGGTTTTTTTTTTTTVN-3′).
4. 5′ RNA ligase-mediated (RLM)-Rapid Amplification of cDNA Ends (RACE) kit (e.g., use FirstChoice™ RLM-RACE).
5. miRNA specific primers.
6. Primer pair flanking the miRNA binding site in the target genes.
7. Primer pair positioned downstream or upstream of the miRNA binding site.
8. Binary vectors for overexpression and promoter analyses.
9. Binary vector for CRISPR/Cas9-mediated mutagenesis.
10. DNA fragment coding for miRNA primary transcript.

11. mRNA of *Arabidopsis IPS1* (*induced by phosphate starvation1*) gene.
12. Reagents for reverse transcription-quantitative PCR (RT-qPCR).
13. Reagents for gene cloning.

2.2 DNA Methylation

For constructing methylC-seq libraries, the following materials are needed:

1. About 1 μg high-quality DNA from experimental and control samples.
2. Unmethylated λ DNA (Promega).
3. Sodium bisulfite conversion kit (e.g., Zymo EZ DNA methylation kit).
4. TruSeq DNA methylation kit (Illumina).
5. Agilent Bioanalyzer high-sensitivity DNA chips (Agilent Technologies).
6. Paramagnetic DNA purification beads (e.g., AgencourtAMPure XP beads from Beckman Coulter).

2.3 Histone Modifications

For constructing ChIP-seq libraries, the following materials are needed:

1. Formaldehyde solution (1%).
2. 5 M NaCl solution.
3. Phenol.
4. Chloroform.
5. Nuclease-free water.
6. M220 focused ultrasonicator (Covaris) or similar sonicator.
7. Nuclei extraction kit.
8. ChIP grade antibodies.
9. Protein A agarose beads.
10. Illumina NEBNext Ultra II DNA Library Prep Kit.

3 Methods

3.1 miRNA

Determining the spatial and temporal expression profiles of miRNAs in root tissues following infection by parasitic nematodes is the first step toward understanding the functional roles of miRNAs in plant–nematode interactions. miRNAs function through post-transcriptional regulation of the target genes, which contain complementary binding motifs. Various methods to examine miRNA-mediated post-transcriptional silencing of target genes are highlighted in this section. Methods to manipulate miRNA expression levels are also discussed.

1. RT-qPCR is frequently used to quantify the abundances of miRNA primary transcripts and mature molecules (*see* **Note 1**). Total RNA enriched with small RNAs is polyadenylated using poly(A) polymerase and then reverse-transcribed using oligo(dT) adapter (5′-GCGAGCACAGAATTAATACGAC T CACTATAGGTTTTTTTTTTTTVN -3′) as described by Hewezi et al. [6]. PCR quantification of mature miRNA expression is conducted using a non-specific reverse primer complementary to poly(T) adapter and a forward primer specific to the mature miRNA sequences. Two adenine residues should be added to the 3′ end of the miRNA-specific primer in order to increase the binding ability of the primer to poly (T) region of the mature miRNA cDNA and to simultaneously reduce its ability to bind to the precursor region [7, 8]. Quantification of the primary transcripts of miRNAs is conducted using the universal reverse primer and a forward primer specific to miRNA primary transcripts or precursors. The specificity of the amplification can be confirmed by running the amplified products on a 2% agarose gel.
2. Generation of transgenic plants expressing a reporter gene under the control of miRNA promoter is an efficient approach to determine miRNA activity in the nematode feeding sites [8–15]. The advantage of determining miRNA promoter activity over PCR quantification is that the activity of miRNA can be visualized specifically in the initial feeding cells as early as 1 day after infection as well as in the developing feeding sites and surrounding root tissues [9]. It is important to take into consideration that many miRNAs are encoded by a gene family containing several members that produce the same mature miRNA molecules. Thus, assaying miRNA promoter activity upon nematode infection would facilitate differentiating between various miRNA family members as previously described [8, 12].
3. The 5′ RLM-RACE assay is frequently used to confirm transcript cleavage of miRNA targeted genes. Total RNA isolated from infected root tissues is ligated to 5′-RACE RNA adaptor without phosphatase treatment and then used for cDNA synthesis using reverse primers specific to the target genes. Amplified PCR products using 5′-RACE forward primer and gene-specific reverse primers are gel purified and sequenced. Sequence analysis of cloned PCR products will indicate the cleavage sites of the targeted transcripts.
4. Degradome sequencing is another method developed for genome-wide identification and confirmation of the miRNA–

mRNA target pair [16]. Degradome sequencing is a transcriptome-based approach in which uncapped 5′ ends of cleaved mRNA transcripts containing poly-A tails are sequenced and then mapped to mRNAs to identify miRNA target genes at the genome level. Analytical tools such as sPARTA [17], PAREsnip2 [18], and PAREfirst [19] can be used for high-throughput identification of miRNA cleavage sites in target genes. All of these software programs are publicly available and can be run in the command line with a Linux-based operating system.

5. RT-qPCR can be used to provide measurable confirmatory data for miRNA-mediated cleavage of target genes during nematode infection. This is done by quantifying the abundance of both cleaved and total transcripts of target genes in RNA samples isolated from nematode-infected root tissues. The level of uncleaved transcript can be measured using a primer pair flanking the miRNA binding sequence, whereas a primer pair positioned downstream or upstream of the miRNA cleavage site would indicate the levels of total transcripts (uncleaved and cleaved). A significant decrease in the levels of uncleaved transcript compared to the levels of total transcripts is an indication of post-transcriptional degradation of miRNA-targeted transcripts [9, 11, 20].

6. A fascinating approach to visualize post-transcriptional regulation of miRNA targets in nematode feeding sites is to generate promoter–reporter constructs that reflect transcriptional and post-transactional regulation of miRNA target genes. Expressing wild-type variant of a reporter gene (GUS or GFP) under the control of target gene promoter would reflect the transcriptional activity of the target gene independent of miRNA activity. In contrast, expressing a modified variant of the reporter gene containing the miRNA binding site at the N-terminus or C-terminus using target gene promoter would reflect miRNA-mediated post-transcriptional regulation of the target gene. This approach was successfully used to show post-transcriptional regulation of *Nitrogen Limitation Adaptation* (*NLA*) gene by miR827 in the syncytium inducted by *Heterodera schachtii* in *Arabidopsis* [9].

7. Constitutive over-expression of miRNA precursor sequences using the 35S promoter frequently induces significant down-regulation of the target genes, allowing the impacts on nematode infection to be assayed [8, 9, 11–15]. In some cases, over-expression of miRNA genes using the strong 35S promoter induces severe developmental irregularities particularly in the root system, preventing such assessment. In this case, over-expression of miRNA precursors can be achieved using a

much less active promoter such as the nopaline synthase (NOS) promoter.

8. Suppressing miRNA activity is another approach to further validate the involvement of miRNAs in establishing plant–nematode interactions. Suppressing miRNA activity can be achieved via different methods. For example, miRNA target mimicry approach can be employed to efficiently inhibit the expression of mature miRNAs [21]. In this method, the 21-nt miR399 binding motif in *Arabidopsis IPS1* (*induced by phosphate starvation1*) gene is substituted by a sequence that mimics the miRNA target site. The mimic sequence generally contains a mismatched loop of 3–4 nucleotides at the miRNA cleavage site that sequesters the activity of miRNA and prevents miRNA-guided cleavage of the target transcripts [21]. This approach has been proven very efficient in inhibiting the activity of miR827, miR858, miR172, and miR778 and hence determining the impacts on plant response to nematode infection [9, 11, 15, 20].
9. miRNA activity can be effectively suppressed by over-expressing non-cleavable coding sequences of miRNA target genes. The non-degradable variants of the target genes are constructed by generating synonymous mutations in the miRNA target sites without changing the encoded protein sequences [8, 9, 11, 13, 20]. For example, the miR858 binding site located in *MYB83* transcript (CCA GGT AGA ACA GAC AAC GAG) was modified to (CCC GGA AGG ACT GAT AAT GAG) without changing the amino acid sequences (PGRTDNE) [11]. The impact of impeding miRNA activity would be more understandable by comparing the transgenic lines over-expressing miRNA-resistant non-cleavable mRNA variants of the target genes with those over-expressing wild-type cleavable variants as described by Hewezi et al. [9].
10. T-DNA insertional mutants can also be used to investigate the function of miRNAs in plant–nematode interactions [14]. However, the frequency of T-DNA insertions in the short miRNA precursor coding regions is low. Additionally, T-DNA insertions in the miRNA promoter region are frequently less effective. With the recent advances of the genome editing approach, Clustered Regularly Interspaced Short Palindromic Repeats (CRISPR)/Cas9-mediated gene editing can be applied to generate mutations directed to the miRNA precursor regions that interrupt precursor formation and hence miRNA maturation and function.

3.2 DNA Methylation

Recently, bisulfite sequencing became the gold standard for genome-wide DNA methylation analysis at a single-nucleotide level. Treatment of DNA by sodium bisulfite specifically converts unmethylated cytosines to uracils via deamination, whereas methylated cytosines remain unaffected. After PCR amplification of treated DNA samples, these uracils are further converted to thymines. By sequencing bisulfite-treated DNA and further mapping sequenced reads to the reference genome, a definite nucleotide distinction between unmethylated and methylated cytosines can be established at the genome level. DNA extracted from non-infected and nematode-infected roots as well as nematode feeding sites are treated by sodium bisulfite and used for library preparation, sequencing, and data analysis as detailed below (*see* **Note 2**).

1. Extract high-quality DNA from nematode-infected samples and the corresponding control root tissues using DNeasy Plant Mini Kit (Qiagen) or similar kits.
2. Add 1 ng of unmethylated λ DNA to 200 ng of sample DNA, and treat with sodium bisulfite using Zymo DNA Methylation-Gold kit following manufacturer's protocol.
3. Quantify single-stranded DNA recovered from previous step using a spectrophotometer with the RNA setting.
4. Use the purified DNA to construct MethylC-seq libraries using TruSeq DNA Methylation kit (Illumina) following the manufacturer's protocol.
5. Purify MethylC-seq libraries using AMPure XP beads (Beckman Coulter) or similar kits following the manufacturer's protocol.
6. Run the DNA samples (MethylC-seq libraries) on an Agilent Bioanalyzer high-sensitivity DNA chip to determine quality and concentration of the libraries.
7. Use Illumina HiSeq platform or similar systems to sequence MethylC-seq libraries with 150 bp paired-end reads.
8. Bisulfite treatment can lead to DNA degradation, and hence optimizing this treatment is of utmost importance to ensure complete conversion of non-methylated cytosines without DNA degradation. Therefore, it is important to include an indicator for bisulfite conversion rate. This can be achieved by adding the nonmethylated λ phage genome to the DNA samples before bisulfite treatment. Bisulfite conversion efficiency of at least 99% is required to ensure accurate identification of differentially methylated cytosines.
9. High-quality sequence reads representing about 20× genome coverage of at least three biological samples of experimental

and control treatments are needed. Only high-quality reads are mapped to the reference genome using Bismark [22].

10. Report files of cytosine methylation generated by Bismark can be used to visualize global DNA methylation levels across protein-coding genes, transposable elements (TEs), and miRNA genes using the ViewBS package [23]. Global DNA methylation level is determined by dividing the number of methylated cytosines at each position by total read coverage. Comparing global DNA methylation levels between experimental and control samples will demonstrate the distribution of differential DNA methylation (hyper- and hypomethylation) over the body and flanking regions of protein-coding genes, TEs, and miRNAs [24, 25] (*see* **Note 2**).
11. For identifying differentially methylated cytosines, only cytosines covered by at least ten reads should be considered. Despite various methods have been used to call differentially methylated regions (DMRs) in various methylome studies, we found that dividing plant genomes into 200-bp non-overlapping sliding windows is the most efficient approach to determine differentially hyper- and hypomethylated regions with a minimum of 25% methylation differences using the R package methyl kit with logistic regression [26].
12. Level, sequence context, and location of cytosine methylation are known to impact gene expression in a similar and unique manner [27, 28]. Therefore, it is important to map DMRs to various annotated regions of protein-coding genes, including promoter, exon, intron, and untranslated regions. This can be performed using the Bioconductor package RTRACKLAYERS [29] and GenomicRanges [30] or similar packages.
13. It is equally important to map DMRs to various families of class I and class II TEs. This will enable determining whether certain TE families are preferentially targeted by parasitic nematodes and whether these families are linked to a specific methylation context of hyper- or hypomethylation [24, 28, 31].
14. To examine the impact of DNA methylation on gene expression accurately, it is imperative to analyze gene expression changes, via RNA-seq, using the same tissue samples used for DNA methylation analysis.
15. It is well known that DNA methylation is a repressive epigenetic mark that leads to gene downregulation. However, several recent reports including ours indicate that DNA methylation can also lead to gene activation [27, 28, 32–34]. For example, hypermethylation of *cis* regulatory element of transcriptional repressors may result in gene upregulation. Also, methylation status of TEs has been shown to impact the

transcriptional activity of nearby protein-coding and miRNA genes [25, 28, 35]. Therefore, it may be of interest to determine the distance between DMR-associated TEs and the nearest genes. This can be done using bedtools closest [36]. As a result, the associations between methylation status, levels, and sequence contexts of TEs and expression levels of closest genes can be determined, providing additional dimension to the impact of DNA methylation on gene expression.

16. Until very recently, the regulatory function of DNA methylation in modulating miRNA gene expression and biogenesis was understudied (*see* **Note 1**). Changes in DNA methylation levels upon soybean infection by soybean cyst nematode (SCN, *Heterodera glycines*) have been recently identified at miRNA loci. Hyper- and hypomethylation were detected in the miRNA promoters and to a lesser extent in the primary transcript-coding regions [25]. DNA methylation at the promoter region could interfere with the binding affinity of transcriptional regulators, thereby impacting miRNA gene transcription. Emerging experimental evidence both in animal and plant systems indicates that miRNA biogenesis could be regulated by hyper- and hypomethylation at regions coding for the primary transcripts [25, 35, 37] (*see* **Note 1**). Hypermethylation at the primary transcript regions may decrease the elongation rate of RNA polymerase II, thereby allowing the newly produced miRNA transcripts to be processed into mature 21-nt molecules very effectively. Thus, DNA methylation status at miRNA loci should be comprehensively analyzed by including both promoter and primary transcript regions in order to examine any potential correlation between DNA methylation and the abundance of primary transcripts and mature miRNA molecules.

3.3 Histone Modifications

Histone modifications are common post-translational modifications to histone proteins that impact chromatin structure and gene transcription. Methylation, acetylation, phosphorylation, and ubiquitylation of histone proteins are the most well-studied modifications. Histone modifications also include ADP-ribosylation, SUMOylation, deamination, and proline isomerization; however, these modifications are yet to be thoroughly studied. The most efficient approach to identify various modifications to histone proteins is the chromatin immunoprecipitation (ChIP)-seq method. In this method, histone-DNA complexes are immuno-precipitated using an antibody that recognizes specific histone modification. Purified DNA is then used for ChIP-seq library construction and sequencing to identify genomic regions associated with such modification as detailed below.

1. Collect root tissues from experimental and control samples, and immediately treat them with 1% formaldehyde solution under vacuum for 25 min for DNA–protein crosslinking.
2. Isolate nuclei using standard protocol.
3. Shear isolated chromatin to an average size of 400 bp using M220 Focused Ultrasonicator (Covaris) using the following settings: 10% duty cycle, 75 W intensity peak incident power, and 200 cycles/burst for 10 min.
4. Immuno-precipitate sonicated DNA using Abcam ChIP grade antibodies (Abcam, Cambridge, UK) that recognize specific histone modification marks.
5. Add the immune complexes to protein A agarose beads and rotate for 12 h at 4 °C.
6. Wash the beads using low- and high-slat washing buffers, and then elute the immune complexes in 500 μL of elution buffer following the manufacturer's instruction.
7. Add 20 μL of 5 M NaCl to the eluted immune complexes, and inoculate at 65 °C for 12 h for reverse crosslinking.
8. Purify DNA use standard phenol–chloroform extraction method, and suspend the purified DNA in nuclease-free water.
9. Use purified ChIP-DNA to construct ChIP-seq libraries using Illumina NEBNext Ultra II DNA Library Prep Kit according to the manufacturer's instructions.
10. Use Illumina HiSeq platform or similar systems to sequence ChIP-seq libraries with 150 bp paired-end reads.
11. ChIP-seq reads are mapped to the reference genome using Bowtie2 (version 2.2.9) [38]. The mapping output files of experimental and control libraries are used to identify statistically significantly enriched peaks of histone-modified marks using the broad peak function of MACS2 tool (version 2.1.1.2) [39]. GoPeaks, a newly developed method for profiling histone modifications, can also be used to identify narrow and broad peaks of aligned reads [40]. Statistically significantly enriched peaks are determined using a q-value <0.05 and a 1.5-fold change cut-off. Final output files include a list of peaks and their lengths, genomic coordinates, fold changes, and p and q-values.
12. Peaks are mapped to various genic features and TEs using the closest function of bedtools2 package (version 2.29.2) [36]. Similar to DNA methylation, histone modifications of TEs can influence the expression of nearby genes.
13. Sequencing depth is vital for distinguishing between immunoprecipitated signals from nonspecific background noise. A

sequencing depth of about 50 million reads per replicate for large plant genomes is adequate.

14. The reproducibility of peak detection across independent biological replicates should be examined. Only peaks identified in at least two independent replicates should be considered as previously described [41].
15. To determine the impact of histone modifications on gene expression precisely, it is highly recommended to analyze gene expression changes, via RNA-seq, using the same tissue samples used for histone modification analysis.
16. In some cases, the impact of histone modifications on gene expression can be insignificant [42]. This can be explained by the fact that specific histone modification marks induce small changes in gene transcription below the level of significance. It is also conceivable that additional histone modification or epigenetic marks such as DNA methylation are simultaneously needed to induce detectable changes in gene expression (*see* **Note 3**). Additionally, peak characteristics, including size, location, and signal intensity can affect gene transcription to various degrees. These factors should be considered when such associations between histone modifications and gene expression are investigated. Using the recently developed single-cell sequencing technology is expected to precisely determine the impact of histone modifications on gene expression (*see* **Note 4**).
17. Because histones undergo numerous changes mediated via the activity of many enzymes, identifying the most influential histone-modifying marks during plant–nematode interactions is very challenging. This can be facilitated by identifying the most significantly differentially expressed genes coding for key enzymes involved in these modifications. For example, the identification of a significant number of histone-modifying genes with altered expression in the galls induced by *Meloidogyne graminicola* in rice [43] was the foundation to study three histone modification marks (H3K9me2, H3K27me3, and H3K9ac,) during the early stage of gall formation [42].
18. Nematode bioassays of mutant lines of histone-modifying genes would also point to specific histone modifying marks to be profiled at the genome level upon nematode infection. For instance, reduced nematode susceptibility of mutant lines of the *SU(var)3-9 homolog 5* (*SUVH5*) and *SUVH6* methyltransferases underscored the importance of H3K9me2 establishment upon *Arabidopsis* infection with *H. schachtii* [20].
19. Effectors from various phytopathogens including plant-parasitic nematodes have been shown to target key histone-modifying enzymes [3, 44]. Histone-modifying enzymes

targeted by nematode effectors could also provide strong indication to such modifications. In this context, it has been reported that cyst nematode effector 32E03 physically interacts with *Arabidopsis* histone deacetylase 1 and alters histone H3 acetylation level at the rDNA chromatin regions [44].

4 Notes

1. Recent experimental data indicate that miRNA-encoding genes are highly vulnerable to DNA methylation changes, which impact both miRNA transcription and biogenesis [25, 35]. Thus, DNA methylation status of miRNA genes should be carefully examined in epigenetic studies in order to better understand the functional consequences associated with transcriptome programming during plant–nematode interactions.
2. Genetic mechanisms generally function in coordination with epigenetic pathways to regulate gene transcriptional activity. One of the major challenges in comparative epigenetic analyses of compatible and incompatible interactions is to define whether epigenetic modifications associated with specific phenotypes are independent of genetic variations. In this context, using resistant and susceptible isogenic lines [24] would enable determining the extent to which epigenetic variations contribute to phenotypic differences between susceptible and resistant interactions without genetic background effects.
3. DNA methylation and histone modifications are highly linked epigenetic marks that together determine chromatin structure and patterns of gene expression, despite the fact that these modifications are established by different enzymatic reactions. DNA methylation can act as a template to recruit various histone modifiers, and correspondingly histone modifications may determine the patterns of DNA methylation [20, 45, 46]. Therefore, it is crucial for histone modification studies to be accompanied with DNA methylation analysis to understand how both epigenetic mechanisms interact to coordinately regulate gene expression in response to nematode infection.
4. Emerging sequencing approaches have proven efficient in profiling DNA methylation and histone modifications at the single-cell level in animal systems [47, 48]. However, single-cell epigenomic methodologies remain largely unexplored in plants. It is likely only a matter of time until these methods are widely adopted. For example, the use of nanopore sequencing for long-read methylation profiling, especially with trained base-calling models specified to a particular plant species [49, 50], could provide information about epigenetic changes

at the cellular level. This could particularly be of interest when studying plant–nematode interactions as parallel transcriptome and epigenome analyses of nematode-induced feeding sites using single-cell approach will reduce cellular heterogeneity, leading to a better understanding of the multifaceted relationship between cellular epigenomes and gene expression.

Acknowledgments

Work in the Hewezi laboratory on epigenetics was supported by funds from the National Science Foundation (Award IOS-1145053), Tennessee Soybean Promotion Board, University of Tennessee Research Foundation, and University of Tennessee Institute of Agriculture.

References

1. Hewezi T (2015) Cellular signaling pathways and posttranslational modifications mediated by nematode effector proteins. Plant Physiol 169:1018–1026
2. Jaubert-Possamai S, Noureddine Y, Favery B (2019) MicroRNAs, new players in the plant–nematode interaction. Front Plant Sci 10:1180
3. Hewezi T (2020) Epigenetic mechanisms in nematode–plant interactions. Ann Rev Phytopathol 58:119–138
4. Law JA, Jacobsen SE (2010) Establishing, maintaining and modifying DNA methylation patterns in plants and animals. Nat Rev Genet 11:204–220
5. Berr A, Shafiq S, Shen W-H (2011) Histone modifications in transcriptional activation during plant development. Biochim Biophysica Acta 1809:567–576
6. Hewezi T, Howe P, Maier TR, Baum TJ (2008) Arabidopsis small RNAs and their targets during cyst nematode parasitism. MPMI 21:1622–1634
7. Gutierrez L, Bussell JD, Pacurar DI, Schwambach J, Pacurar M, Bellini C (2009) Phenotypic plasticity of adventitious rooting in Arabidopsis is controlled by complex regulation of AUXIN RESPONSE FACTOR transcripts and microRNA abundance. The Plant Cell 21:3119–3132
8. Hewezi T, Maier TR, Nettleton D, Baum TJ (2012) The Arabidopsis microRNA396-GRF1/GRF3 regulatory module acts as a developmental regulator in the reprogramming of root cells during cyst nematode infection. Plant Physiol 159:321–335
9. Hewezi T, Piya S, Qi M, Balasubramaniam M, Rice JH, Baum TJ (2016) Arabidopsis miR827 mediates post-transcriptional gene silencing of its ubiquitin E3 ligase target gene in the syncytium of the cyst nematode *Heterodera schachtii* to enhance susceptibility. Plant J 88:179–192
10. Díaz-Manzano FE, Barcala M, Engler G, Fenoll C, de Almeida-Engler J, Escobar C (2016) A reliable protocol for *in situ* microRNAs detection in feeding sites induced by root-knot nematodes. Front Plant Sci 7:966
11. Piya S, Kihm C, Rice JH, Baum TJ, Hewezi T (2017) Cooperative regulatory functions of miR858 and MYB83 during cyst nematode parasitism. Plant Physiol 174:1897–1912
12. Noon JB, Hewezi T, Baum TJ (2019) Homeostasis in the soybean miRNA396–GRF network is essential for productive soybean cyst nematode infections. J Exp Bot 70:1653–1166
13. Noureddine Y, Mejias J, da Rocha M, Thomine S, Quentin M, Abad P et al (2022) Copper microRNAs modulate the formation of giant feeding cells induced by the root knot nematode *Meloidogyne incognita* in *Arabidopsis thaliana*. New Phytol 236:283–295. https://doi.org/10.1111/nph.18362
14. Medina C, da Rocha M, Magliano M, Ratpopoulo A, Revel B, Marteu N et al (2017) Characterization of microRNAs from Arabidopsis galls highlights a role for miR159 in the plant response to the root-knot nematode Meloidogyne incognita. New Phytol 216: 882–896
15. Díaz Manzano FE, Cabrera J, Ripoll JJ, Del Olmo I, Andrés MF, Silva AC et al (2018) A role for the gene regulatory module

microRNA172/TARGET OF EARLY ACTIVATION TAGGED 1/FLOWERING LOCUS T (mi RNA 172/TOE 1/FT) in the feeding sites induced by Meloidogyne javanica in Arabidopsis thaliana. New Phytol 217:813–827

16. Addo-Quaye C, Eshoo TW, Bartel DP, Axtell MJ (2008) Endogenous siRNA and miRNA targets identified by sequencing of the Arabidopsis degradome. Curr Biol 18:758–762
17. Kakrana A, Hammond R, Patel P, Nakano M, Meyers BC (2014) sPARTA: a parallelized pipeline for integrated analysis of plant miRNA and cleaved mRNA data sets, including new miRNA target-identification software. Nucleic Acids Res 42:e139
18. Thody J, Folkes L, Medina-Calzada Z, Xu P, Dalmay T, Moulton V (2018) PAREsnip2: a tool for high-throughput prediction of small RNA targets from degradome sequencing data using configurable targeting rules. Nucleic Acids Res 46:8730–8739
19. Alzahrani S, Applegate C, Swarbreck D, Dalmay T, Folkes L, Moulton V (2021) Degradome assisted plant microRNA prediction under alternative annotation criteria. IEEE/ACM Trans. Comput Biol Bioinform 19: 3374–3383. https://doi.org/10.1109/TCBB.2021.3115023
20. Bennett M, Piya S, Baum TJ, Hewezi T (2022) miR778 mediates gene expression, histone modification, and DNA methylation during cyst nematode parasitism. Plant Physiol. 189: 2432–2453
21. Franco-Zorrilla JM, Valli A, Todesco M, Mateos I, Puga MI, Rubio-Somoza I et al (2007) Target mimicry provides a new mechanism for regulation of microRNA activity. Nat Genet 39:1033–1037
22. Krueger F, Andrews SR (2011) Bismark: a flexible aligner and methylation caller for Bisulfite-Seq applications. Bioinformatics 27:1571–1572
23. Huang X, Zhang S, Li K, Thimmapuram J, Xie S (2018) ViewBS: a powerful toolkit for visualization of high-throughput bisulfite sequencing data. Bioinformatics 34:708–709
24. Rambani A, Pantalone V, Yang S, Rice JH, Song Q, Mazarei M et al (2020) Identification of introduced and stably inherited DNA methylation variants in soybean associated with soybean cyst nematode parasitism. New Phytol 227:168–184
25. Rambani A, Hu Y, Piya S, Long M, Rice JH, Pantalone V et al (2020) Identification of differentially methylated miRNA genes during compatible and incompatible interactions between soybean and soybean cyst nematode. MPMI 33:1340–1352
26. Akalin A, Kormaksson M, Li S, Garrett-Bakelman FE, Figueroa ME, Melnick A et al (2012) methylKit: a comprehensive R package for the analysis of genome-wide DNA methylation profiles. Genome Biol 13:R87
27. Rambani A, Rice JH, Liu J, Lane T, Ranjan P, Mazarei M et al (2015) The methylome of soybean roots during the compatible interaction with the soybean cyst nematode. Plant Physiol. 168:1364–1377
28. Hewezi T, Lane T, Piya S, Rambani A, Rice JH, Staton M (2017) Cyst nematode parasitism induces dynamic changes in the root epigenome. Plant Physiol 174:405–420
29. Lawrence M, Gentleman R, Carey V (2009) rtracklayer: an R package for interfacing with genome browsers. Bioinformatics 25:1841–1842
30. Lawrence M, Huber W, Pages H, Aboyoun P, Carlson M, Gentleman R et al (2013) Software for computing and annotating genomic ranges. PLoSComput Biol 9:e1003118
31. Silva AC, Ruiz-Ferrer V, Müller SY, Pellegrin C, Abril-Urías P, Martínez-Gómez Á et al (2022) The DNA methylation landscape of the root-knot nematodes induced pseudo-organ, the gall, in Arabidopsis, is dynamic, contrasting over time, and critically important for successful parasitism. New Phytol 236:1888–1907. https://doi.org/10.1111/nph.18395
32. Gent JI, Ellis NA, Guo L, Harkess AE, Yao Y, Zhang X et al (2013) CHH islands: *de novo* DNA methylation in near-gene chromatin regulation in maize. Genome Res 23:628–6237
33. Song Q-X, Lu X, Li Q-T, Chen H, Hu X-Y, Ma B et al (2013) Genome-wide analysis of DNA methylation in soybean. Mol Plant 6:1961–1974
34. Niyikiza D, Piya S, Routray P, Miao L, Kim WS, Burch-Smith T et al (2020) Interactions of gene expression, alternative splicing, and DNA methylation in determining nodule identity. Plant J 103:1744–1766
35. Piya S, Lopes-Caitar VS, Kim WS, Pantalone V, Krishnan HB, Hewezi T (2021) Hypermethylation of miRNA genes during nodule development. Front Mol Biosci 8:616623
36. Quinlan AR, Hall IM (2010) BEDTools: a flexible suite of utilities for comparing genomic features. Bioinformatics 26:841–842
37. Glaich O, Parikh S, Bell RE, Mekahel K, Donyo M, Leader Y et al (2019) DNA methylation directs microRNA biogenesis in mammalian cells. Nat Commun 10:5657

38. Langmead B, Salzberg SL (2012) Fast gapped-read alignment with Bowtie 2. Nat Methods 9: 357–359
39. Zhang Y, Liu T, Meyer CA, Eeckhoute J, Johnson DS, Bernstein BE et al (2008) Model-based analysis of ChIP-Seq (MACS). Genome Biol 9:R137
40. Yashar WM, Kong G, VanCampen J, Curtiss BM, Coleman DJ, Carbone L et al (2022) GoPeaks: histone modification peak calling for CUT&Tag. Genome Biol 23:144
41. Piya S, Liu J, Burch-Smith T, Baum TJ, Hewezi T (2020) A role for Arabidopsis growth-regulating factors 1 and 3 in growth–stress antagonism. J Exp Bot 71:1402–1417
42. Atighi MR, Verstraeten B, De Meyer T, Kyndt T (2021) Genome wide shifts in histone modifications at early stage of rice infection with *Meloidogyne graminicola*. Mol Plant Pathol 22:440–455
43. Kyndt T, Denil S, Haegeman A, Trooskens G, Bauters L, Van Criekinge W et al (2012) Transcriptional reprogramming by root knot and migratory nematode infection in rice. New Phytol 196:887–900
44. Vijayapalani P, Hewezi T, Pontvianne F, Baum TJ (2018) An effector from the cyst nematode *Heterodera schachtii* derepresses host rRNA genes by altering histone acetylation. Plant Cell 30:2795–2812
45. Vaillant I, Paszkowski J (2007) Role of histone and DNA methylation in gene regulation. Curr Opin Plant Biol 10:528–533
46. Du J, Johnson LM, Jacobsen SE, Patel DJ (2015) DNA methylation pathways and their crosstalk with histone methylation. Nat Rev Mol Cell Biol 16:519–532
47. Clark SJ, Lee HJ, Smallwood SA, Kelsey G, Reik W (2016) Single-cell epigenomics: powerful new methods for understanding gene regulation and cell identity. Genome Biol 17:72
48. Casado-Pelaez M, Bueno-Costa A, Esteller M (2022) Single cell cancer epigenetics Trends. Cancer. https://doi.org/10.1016/j.trecan.2022.06.005
49. Wick RR, Judd LM, Holt KE (2019) Performance of neural network base calling tools for Oxford Nanopore sequencing. Genome Biol 20:129
50. Ferguson S, McLay T, Andrew R, Bruhl J, Schwessinger B, Borevitz J et al (2022) Plant species-specific basecaller improves actual accuracy of nanopore sequencing. Research Square, Preprint. https://doi.org/10.21203/rs.3.rs-1919465/v1

Chapter 13

Heat Shock Proteins as Potential Indicators of Induced Stress in Nematodes

Monica Celi, Mirella Vazzana, and Barbara Manachini

Abstract

Heat shock proteins (HSPs) in all animals studied to date constitute potential indicators of stress, under various environmental conditions. The goal of this chapter is to show, for the first time, the suitability of the approach based on evaluation of the expression levels of heat shock proteins, as good indicators of stress induced in nematodes by the cultivation of resistant plant varieties or by other potential stressors.

Key words Heat shock protein, Plant feeder nematodes, Resistant plants, Bioindicators, Stress

1 Introduction

All animal cells express a suite of proteins known as heat shock proteins (Hsps), which act as molecular chaperones, participating in the folding/refolding of newly synthesized or denatured proteins. HSPs are constitutively expressed and can be induced after exposure to specific stressors [1, 2]. Named according to their molecular weight (e.g., HSP100, HSP90, HSP70, HSP60, HSP40, and the smaller HSPs such as HSP 27 and HSP10), these proteins are induced by multiple environmental stresses including exposure to trace metals, organic pollutants, pesticides, changes in temperature or osmolarity, hypoxia/anoxia, and exposure to ultraviolet radiation in many animals, both vertebrates and invertebrates such as insects and nematodes [3, 4].

The value of HSPs as suitable biomarkers in ecotoxicology has already been described in regard to different animal species and/or stressors [5–8]. In particular, with regard to nematodes, it was shown that after specific stress condition, HSP 90 levels increased relative to the respective control [9–11]. HSP expression can be also influenced by ecological interactions (e.g., competition and habitat suitability), resource utilization, and many other parameters [12, 13].

Sergio Molinari (ed.), *Plant-Nematode Interactions*, Methods in Molecular Biology, vol. 2756,
https://doi.org/10.1007/978-1-0716-3638-1_13,

The use of HSPS to study the impact of a stressor such as resistant varieties or different practices of crop cultivations was based on observations showing that the sensitivity of HSP60 to induction in certain organisms (e.g., in *Plectus acuminatus*) is significantly greater than the use of other comparable parameters for quantifying adverse effects on biomass or reproduction [13–15]. Indeed, differences in the synthesis of HSP60 would not only indicate a differential response to stress but also a way to cope with pathogen attacks or host resistance [15]. HSP70 has already been identified and studied in the bacterial feeder nematode, *Caenorhabditis elegans* [16], and it has also been characterized in other nematode species such as *Brugia malayi* [17] and *Parastrongyloides trichosuri* [18] and in free living nematodes [19]. Hsp90s are essential for viability under all conditions in eukaryotes; they are expressed in response to adverse environmental or chemical stresses and patho-physiological situations and play an important role in cell survival [20–23]. HSPs play also a great role in the success of plant-parasitic nematode, like *Globodera* spp., under climate change [24]. Here we used the expression of stress proteins to detect the potential influence of host plants or agricultural management on the nematode specie and on nematode community. We believe that this protocol could give a quick and standardized method to exemplify the answer of nematode to a potential stressor.

2 Materials

2.1 Nematode Extraction and Collection

According to the nematode species or community that need to be studied, the extraction of the alive specimens would be performed by a variety of techniques to keep them alive and in unmodified conditions. These, well-known, methods based on the motility of the nematodes can be:

- Baermann funnel
- Mistifier, incubation
- Ostenbrink dish, flotation, and sieving
- Flegg modified Cobb technique
- Oostenbrink elutriator

2.2 Nematode Lysate

- TNES buffer (50 mM Tris–HCl; 100 mM NaCl; 2 mM EDTA; 1% NP40; 2 μg/μL antipain, leupeptin, and bestatin; 1 μg/μL aprotinin and pepstatin; 1 mM benzamidine; and 0.1 mM AEBSF, pH 7.4).

2.3 Detection and Quantification of HSP Proteins

- TBS-T (20 mM Trizma base, pH 7.5, 300 mM NaCl, 0.1% (v/v), Tween 20 with 0.02% sodium azide).
- Protease inhibitors.

- SDS-polyacrylamide mini-gels.
- Nitrocellulose membranes.
- BCPI/NBT (5-bromo-4-chloro-3-indolyl-phosphate/nitro blue tetrazolium).
- Antibodies*:* Preliminary studies were carried out to ascertain the specificity of mammalian antibodies (primary and secondary, Sigma Aldrich, St. Louis, MO USA) with the specific HSP proteins with molecular weights of 27 KDa, 60 KDa, 70 KDa, and 90 KDa, respectively. Antibody selection is based on information collected from a review of available literature and preliminary assessments [25, 26]. As the antibody specificity is confirmed, the analyses is conducted with all samples using the following antibodies:
- Anti-HSP90 monoclonal antibody produced in mice (H1775 Sigma Aldrich) 1:1000.
- Anti-HSP70 monoclonal antibody produced in mice (H-5147, Sigma Aldrich) 1:2500.
- Anti-HSP60 monoclonal antibody produced in mice (H3524, Sigma Aldrich) 1:500.
- Anti-HSP27 polyclonal antibody produced in rabbits (SAB4501457, Sigma Aldrich) 1:1000.
- BSA (bovine serum albumin).

3 Methods

3.1 Nematode Collection and Extraction

After sampling of soil or plant root, extract the nematodes. The extraction of nematodes should be done according to the nematode species or community to be studies but considering that the extraction must keep the specimens alive. A variety of techniques based on motility of nematodes are available. The nematodes extracted from soil or roots should be concentrated alive (*see* **Note 1**), and the upper aqueous part is discarded (*see* **Note 2**). Check on the supernatant is performed to control if nematodes are lost. This procedure is repeated different times, to finish the volume of water obtained from the extraction method. Up to 250 nematodes should be collected and put in 0.5 mL Eppendorf tube.

3.2 Preparation of Nematode Lysates

- Extracts of nematode are prepared in ice-cold TNES buffer added with protease inhibitors (Complete Mini Protease Inhibitor Cocktail, Sigma).
- Homogenize in liquid nitrogen using a mortar and a pestle. Fine powder is then re-suspended in 1.0 mL TNES.
- Sonicate for a total of 15 min. Thereafter, centrifuge at 15,000× *g* for 30 min at 4 °C.

- Collect the supernatant and dialyse against 50 mM Tris–HCl (pH 7.5) overnight.
- Protein content is estimated using a Qubit® 2.0 Fluorometer (Invitrogen by Life Technologies, Singapore).

3.3 Detection and Quantification of HSP Proteins

Equal amounts of cellular lysate samples (15 μg) after incubation at 100 °C for 5 min (reducing conditions), per treatment/replicate analyzed on 10% acrylamide gels for SDS-PAGE.

- Transfer SDS-polyacrylamide mini-gels to nitrocellulose membranes (*see* **Note 3**) using a semi-dry transfer apparatus (Trans-Blot SD, BioRad, Hercules, CA), and add 5% bovine serum albumin in TBS-T for 60 min at room temperature (RT) for saturating excess protein-binding sites on membranes.
- Incubate the membranes overnight at 4 °C with gentle stirring, in a solution of various anti-HSP antibodies: anti-HSP90 monoclonal antibody produced in mice (H1775 Sigma Aldrich) 1: 1000 in 3% BSA (bovine serum albumin); mouse monoclonal in TBS-T; anti- HSP70 monoclonal antibody produced in mice (H-5147, Sigma Aldrich) 1:2500 in 3% BSA in TBS-T; anti-HSP60 monoclonal antibody produced in mice (H3524, Sigma Aldrich) 1:500 in 3% BSA in TBS-T; and anti- HSP27 polyclonal antibody produced in rabbits (SAB4501457, Sigma Aldrich) 1: 1000 in 3% BSA in TBS-T as the primary antibody.
- Incubate for 2 h at 20 °C with a diluted solution of the secondary antibody anti-mouse IgG (A5324, Sigma Aldrich, 1:10000 anti-mouse IgG in TBS-T) conjugated with alkaline phosphatase to highlight the presence of HSP70 and HSP90 and with a diluted solution of the secondary antibody anti-rabbit IgG (A3937, Sigma Aldrich, 1:7500 anti-rabbit IgG in TBS-T) conjugated with alkaline phosphatase to mark HSP27.
- Use the colorimetric detection of alkaline phosphatase activity BCPI/NBT (5-bromo-4-chloro-3-indolyl-phosphate/nitro blue tetrazolium).
- Molecular weights and the densitometric analysis of the band intensities are performed using AlphaEaseFC software (*see* **Notes 4** and **5**) after image acquisition using a flatbed scanner (Epson Perfection V700 Photo, Dual Lens System).
- Evaluate the data as integrated optical density value (IDV) (*see* **Note 6**).
- For each biological sample, three technical replicates are suggested.

In Fig. 1 is reported a typical result of integrated optical density value analysis (IDV) after Western blotting. In this case the expression of HSP90, so its amount, is significantly higher ($p < 0.05$) in

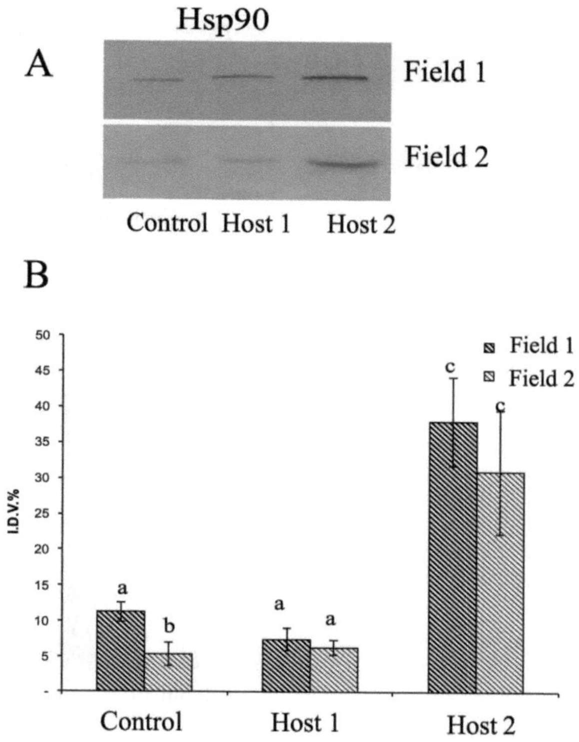

Fig. 1 Effect of different host plants on levels of HSP90 protein in a cyst nematode species (*Globodera pallida* (Stone)) Behrens collected from two different fields. Representative Western blot of HSP90 (**a**) levels across nematode samples taken from fields cultivated with two different host plants (Host 1 and Host 2) and the control as susceptible variety to plant feeder nematodes. Results are given as means ($n = 3$) and standard deviations. Histogram of integrated optical density (IDV) of proteins bands and HSP90 (**b**). Bars with different letters indicate statistical differences as revealed by post hoc Tukey's t-test for multiple comparisons ($p < 0.05$)

3.3.1 Expression of HSPs by a Single Plant-Parasitic Nematode Species in Response to Hosts

Globodera pallida collected from host 2 (*see* Fig. 1a, b) and then from control and host 1 (susceptible hosts) collected in both experimental locations (Field 1 and 2).

3.3.2 Expression of HSPs by Nematode Community in Response to Different Host Plants and to Nematocides

Integrated optical density value analysis (IDV) after Western blotting is represented in Fig. 1.

The production of HSP90 is greater in the cultivation of host 2 in samples from both locations (*see* Fig. 2a). The densitometric analysis of HSP90 protein levels (*see* Fig. 2b) revealed significantly higher amounts produced by nematodes collected from soils cultivated with host 2 ($p < 0.05$) compared to plots grown with the other host. Also, nematodes subjected to the nematocidal treatment showed a higher production of HSP90 compared those from the control and host 1. The amount of HSP90 was apparently

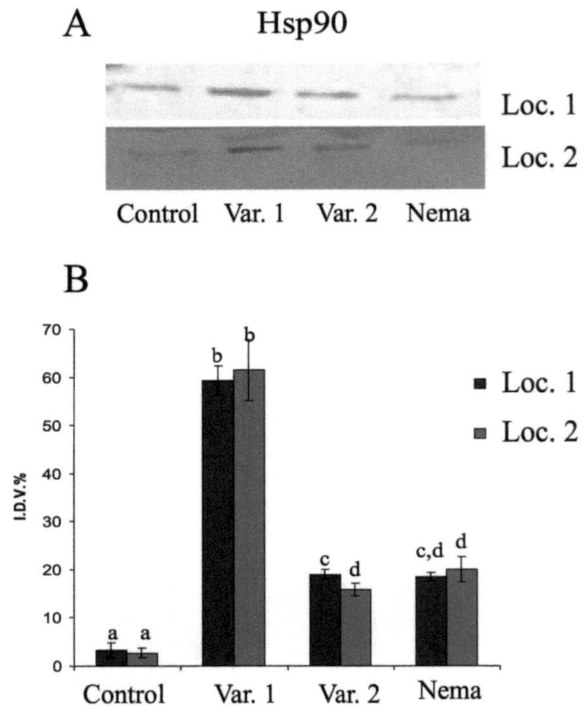

Fig. 2 Impact of different plant varieties on protein expression levels of HSP90 in soil nematode populations collected from experimental field plots in two different locations (loc. 1 and loc. 2). Representative Western blot of HSP27 (A) and of HSP90 (C) in soil taken from Slovakia and from Spain. Histogram of integrated optical density value (IDV) of the HSP27 bands (B) and of the HSP90 bands (D). Results are given as mean of three replicates, and error bars indicate standard deviations. GM, genetically modified maize; ISO, near-isogenic cultivar; I, inter-field zones; Nema, nematocide applied in laboratory on the nematodes collected from the plot ISO. Different letters (a-b) are assigned to statistically different results (ANOVA with post-hoc Tukey's tests, $p < 0.05$)

neither influenced by the locality for the same host ($p < 0.05$). Thus the HSP90 protein, it is significantly more abundant ($p < 0.055$) in nematode community collected from resistant variety plots (*see* Fig. 2a, b) compared to the susceptible control host. Varity 1 modulates more the expression of Hsp90 than variety 2 and nematocide. The effect of nematocide is comparable to the one induced by variety 1.

4 Notes

1. Wait some 5 min to be sure that nematodes fall at the bottom of the collecting glass beaker.
2. Having nematode at the bottom of the beaker helps to concentrate them for counting, and collect them in the Eppendorf.

3. Nitrocellulose is the globally preferred membrane substrate in diagnostic assays where antigen–antibody binding occurs. When choosing a nitrocellulose membrane, quality assurance and membrane consistency are critical to give you confidence.
4. Densitometry data generated are commonly used to compare protein abundance between samples. It should be noted that the intensity of a stained band on a gel diminishes with time due to damaging effects of UV and light illumination. Digital imaging should be done immediately on a freshly stained gel.
5. Gel calibration is an important initial step to undertake to provide the highest precision in the quantification process. So at the beginning of the analysis, it is important to determine the linear capacity of the user's gel system (for a specified agarose gel concentration, buffer, gel dimension, gel thickness, staining) with respect to the amount of a specific ladder.
6. The results presented (in Figs. 1 and 2) are the mean optical density of all the measurements.

Acknowledgment

This work was supported by MUR FONDI PNR D.M. 737/2021 (PRJ-0823).

References

1. Morimoto RI (1998) Regulation of the heat shock transcriptional response: cross talk between a family of heat shock factors, molecular chaperones, and negative regulators. Genes Dev 12:3788–3796
2. Samali A, Orrenius S (1998) Heat shock proteins: regulators of stress response and apoptosis. Cell Stress Chaperones 3:228–236
3. Gupta SC, Saxena DK, Chowdhuri DK (2005) Comparative toxic potential of market formulation soft organophosphate pesticides in transgenic *Drosophila melanogaster* (Hsp70-lacZ). Cell Biol Toxicol 21:149–162
4. Shashikumar S, Rajini PS (2010) Cypermethrin elicited responses in heat shock protein and feeding in *Caenorhabditis elegans*. Ecotoxicol Environ Saf 73:1057–1062
5. Köhler HR, Triebskorn R, Stöcker W, Kloetzel PM, Alberti G (1992) The 70 kD heat shock protein (HSP 70) in soil invertebrates: a possible tool for monitoring environmental toxicants. Arch Environ Contam Toxicol 22:334–338
6. Bierkens JGEA (2000) Applications and pitfalls of stress proteins in biomonitoring. Toxicology 53:1–72
7. Skantar M, Carta LK (2004) Molecular characterization and phylogeny evaluation of the Hsp90 gene from selected nematodes. J Nematol 36:466–480
8. Siamba DN, Mulambalah CS, Ngeiywa MM, Wamae LW (2012) Stress response in infective larvae (L3) of the parasitic nematode *Haemonchus contortus* is accompanied by enhanced expression of Heat Shock Proteins (HSP70). Curr Res J Biolo Sci 4:345–349
9. Birnby DA, Link EM, Vowels JJ, Tian H, Colacurcio PL, Thomas JH (2000) A transmembrane guanylylcyclase (DAF-11) and Hsp90 (DAF-21) regulate a common set of chemosensory behaviors in *Caenorhabditis elegans*. Genetics 155:85–104
10. Devaney E, O'Neill K, Harnett W, Whitesell L, Kinnaird JH (2005) Hsp90 is essential in the filarial nematode *Brugia pahangi*. Int J Parasitol 35:627–636
11. Fanelli E, Troccoli A, Tarasco E, De Luca F (2021) Molecular characterization and

functional analysis of the Hb-hsp90-1 gene in relation to temperature changes in *Heterorhabditis bacteriophora*. Front Physiol 12:615653

12. Chen B, Feder ME, Kang L (2018) Evolution of heat-shock protein expression underlying adaptive responses to environmental stress. Mol Ecol 27:3040–3054
13. Kammenga JE, Arts MSJ, Oude-Breuil WJM (1998) HSP60 as a potential biomarker of toxic stress in the nematode *Plectus acuminatus*. Arch Environ Contam Toxicol 34:253–258
14. Kammenga JE, Dallinger R, Donker MH (2000) Biomarkers in terrestrial invertebrates for ecotoxicological soil risk assessment. Rev Environ Contam Toxicol 164:93–147
15. Vabulas RM, Ahmad-Nejad P, Costa C, Miethke T, Kirschning CJ, Haucker H, Wagner H (2001) Endocytosed HSP60s use toll-like receptor 2 (TLR2) and TLR4 to activate the toll/interleukin-1 receptor signaling pathway in innate immune cells. J Biol Chem 276: 31332–31339
16. Heschl MF, Baillie DL (1989) Characterization of the hsp70 multigene family of *Caenorhabditis elegans*. DNA 8:233–243
17. Selkirk ME, Denham DA, Partono F, Maizels RM (1989) Heat shockcognate 70 is a prominent immunogen in *Brugian filariasis*. J Immunol 143:299–308
18. Newton-Howes J, Heath DD, Shoemaker CB, Grant WN (2006) Characterisation and expression of an Hsp70 gene from *Parastrongyloides trichosuri*. Int J Parasitol 36:467–474
19. Arts MJ, Schill RO, Knigge T, Eckwert H, Kammenga JE, Kohler HR (2004) Stress protein (hsp70, hsp60) induced in isopods and nematodes by field exposure to metals in a gradient near Avonmouth, UK. Ecotoxicology 13:739–755
20. Pantzartzi CN, Drosopoulou E, Scouras ZG (2013) Assessment and reconstruction of novel HSP90 genes: duplication, gains and losses in fungal and animal lineages. PLoS One 8:e73217
21. Queitsch C, Sangster TA, Lindquist S (2002) Hsp90 as a capacitor of phenotypic variation. Nature 417:618–624
22. Eckl J, Sima S, Marcus K, Lindemann C, Richter K (2017) Hsp90-downregulation influences the heat-shock response, innate immune response and onset of oocyte development in nematodes. PLoS One 12:e0186386
23. Gillan V, Maitland K, McCormack G, Him NA, Devaney E (2009) Functional genomics of hsp-90 in parasitic and free-living nematodes. Int J Parasitol 39:1071–1081
24. Jones LM, Eves-van den Akker S, van Oosten Hawle P, Atkinson HJ, Urwi PE (2018) Duplication of hsp-110 Is Implicated in Differential Success of *Globodera* species under climate change. Mol Biol Evol 35:2401–2413
25. David CL, Smith HE, Raynes DA, Pulcini EJ, Whitesell L (2003) Expression of a unique drug resistant Hsp90 ortholog by the nematode *Caenorhabditis elegans*. Cell Stress Chaperones 8:93–104
26. Celi M, Vazzana M, Sanfratello MA, Parrinello N (2012) Elevated cortisol modulates Hsp70 and Hsp90 gene expression and protein in sea bass head kidney and isolated leukocytes. Gen Comp Endocrinol 175:424–431

Chapter 14

Reproduction of Entomopathogenic Nematodes for Use in Pest Control

Mahfouz M. M. Abd-Elgawad

Abstract

The growing interest in the use of entomopathogenic nematodes and their symbiotic bacteria as promising biocontrol agents of many arthropod pests and pathogens has created running technologies to expand their use globally. The related laboratory procedures and tests on these nematodes such as their isolation, count, culture, identification, pathogenicity, virulence, and environmental tolerance should form the solid basis for such an expansion with reliable uses. Extensive practical details of such procedures and tests as well as how to identify and overcome the problems associated with these aspects are addressed in this chapter.

Key words Entomopathogenic nematode isolation, Characterization, Culture, Identification, Laboratory techniques

1 Introduction

It is clear that the way out to avoid the current excessive use of toxic chemicals and the consequent serious health and environmental damage they cause is to follow clean and secure agricultural approaches. Being eco-friendly, entomopathogenic nematodes (EPNs) and their symbiotic bacteria and bioactive compounds rank high among such approaches. References are proving that the complex of nematode–bacterium as a dyad [1] or even their symbionts [2, 3] possess virtually all the attributes of an ideal biological control agent. They are extraordinarily lethal with most insects being killed within 36 h. Dozens of important insect pests are susceptible, yet non-target populations are mostly undisturbed [4]. Their various types of insect hosts include butterflies, moths, flies, grasshoppers, beetles, and crickets especially their subterranean life stages [5]. Thus, EPN species of the two main genera, *Heterorhabditis* and *Steinernema*, are currently commercially released, while others are under experimentation with a clear aim at preserving natural resources from pollution and offering clean

Sergio Molinari (ed.), *Plant-Nematode Interactions*, Methods in Molecular Biology, vol. 2756,
https://doi.org/10.1007/978-1-0716-3638-1_14,

and benign agricultural products [6]. The EPN life cycle starts with an infective juvenile (IJ), as the only free-living stage. It is responsible to search for and infect new arthropod host. After entering a host insect, IJs release their mutualistic bacteria, i.e., *Photorhabdus* or *Xenorhabdus* for heterorhabditids and steinernematids, respectively. The bacteria multiply rapidly within the insect body causing septicemia and priming the body contents as a diet for EPN development and reproduction. The nematodes reproduce for several generations inside the insect cadaver maturing via the growth stages of second-stage juveniles (J2) to fourth-stage juveniles (J4) into adults. While IJs of steinernematids mature into females or males, heterorhabditids IJs develop into self-fertilizing hermaphrodites, but later generations generate the two sexes. Therefore, while an individual of heterorhabditid nematodes can successfully infect and reproduce within its susceptible host, the two sexes of steinernematids must infect the host in order to reproduce and complete the life cycle. When food resources in the cadaver are depleted, the adults generate new IJs that are generally adapted to the environment, i.e., outside the cadaver. Mostly after 6–10 days, depending on host/EPN type and environmental conditions, hundreds of thousands of IJs, carrying with them mutualistic bacteria, emerge from the cadaver to seek for new hosts [7].

To support typically used procedures in a basic standard format, this chapter highlights their currently used laboratory methodologies. It provides tooling up of EPNs for optimizing conditions necessary for their biology, ecology, and reproduction in basic laboratory techniques especially for consequent use as biopesticides. These tools and procedures include an update of some approaches that can play an important role in pre-field application technologies and resulting data.

2 Materials

2.1 Laboratory Host Insect Rearing: Galleria mellonella

G. mellonella eggs, wax papers, plastic trays, plastic boxes, artificial diet [200 g wheat bran, 400 g corn flour, 350 ml honey, 200 g milk powder, 100 g yeast powder, 350 ml glycerin], ethanol, cotton (*see* **Note 1**).

2.2 Sampling

Soil samplers (e.g., augers, hand trowel, spade, iron core, bladed shovel, shovel), ice box, polyethylene bags, waterproof marker. A measure to determine *X* and *Y* coordinates if spatial analysis by distance indices (SADIE) is needed for examining EPN spatial distribution (*see* **Note 2**).

2.3 EPN Extraction

1. Direct method: Baermann funnel technique.

Soil sample, glass funnel, iron stand to hold the funnel within a ring, rubber tube, clamp/clip to block the tube, tissue paper, tap water, beaker/Petri plate, support under filter paper/paper towel (e.g., plastic/wire sieve of adequate aperture to permit IJ passage (i.e., 250 μm).

2. Indirect methods: insect-baiting techniques.

Insects, soil samples, Petri dishes, plastic cups, Whatman filter paper, White trap [8], tissue culture flask, BOD incubator, dissecting microscope (*see* **Note 3**).

2.4 Collection and Limited Production of EPN-Infective Juveniles (IJs) from Infected Cadavers

White trap method: Petri dish, plastic container, Whatman filter paper, sterile distilled water, beaker, 0.1% formalin solution, tissue culture flask, BOD incubator.

2.5 Koch's Postulates Method

Get the IJs which you doubt its pathogenic ability or want to confirm it (e.g., IJs emerging from infected cadaver), items of White trap method, light microscope.

2.6 Counting of Nematodes

1. Direct count method (small number of nematodes ≤50): Petri dish, nematode suspension, dissecting microscope, micropipette.
2. Dilution method (large number of nematodes<50): beaker, nematode suspension, micropipette, nematode counting slide (eelworm counter), dissecting microscope.

2.7 Storage and Transportation of Nematodes

IJs, tissue culture flask, BOD incubator, aquarium pump, distilled water, Triton X-100, liquid nitrogen, sponge.

2.8 EPN Identification

2.8.1 Study of EPN Morphology: Light Microscopy Studies—Handling, Mounting, and Identification of Nematodes

Nematode culture; glass pipette/suction device; dissecting needle and microscope; nylon bristle/fine wire/small wire loop; stereoscopic microscope; small plastic Petri dish/watch glass; 26–76 mm glass slide; hot plate; water; Ringer's solution [standard 9 g NaCl, 0.42 g KCl, 0.37 g CaCl2·2H2O, 0.20 g NaHCO3, 1000 mL distilled water]; water bath; fixatives, e.g., TAF fixative [7 mL formalin, 2 mL triethanolamine, 91 mL distilled water]; 5% glycerin; incubator; desiccator; concave glass slide; nematode pick; glass wool (5 mm in length); cover glass (18 mm wide); ethanol; glycerol; forceps/needle; filter paper; nail polish; small brush; phase contrast microscope; camera lucida; recent EPN identification key(s) [e.g., 9, 10].

2.8.2 Hemolymph Hanging Drop Technique

G. mellonella larvae, 70% ethanol, serum-free medium, IJs.

2.8.3 Molecular Characterization

1. Extraction of DNA from IJs: Extraction buffer (1% SDS, 0.05 M EDTA, 400 μL proteinase K), TE buffer (0.001 M EDTA, 0.01 M Tris–HCl pH 7.5), chloroform, phenol, isoamylalcohol, ethidium bromide, agarose, pestle with glass mortar, microcentrifuge, IJs, micropipette, other miscellaneous [11].
2. Quantification of deoxyribonucleic acid (DNA): DNA sample from IJs, UV light, PCR machine, electrophoresis apparatus, agarose, ethidium bromide, Eppendorf tubes, double distilled water, micropipettes, ice pack, forward and reverse primers 18S and 26S.
3. Restriction digestion and electrophoresis of PCR: Nematode DNA, PCR machine, electrophoresis unit, UV light lamp, micropipette, digestion enzymes, TBE buffer, agarose, ethidium bromide.
4. Randomly amplified polymorphic DNA (RAPD) analysis: Nematode DNA, pH meter, spectrophotometer, refrigerator, vortex, gel documentation, electrophoresis unit, PCR, PCR mixture, PCR mixture consists of 1.5 U Taq DNA polymerase, 2 μL dNTPs (2.5 mM each), 2.5 μL *Taq* buffer (10X), 2 Mm MgCl2, 0.5 μM primer, 100 ng genomic DNA, double distilled water to make up the volume 25 μL.

2.9 Phylogenetic Analysis

2.9.1 DNA Cloning and Sequencing

Purified ITS and 16S rDNA PCR product for each EPN species/strain, pGEM-T Easy vector, QIAprep Spin Miniprep Kit (Qiagen) for sequencing, Applied Biosystems 3730 DNA Analyzer, adequate primers, Gel Manager software package from Bio Systematica.

2.9.2 Phylogenetic Analysis

Use proper outgroup taxa (e.g., *Xenorhabdus nematophila* and *Caenorhabditis elegans* to root the trees for *Photorhabdus* and *Heterorhabditis*, respectively), available multiple sequence alignment program (MALIGN) [12], relevant experienced taxonomist.

2.9.3 Phylogeny Construction

Related and recent keys of systematics and biogeography, cladistics and vicariance, relevant experienced taxonomist.

2.9.4 Evaluation of Tree Topology

Related software programs such as molecular evolutionary genetics analysis (MEGA) software version 4.0 and the Fitch package of PHYLIP [13].

IJ suspension, *G. mellonella* larvae, Petri dishes, well tissue culture plates, sterilized sand, Baermann funnels, heat–cool incubator, stereomicroscope.

2.10 Ecological Characterization

2.10.1 Heat Tolerance Test

2.10.2 Soil Moisture Test

Plastic tubes, IJs, *G. mellonella* larvae, cellophane tape, distilled water, soil, IJs, well plates, 0.5% pepsin solution.

2.10.3 Desiccation Tolerance Test

IJs, Petri dish, filter paper (55 mm, Whatman No. 1), vacuum filter, potassium chloride (KCl) or potassium sulfate (K_2SO_4), plastic desiccator, dissecting needle.

2.10.4 Soil pH Test

Soil sample, IJ suspension, hydrochloric acid, sodium hydroxide.

2.10.5 Hypoxia Test

Eppendorf tubes, incubator with shaker, IJs, sterile tap water, Petri dishes.

2.10.6 Penetration Bioassay

G. mellonella larvae, stereoscopic microscope, 0.5% pepsin solution, 24-well plate, IJs, filter paper.

2.10.7 Exposure Time Bioassay

G. mellonella larvae, stereoscopic microscope, 24-well tissue culture, IJs, filter paper, Petri dishes.

2.10.8 One-on-One Bioassay

24-well tissue culture, *G. mellonella* larvae, IJs, filter paper/sand.

2.10.9 Foraging Behavior/Sand Bioassay

G. mellonella larvae, 45 mm height and 40 mm diameter plastic tubes, IJs, moist sterile sand.

2.10.10 Dose–Response Bioassay

G. mellonella larvae, Petri plate/well plate, IJs, filter paper.

2.11 Nematode Production

2.11.1 In Vivo methods

1. Applying nematodes on insects for in vivo culture on White trap or its simplified modifications: IJ suspension, *G. mellonella* larvae, plastic petri dish, Whatman No. 1 filter paper.
2. Automated apparatus: shallow perforated holding stainless steel trays for insects, water supply valve, relative humidity chamber (or conditioning apparatus), water tubes, rigid plastic pipes with atomizer spray nozzles, a timer connected to a water supply, pan for collecting IJ suspension.

2.11.2 In Vitro Methods

1. Liquid medium method, WOUTS agar medium: WOUTS agar [14], 16 g nutrient broth, 12 g agar, 5 g sunflower oil, 1000 mL distilled water; YS medium [15], 5 g yeast extract, 5 g NaCl, 0.5 g $NH_4H_2PO_4$, 0.5 g K_2HPO_4, 0.2 g $MgSO_4{\cdot}7H_2O$, 1000 ml distilled water; egg surface

sterilization solution [0.5 ml of 12% NaOCl solution, 1.5 mL NaOH 4 mol, 10 mL distilled water].

2. Solid medium method: 100 g dog food chunks, 1 g agar, 500 ml distilled water, IJs.

2.12 Virulence Bioassays

Petri dish, IJs, Whatman No. 1 filter paper, mostly larvae of test insects.

2.12.1 Petri Plate Bioassay

2.12.2 Sand Column Bioassay

Petri dish, IJs, sterile sand, test insects (mostly larvae).

2.12.3 Leaf Disk Bioassay

Petri dish, IJs, leaves favored by the plant host, distilled water, test insects (mostly larvae).

2.13 Direct Method of Isolating the Symbionts and Their Phase Variants

A sterilization agent for surface sterilization of EPNs [e.g., sodium hypochlorite or merthiolate], media [e.g., yeast salt broth (YS)/ Luria–Bertani (LB) broth/nutrient broth (NB)], sterile Ringer's solution, nutrient bromothymol blue agar (NBTA) media, Mac-Conkey Agar (MAC) media, IJ suspension, *G. mellonella* or *S. litura* larvae, microcentrifuge, Petri dishes, beaker, pipette, glass tissue homogenizer, stereo zoom microscope or phase contrast microscope, incubator.

2.14 Preparation of Whole Cell Suspension and Cell-Free Supernatant

Stock of *Xenorhabdus/Photorhabdus*, NBTA medium (nutrient bromothymol blue agar), LB (Luria–Bertani) broth [11], phosphate-buffered saline solution.

3 Methods

The EPN suspensions usually in the form of IJs are stored in tissue culture flasks within bio-oxygen demand (BOD) incubators at ecological conditions suitable to each species/strain. They should be in dark or away from ultraviolet radiation. The IJs are generally active at ambient temperature of 25 ± 2 °C and in soil with 10–30% soil moisture [7], unless indicated otherwise for definite species/ strain. Laboratory workers/researchers should avoid using their hands after touching insects to rub their bodies, especially the eyes, to prevent any allergic reactions that sometimes happen. Gloves and proper protective clothing should be worn.

The related methodology is best done at 25 ± 1 °C, but all procedures can be done at room temperature unless otherwise specified.

3.1 Culturing EPN–Host Insect Galleria mellonella

1. Put both glycerol and honey in the tray.
2. Add the other ingredients (milk powder, corn flour, wheat bran, and yeast powder), and then mix well.
3. Cover this tray with muslin cloth for 5–6 days before use.
4. Collect *G. mellonella* eggs from primary sources on tissue paper, and put them on the diet (about 200 g) surface in the container with aeration.
5. Eggs will hatch into first instar larvae that appear as tiny thread-like structures, colorless. After 7–15 days, develop to second instars (usually the higher the temperature up to 35 °C, the quicker the insect develops to the next instar/life stage).
6. Transfer the third instar larvae to a new container with diet to continue developing until the fifth instar larvae.
7. Use the fifth or the fourth instar larvae for the intended bioassays/experiments; otherwise they will develop into the pupa stage and then adult.
8. To continue culturing, put the pupa in other trays with pleated wax pieces/paper for the oviposition of the female moths.
9. They will lay eggs at 3–5 days old and continue laying eggs for several days (females' life span is 7–14 days at 26 °C).
10. Use a razor blade/brush to transfer the eggs from the wax to a container to repeat insect culturing (*see* **Note 1**).

3.2 Sampling

1. Insect pests with suspected symptoms of EPN infection may be collected as samples (*see* **Note 2**). Any soil with an insect pest or suspected of its presence is preferable for sampling.
2. For soil sampling, collect about 0.5–1 kg soil samples (often 10–50 cm soil depth, 3–70 cm diameter) at the sampling site via randomly, stratified random, or systematic sampling method by one of the abovementioned sampler tools.
3. Dig out number of samples adequate to the sampling goal and sampled area.
4. Put each sample in a firm plastic bag.
5. Tag samples with a water-resistant marker to record the related information (e.g., locality/site name, date, habitat, encountered insects, allied vegetation, temperature, precipitation, altitude, etc.).
6. Arthropods and other invertebrates found should be gathered with the sample and classified as they may signify probable EPN hosts.
7. Clean sampling tools in-between samples (e.g., via proper water washing or 70% ethanol).

8. Maintain samples at ice box at ≤27 °C during transportation to the laboratory.
9. Analyze a part of the soil sample to know soil composition and structure (silt, sand, and clay content), moisture, organic matter, type, pH, texture, EC (electrical conductivity), etc.
10. Keep soil samples at adequate temperatures (15–25 °C) to avoid desiccation until EPN extraction.
11. Rational or functional sampling may detect more EPNs with better adaptation to control specific pest(s). It is based on combining four factors: *i*) choosing the appropriate sampling method (random, stratified random, and systematic); *ii*) time (a season with greater abundance of susceptible pests), *iii*) site (weeded area under tree canopy where weeds can offer protection, shelter, and food for insect pests); and *iv*) extraction technique, i.e., multiple *Galleria*-baiting cycles (*see* **Note 3**).

3.3 EPN Extraction

3.3.1 Direct Extraction of EPNs by Baermann Funnel Technique

1. Set a glass funnel with clamped hose (tube) into the ring stand.
2. Put a tissue paper over a wire sieve horizontally in the wide open of the funnel.
3. Put soil sample gently onto the tissue paper.
4. Add water delicately over the surface of the tissue paper, and maintain it just above the soil level by adding water to the sample to substitute evaporated water.
5. Discharge the hose clamp, and gather the water suspension daily (for about 3 days) in a beaker/Petri plate to examine its nematodes under a microscope (*see* **Note 3**).

3.3.2 Indirect Extraction of EPNs by Insect-Baiting Technique

1. Place about 500 cm^3 aliquot from each soil sample in a plastic cup (e.g., 10 cm diameter, 7 cm height) with a lid (it is preferable to maintain adequate soil moisture in the range 20–30%).
2. Add four last instar *G. mellonella* larvae evenly on the surface of each soil sample in each cup, and immediately cover the cups to prevent insect escaping.
3. Incubate the cups at 25 °C and 55% relative humidity in darkness while inverting these cups daily to maximize insect migration within the soil.
4. Remove the dead insects, and put those with symptoms of EPN infection on White trap [8] in the incubator to allow EPN reproduction within the infected insects before emergence of EPN-infective juveniles. Multiple *Galleria*-baiting cycles (*see* Table 1) may be followed for possible increase of extraction efficiency (*see* **Note 3**).
5. After 6 days from insect infection, visually observe the suspension daily to examine the presence of IJs in water.

Table 1
Multiple cycles of *Galleria* baiting that could detect new samples positive for *Heterorhabditis indica*. No new samples positive for the nematodes were detected after cycle 5

Cycle no. Positive samples	Cycle 1	Cycle 2	Cycle 3	Cycle 4	Cycle 5
Total positive samples at the cycle end	20	29	32	36	37
New positive samples in the cycle	20	9	3	4	1

6. On observing IJs in water, use a proper beaker to collect the IJ suspension, and add water to replace the collected suspension. Repeat this step daily (usually for 3–5 days) until few or no IJs are found in the suspension (*see* **Note 3**).
7. Use the IJ suspension for the required tests or store it till needed.

3.4 Collection of EPN-Infective Juveniles (IJs) from Infected Cadavers

3.4.1 White Trap

1. Put the bottom of an inverted Petri dish inside a larger one or plastic jar (*see* Fig. 1).
2. Add about 200 mL of sterile distilled water or 0.1% formalin solution into the plastic jar.
3. Put a circular Whatman filter paper inside the inner Petri dish, and set the periphery of filter paper to be in contact with water/formalin solution in the larger dish or plastic jar.
4. Gently transfer EPN-infected cadavers (about 4–15 per trap according to the dish diameter) on the filter paper inside the inner dish (don't let the cadavers touch, but beside, each other to avoid contamination).
5. Close the larger dish or jar with the lid, and incubate it at 20–27 °C or temperature suitable to a definite species/strain.
6. The nematode IJs will begin to emerge from cadavers usually after 1–3 weeks according to the used nematode species/strain and move into water/formalin solution in the trap.
7. Transfer IJ suspension daily from the jar/dish into a beaker with a proper volume until no emergence of IJs from cadavers is seen.
8. Let the nematode IJs sink and settle in the beaker for 30 min, and discard the supernatant.
9. Rinse the IJs by adding 0.1% formalin solution/sterile distilled water (SDW), and allow the IJs to decant.
10. Steps 8 and 9 may be repeated 2–3 times to get rid of any cadaver remains (clear suspensions).

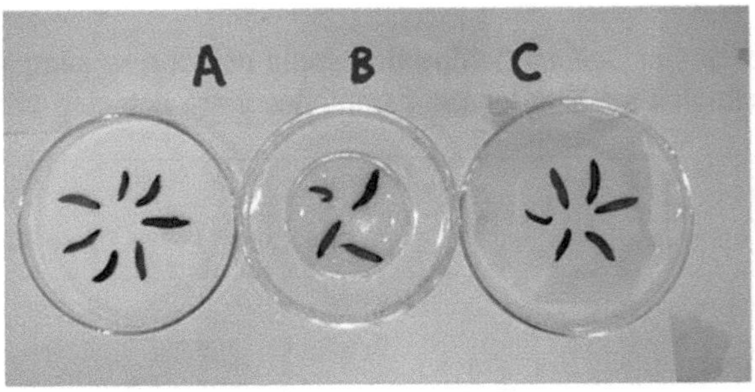

Fig. 1 Symptoms (glowing reddish color) of *Heterorhabditis* indica-infected *G. mellonella* larvae (**a**). A typical White trap has the insects within the inner Petri dish, and the area around it is filled with water below the level of dish (**b**). Emerging nematodes crawl over the inner dish into the water and are collected daily. The inner dish may be upside down to facilitate nematode migration by gravity (**c**)

11. Transfer IJ suspension to clean transparent tissue culture flask, and label it with date of IJ collection, nematode species/strain, and any other necessary data.
12. Store the flask with nematode suspension at temperature proper to the nematode species/strain (*see* **Note 4**).

3.4.2 Modified White Trap Method

Several modifications have been made. Mainly, they intended either to facilitate IJ migration into the water/formalin solution (*see* Fig. 1) or to scale up by simply providing larger/more White traps, usually by placing several dishes of insect cadavers within a large tray (*see* **Note 4**).

3.5 Koch's Postulates: Inoculation of IJs to a Susceptible Host

This test confirms the pathogenicity of IJs via their ability to infect and reproduce on fresh *G. mellonella* larvae.

1. Add the target IJs onto moist filter paper in Petri dishes.
2. Add two to three larvae of live *G. mellonella* (about 20 IJs per insect).
3. Incubate the dishes for 1–3 days, in dark condition.
4. Move the infected (dead) larvae to White trap to confirm IJ emergence from the insects on the trap as aforementioned.
5. Use microscope to check the EPN-IJs which may be stored for further use.

3.6 Counting of Nematodes

1. If the sample has few nematodes ($n \leq 50$), transfer them into a Petri dish via a micropipette, and sum them in the suspension independently under a dissecting microscope (magnification 10–20 × for adults and 40 × for infective juveniles).

2. If the sample has more nematodes ($n \geq 50$), prepare an initial homogeneous diluted suspension in a beaker of about 100 IJs/mL to facilitate counting via adding water.
3. Take 1 mL of nematode initial suspension, and add to L ml of water (L is utilized to have another suspension of about 100 IJs/ml, i.e., 1: L + 1 dilution).
4. Take at least three 1 mL samples by Pasteur pipette from the well-mixed, diluted suspension to count the nematodes per sample using Hawksley slide under light microscope or using eelworm counter under a dissecting microscope; take fresh sample from homogeneous suspension each time.
5. Calculate the nematode concentration in the original suspension via the following formula:

   S = [N X 1 (L + 1)]/ M where S is concentration (nematode/ml) in initial suspension, (L + 1) is dilution, N is average number of nematodes per counted sample, and M is number of ml per counted sample.
6. Any further or needed dilutions can consequently be made via the following formula:

   A = D X C/ B where A is volume in ml of the suspension to be diluted, D is the needed concentration in new dilution, C is volume in ml of the new dilution, and B is number of nematodes per ml in suspension before dilution.
7. Thus, C – A = volume in ml of water to be added to get the needed concentration.

3.7 Storage and Transportation

1. The EPNs are stored (about 2000–3000 IJs/ml) in distilled water with a drop of Triton X-100 to preclude IJs from sticking to the container wall or in 0.1% formalin if contamination is a problem.
2. When IJs are stored without aeration, IJs should be in a shallow water suspension ($\leq$ 1 cm depth); sterilized tissue culture flask is a model for such a type of storage as it has large surface-to-volume ratio.
3. On the contrary, aerated suspension can favor storage of up to 100,000 IJs/ml (e.g., an aquarium pump may be attached to the container to push fine bubbles into the EPN suspension).
4. The IJs are usually stored at 10–15 °C for *Heterorhabditis* spp. and 4–10 °C for *Steinernema* spp. for 2–12 months according to the nematode species/strain.
5. The IJs may be coated on autoclaved polyether-polyurethane sponge as alternative method for storing and transporting large quantities of IJs (*see* **Note 5**).

3.8 EPN Identification

3.8.1 Study of EPN Morphology

1. To handle or pick single nematodes in the suspension, use a dissecting needle with fine wire, toothbrush bristle, or small wire loop attached to its end.
2. Set the targeted nematodes to be near the center of the dish, in shallow water.
3. Lift the nematode to the surface of the water via one of these handling devices (e.g., the needle) while viewing with the stereoscopic microscope, and rapidly draw the nematode out of the water suspension, and put it on the glass slide on the binocular/microscope stage to initially ensure that you have the target EPN.
4. For EPN identification to the species level and permanent storage, EPNs must first be either killed in hot water (at 80 °C for 4–6 s) or killed and fixed in hot fixatives before properly mounted on glass slides (*see* **Note 6**).
5. Set the light microscope to optimize the depth of focus, working distance, and magnification to properly examine EPN morphological characters and morphometrics.
6. Use camera lucida if you need to draw by hand or take morphometrics, not to photograph.
7. Consult recent/conclusive identification keys of EPNs to determine the EPN species. Use also molecular tool for sound identification (*see* **Note 6**).

3.8.2 Cross-hybridization/Breeding Assays via Hemolymph Hanging Drop Technique

1. Bleed, by sterilized needle, surface-sterilized greater wax moth larva (via 70% ethanol for few seconds) from the point between the head and the first prothoracic segment.
2. Transfer a drop of the emerging insect hemolymph (not fat) on a coverslip, and add 10 mL serum-free medium (SF-900 II SFM Gibco) to it to avoid desiccation of the hemolymph.
3. Add 30–50 surface-sterilized IJs to the drop, and swap the coverslip upside down to smoothly fix it on concave slide.
4. Put the slide in a 100 × 15 mm Petri dish having a filter paper disk drenched in water to preclude the sample from drying.
5. Incubate the dish after covering it by the lid inside a plastic carrier at 25–27 °C (to prevent water loss).
6. Check nematode development daily to transfer EPN developed males and females to new hanging drop slides with adults of the opposite sex of the other EPN isolate; the ratio 5:5 males to females is preferred (*see* **Note 7**).

3.8.3 Extraction of DNA from IJs

1. Transfer 1 mL suspension of IJs via micropipette to a mortar, and grind them (using a pestle) by pouring liquid nitrogen to get a fine white powder.

2. Use 1 mL of extraction buffer for more crushing the IJs, and then put them at water bath (65 °C for 1 h).
3. Centrifuge the sample (15,000 *g* for 10–15 min).
4. Get the supernatant, and add to it similar volume of the mixture phenol/chloroform/isoamylalcohol (ratio 25:24:1), and mix the solution well.
5. Centrifuge again (15,000 g for 10 min), and collect the upper layer to add to it similar volume of another mixture (chloroform/isoamylalcohol, 24:1), and mix the solution well before centrifuging at 15,000 g for 10 min.
6. Separate the upper phase, and pour double volume of ethanol and 3 M ammonium precipitate for better DNA precipitation.
7. Freeze the mixture for 2 h, and then centrifuge (15,000 g for 30 min).
8. The DNA precipitated from aqueous phase should be dried before re-suspending in TE buffer.
9. Dry the pellet DNA precipitate within a microcentrifuge at ambient temperature in vacuum.
10. Re-suspend the obtained pellet in TE buffer (50 μL).
11. The extracted DNA could be stored at 4 °C if not needed soon.
12. Run agarose gel electrophoresis using the DNA sample and 1 kb markers of lambda DNA as a standard.
13. A 5 μL of DNA sample is loaded on 0.8% agarose gel (100 mA, for 30 min), while the gel is watched under UV light (*see* **Note 8**).

3.8.4 Quantification of DNA and PCR Amplification of ITS Region

1. Use UV spectrophotometer to quantify the genomic DNA and amplified products (the sample should be pure to accurately measure the amount of UV irradiation absorbed by the bases).
2. Record the diluted DNA sample (1:250) at 260 nm using distilled water as blank.
3. Estimate the DNA amount via the following formula:

   Amount of DNA (μg/μL) = [the optical density at 260 nm (OD_{260}) × 50 × dilution factor]/1000.
4. Amplify, via a DNA thermal cycler, the ITS region in a reaction volume of 100 μL for individual strains comprising 75.6 μL H_2O, 10 μL of 10 X PCR buffer, 10 μL of purified DNA, 1 μL of primer forward (200 μM), 1 μL of primer reverse (200 μM), 0.4 μL of Taq polymerase (5 U/μL), and 2 μL of dNTPs (2 mM).
5. The abovementioned samples are put in the thermal cycler, which is pre-heated to 95 °C and incubated at 94 °C for 3 min. Start the amplification at an annealing temperature of

55 °C for 1 min and then for 1 min 30 s at 72 °C as an extension period and a denaturation for 30 s at 94 °C.

6. Terminate the last round after 35 cycles, of 1 min at 94 °C, 3 min at the annealing temperature of 55 °C, and 5 min at 72 °C for extension, to assure that all the obtained amplified products have full length.
7. Store the amplified samples at 4 °C till needed, or run agarose gel electrophoresis via mixing 10 μL of the amplified product with 2 μL loading buffer (0.1% bromophenol blue, 0.25% xylene cyanol, 40% w/v sucrose in water).
8. The 2.5% agarose gel comprising 1 μg/mL ethidium bromide is run at room temperature and 60 volts for 3 h in TBE running buffer.
9. Apply a 100 bp DNA marker as a check, and view the gel under gel documentation.
10. For sequencing the genomic material, purify first the PCR products with a PCR purification kit, and then sequence the purified DNA in both directions on automated sequencers.
11. Apply DNA sequencer for sequence editing and authenticating base data (*see* **Note 8**).

3.8.5 Restriction Digestion and Electrophoresis of PCR Products

1. A 0.5 μL of chosen restriction endonucleases (e.g., *Sau* 3 AI, *Pst* I; *Pvu* II) is digested with 5 μL of ITS PCR products (all digestions should be done at 37 °C for 3 h).
2. Use electrophoresis to get the bands of the digestion products in 2% agarose gels (w/v), in a 1 × TBE buffer (after 2 h at 100 volts).
3. Photo the bands on the gel using UV light, and store the photos digitally by scanning.
4. The obtained restriction fragment length polymorphism (RFLP) patterns of the *Steinernema* or *Heterorhabditis* isolates are compared to profiles listed in the references.
5. These fragments are separated by electrophoresis on a 2% agarose gel comprising ethidium bromide (0.2%) in 0.5 × TBE buffer (27.5 g boric acid +54 g Tris base +20 mL 0.5 M EDTA (pH 8.0) to 1 l aqua test).
6. Examine the photographs of the fragments under UV light, and compare RFLP patterns with those of other described species (*see* **Note 8**).

3.8.6 Randomly Amplified Polymorphic DNA (RAPD) Analysis

1. Use the thermal cycler for 25 μL volume having PCR mixture and nucleotide sequences of RAPD primers (*see* **Note 9**).
2. Fix the reaction mixture at 95 °C and soon maintain at 94 °C for 5 min.

3. Incubate again at 94 °C for 1 min, 37 °C for 1 min, and 72 °C for 2 min and lastly at 72 °C for 5 min to assure full lengths of all the final amplified products.
4. Mix the samples (20 μL) with 2 μL loading buffer (0.25% xylene cyanol, 0.1% bromophenol blue, 40% w/v sucrose in water), and electrophorese on a 1.6% (w/v) agarose gel having 1 μg/mL ethidium bromide at ambient temperature (60 V for 3 h in TBE running buffer; employ 100 bp DNA marker as a check).
5. Photograph gels with amplification fragments under UV light.
6. Assure reproducibility of DNA profiles via repeating all RAPD reactions at least twice.
7. To analyze RAPD data, apply an average percentage similarity in band sharing in order to detect the magnitude of genetic variation among EPN strains (*see* **Note 9**).

3.9 Phylogenetic Analysis

3.9.1 DNA Cloning and Sequencing

1. The purified ITS and 16S rDNA PCR generated per each strain are cloned into pGEM-T Easy vector.
2. Extract the plasmids having ITS and 16S rDNA PCR products using QIAprep Spin Miniprep Kit (Qiagen) to sequence the plasmids from both ends by the universal primers SP6 and T7 (use an Applied Biosystems 3730 DNA Analyzer).
3. Sequence the plasmids from both ends via the universal primers SP6 and T7 with an Applied Biosystems 3730 DNA Analyzer.
4. Sequence the purified *gyrB* gene PCR products at both ends using the oligonucleotides with the same sequences (*see* **Note 10**).
5. Apply single DNA sequences to get the related map of the ITS1 location for the expected endonuclease restriction site.
6. Use the Gel Manager software (from Bio Systematica) to analyze single gel graphs (images) of restriction enzymes.
7. Produce alignments of ITS or *gyrB* with previously known EPNs or bacteria using the obtained high-quality sequences for comparisons.

3.9.2 Phylogenetic Analysis

1. Omit the unaligned sequence ends and unreliable sequences via manual inspection of every aligned sequence.
2. Set the gaps in the aligned domains as either a fifth base or missing data.
3. Apply *Caenorhabditis elegans* and *Xenorhabdus nematophila* as outgroup taxa in order to set the tree roots for *Heterorhabditis* and *Photorhabdus*.
4. The outgroup sequences are aligned separately with MALIGN to examine if the absence or presence of different output taxa

had an impact on the alignment and the generated phylogenetic trees (*see* **Note 10**).

5. Apply parsimony analyses on phylogenetically revealing characters.
6. Use insertion or deletion event (indel) coding to consider the gaps that may result from single indels, comprising all gaps as independent characters.
7. Each clear single indel is considered an individual character.
8. Attach the matrix of the indels to the dataset.

3.9.3 Phylogeny Construction

1. Treat all variable characters in the outgroup section (taxa) as plesiomorphic.
2. Treat homologous characters in the ingroup as apomorphic.
3. Delete multi-state characters from any computation as they form difficulty.
4. Delete also monomorphic/autapomorphic characters from phylogenetic analysis because they offer no hierarchical data.
5. Set a tree of relationships for each of the remained characters, and establish a conclusive consensus tree from these characters using combinable components (*see* **Note 10**).

3.9.4 Evaluation of Tree Topology

1. Calculate phylogenetic figures using a neighbor joining algorithm via an "evolutionary model" that finds out various evolutionary rates of different sites (*see* **Note 11**).
2. Set bootstrap analyses with 1000 re-samplings to assess phylogenetic tree topologies of all techniques.
3. For the significance of the DNA sequence alignment, apply RNA secondary structure and minimum energy estimates (*see* **Note 11**).
4. Set the parameters as follows: conservativity, 0.8; conserved factor, 2; compensated factor, 4; cluster factor, – 2; and energy threshold, −4.
5. Apply the Fitch package of PHYLIP (*see* **Note 11**).

3.10 Ecological Characterization (See Note 12)

3.10.1 Heat Tolerance Test

1. Use 50 mm diameter Petri dish to put 8 g of wet (15% water w/w) and sterilized sand.
2. Add 100 EPN-IJs in 100 μL suspension to the dish bottom. Set proper number of similar dishes as replicates.
3. Use parafilm to seal the Petri dish.
4. Incubate each set (treatment) in the dark at specific temperature for different hours (e.g., 1–48 h). The range of temperature (as different treatments) to be tested varies according to

basic information about the intended EPN species/strain (e.g., 5–30 °C or 20–42 °C).

5. Then, the dishes are transferred to Baermann funnels for definite time (several hours to 1 day).
6. The extracted living nematodes via Baermann funnels are counted under a stereomicroscope.
7. To assure that IJs are still living, probe with a fine needle, and check the nematode mobility.
8. The live EPNs may be further tested for pathogenicity by confining them with *G. mellonella* larvae on a filter paper within Petri dish to be covered, and then check their pathogenicity via Koch's postulates.

3.10.2 Soil Moisture Tests

1. Prepare sets of same soil textures but with various water/moisture content (e.g., 4, 6, 7, 8, 10, 12, 13, 14, 16, 19, 18, 20, 21, 22, 24, and 26% w/w).
2. Put specific number of IJs (e.g., 50, 100, or 150) in 25 μL distilled water on the lowest part of each well with proper replicates (5–10), filled with pre-wetted soil to set the required soil moisture.
3. Add *G. mellonella* larva to the top of soil in each well. Untreated wells (without IJs) are used as the control to record their insect mortality, and apply the corrected mortality (*see* **Note 12**).
4. Incubate at a fixed temperature (usually between 22 and 28 °C) for 48–96 h after the plate is covered and sealed by a cellophane tape and turned upside down.
5. The plates should be turned upside down every 6–8 h during the incubation, to watch the upward movement of the insect and to set a regular IJs distribution.
6. Examine the insects recovered from the individual wells after the fixed exposure period (e.g., 48, 72, or 96 h) to report the number of dead and live insects (or take periodic samples at each of these times to examine insect mortality and Koch's postulates).
7. Otherwise, the number of EPN-IJs/insect host may be assessed under a dissecting microscope via washing larva in distilled water before dissecting in a solution of 0.5% pepsin and incubating at 37 °C for 2 h to break down the insect tissues to facilitate IJ assessing.

3.10.3 Desiccation Tolerance Test

1. Within a Petri dish, put 2000 IJs suspended in 100 μL on filter paper disk.
2. Use vacuum filter to get rid of excess moisture.

3. Transfer the disk on a 50 mm diameter Petri dish to standard desiccators set at different relative humidity (RH) levels (e.g., 70, 80, 85, 90, 95, 98% RH) via utilizing saturate salt solutions [K_2SO_4 or KCl]).
4. Incubate for specific range of different hours (e.g., 24, 36, 48, 60 or 72 h) at 25 °C.
5. Rehydrate the IJs after each of these periods via immersion in Petri dishes containing 10 mL distilled water for 24 h.
6. Examine IJ viability by observing motility; negative response to probe with a fine needle is deemed dead.
7. Examine samples from each plate to count the number of dead/live IJs under a dissecting microscope, and estimate survival percentage.

3.10.4 Soil pH Test

1. Use tissue culture flasks containing 500 or 1000 IJs in distilled water with soil incubated at 25 ± 2 °C.
2. Set the chosen pH range (e.g., 3.5–9.5) for the test in the flasks using NaOH for alkalinity and HCl for acidity.
3. List the dead/living IJs at daily or weekly intervals for the tested period (e.g., 2–4 weeks) to estimate the survival percentage.

3.10.5 Hypoxia Test

1. Put 10,000 IJs suspended in 1.5 mL distilled water within 100 μL Eppendorf tubes.
2. Put these tubes in a shaker within incubator (150 rpm at 25 °C) for a definite time (e.g., 24, 36, 48, 60, 72, 84, or 96 h).
3. Transfer the IJs into 50 mm diameter Petri dishes with 8.5 mL distilled water.
4. Count the IJs number (living and dead) after incubating at 25 ° C for 24 h to estimate the survival percentage.

3.10.6 Penetration Bioassay

1. Put a 100 μL sterile water suspension containing 200 IJs in each replicate (well) of plates.
2. Put a single *G. mellonella* larva in each well, and use a check (well plate with sterile water and the insect).
3. Score the daily mortality of insects (e.g., 1–5 days).
4. Rinse the dead insects in tap water to remove IJs found on their body, and save them individually on moist filter paper within 5 cm diameter Petri dishes.
5. Incubate at 25 ± 2 °C for additional 2 days before dissecting the larvae under a dissecting microscope.
6. Compute the number of males/females in each insect (use 0.5% pepsin as aforementioned).

7. Express the rate of nematode penetration as the percentage of the IJs that entered the host insect daily.

3.10.7 Exposure Time Bioassay

1. To estimate the time needed by the IJs to cause the host mortality, put 400 IJs suspended in 100 μL water, and add a single *G. mellonella* larva per well.
2. Take out insects randomly after definite times (e.g., 1, 3, 6, and 8 h).
3. Wash the larvae, and put them singly on moist filter paper in 5 cm diameter Petri dishes to report mortality at intervals (e.g., 6, 12, 24, 36, and 48 h) post-exposure to the IJs.
4. Compute ET_{50} value (IJ exposure time in minutes/hours needed to achieve 50% larval mortality).

3.10.8 One-on-One Bioassay

1. Use a proper well plate (e.g., 24-well tissue culture plate with lid) to cut filter paper into circles to fit into the diameter of the well, or put sand (1 g) in each well.
2. Transfer a single IJ (suspended in SDW) via a 25 μL micropipette to each well (replicate).
3. Add a single *G. mellonella* larva into each well, and cover with the lid, and incubate.
4. Report mortality at definite intervals (usually 1, 2, and 3 days) to estimate the mortality percentage.

3.10.9 Foraging Behavior/Sand Bioassay

1. Put 100 IJs suspended in 100 μL sterile water into single tubes, and then gently add moist sterile sand to fill them.
2. Add a single *G. mellonella* larva to the top of sand in each tube, and close it by the lid (it is advisable to place the insect at different dimensions—different treatments—from the bottom of the tube where the nematodes are).
3. Put the tubes upside down to allow the IJs to move toward the host insect.
4. Take the larvae out of the tube after 24 h and put them into Petri dishes.
5. Incubate for about 2–3 days before reporting the foraging ability (percentage of penetrating IJs, developed males/females, and insect mortality).

3.10.10 Dose–Response Bioassay

1. To assess the IJ dose or concentration needed to induce host lethality, use different doses of IJ suspensions (e.g., 5, 10, 20, 40, 60, 80, 100, or 300 IJs/one insect of intended species/stage).
2. Wet a filter paper by one of the IJ suspensions (a replicate) at the bottom of 40 mm Petri plate.

3. Put a single insect on each filter paper, and cover with lid followed by incubation at 25 ± 2 °C.
4. Set an untreated control/check (insects without nematodes).
5. Report the lethal effect of IJ dose via counting dead insects (e.g., at daily intervals for 3 days).

3.11 Nematode Production

3.11.1 In Vivo Methods: Applying Nematodes on Insects for In Vivo Culture on White Trap or Its Simplified Modifications

1. Set the IJ suspension at 200 IJs/mL.
2. Distribute regularly 1 mL of this suspension on a 10 cm diameter filter paper within the top (cover dish) of a 100 × 15 mm Petri dish.
3. Place ten larvae (mostly *G. mellonella*) to the dish so that an average of 20 IJs are allocated to each insect (many IJs per larva generate few progenies due to competition and/or contamination).
4. Close this top (cover dish) with the Petri bottom inverted, label (e.g., nematode species/strain, isolate code/name, date of infection trap), and incubate the dish at 25 ± 2 °C in the dark.
5. Transfer the cadavers with symptoms of EPN infection (discard the cadavers if they smell putrid) to White trap after about 3–5 days (depending on the insect and nematode species/strain) to gather the emerging IJs as aforementioned.

3.11.2 Automated Apparatus (LOTEK System)

1. Use a proper number of shallow perforated (e.g., 30 × 26 × 4 cm) trays (perforated 20 gauge aluminum [30% open area] for the walls) with perforation size of 1.6 mm diameter to retain insects but allow IJs to pass for inoculation, incubation, and then harvest (*see* Fig. 2). Such tray dimensions can have about 500 insect hosts (mostly *G. mellonella*).
2. Each tray is dipped into a concentrated IJ suspension (about 1–2 × 10^4 IJs $ml^{-}1$ for 2 s and then stacked in a high humidity environment (conditioning apparatus) for 3 days to allow optimum IJ development (*see* **Note 13**).
3. The trays are transferred to the harvester, which conceptually resembles the mist chambers used to extract nematodes from soil. The tray is suspended under two rigid plastic pipes, each equipped with an atomizer nozzle (*see* Fig. 2).
4. A timer is connected to a water supply to provide periodic misting cycles for the tray (i.e., a continuous 3 min misting cycle at 6 h intervals) so that IJs emerging from cadaver are exposed to free water; the nematodes are rinsed with the runoff water into an angled drip tray. Each drip tray serves ten or more holding trays arrayed overhead and simply directs nematodes by gravity flow into a collecting pipe and then into a central storage tank.

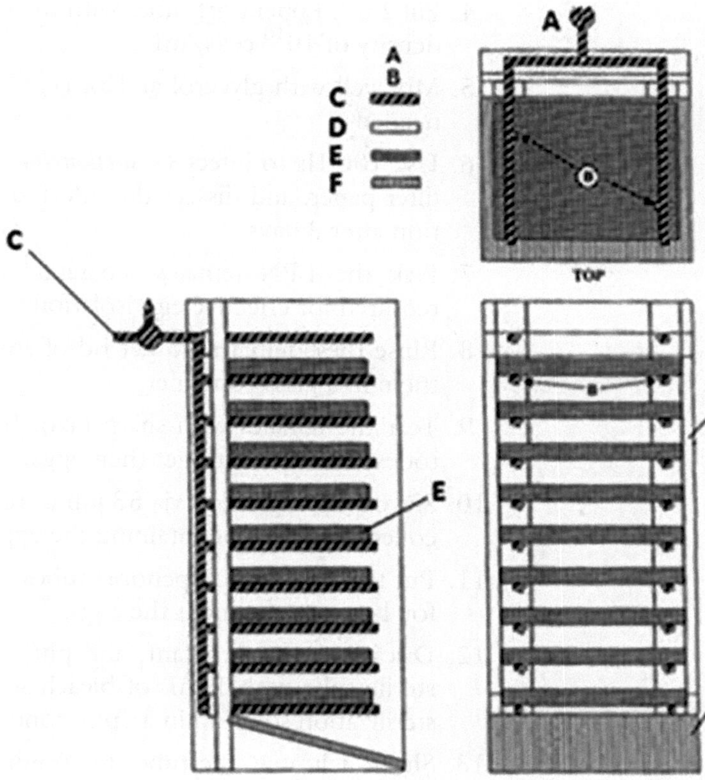

Fig. 2 An automated system for in vivo mass production of entomopathogenic nematodes and their symbiotic bacteria [38]. (**a**) Water supply valve, (**b**) spray nozzles, (**c**) water tubes and tray holders, (**d**) device support structure, (**e**) insect trays, (**f**) nematode suspension collection trays. (Courtesy of Randy Gaugler)

5. Harvest is completed in 2–3 days from IJ emergence time. Cadaver remains (mostly cuticle) are discarded, and the holding tray replaced for in situ cleaning via detergents and disinfectants delivered via the misting nozzles.

3.11.3 In Vitro Methods: Liquid Medium Method Pre-culturing the Symbiotic Bacteria

1. Use 24-h-infected *G. mellonella* larvae to extract the symbiotic bacteria by removing a larval leg and streaking the released hemolymph on nutrient agar (NA) plates (e.g., 1.2% agar and 0.8% Difco Bacto nutrient broth) (*see* **Note 14**).
2. After incubating plates at 25 °C for 3 days, examine the bacterial characteristics of the developed colonies, and repeatedly subculture the required one until a pure culture is obtained.
3. Culture the bacteria (symbionts) in a 100 mL flask with 30 mL yeast salt broth (YS) to get a homogenous inoculum of the symbionts for the nematode culture. Hold culture in shaker at 25 °C (180 rpm).

4. Fill 2 mL Eppendorf tubes with the bacterial suspension at cell density of 10^{10} cells/mL.
5. Mix well with glycerol at 15% (v/v) to store at −80 °C until needed.
6. Use 100 IJs to infect *G. mellonella*-last instar larva in a moist filter paper, and dissect the infected cadaver in Ringer's solution after 3 days.
7. Pick the EPN females separately (≥ 100 big females are required for effective egg isolation).
8. Rinse these females, to get rid of any cadaver tissues, and put them in a glass container.
9. Tear the females with sharp razor blades to shred the nematodes, and vortex to get their eggs.
10. Sieve this suspension via 53 μm to retain the adult tissues, and collect the filtrate containing the eggs.
11. Put the filtrate in Eppendorf tubes, and centrifuge at 450 × g for 1 min to sediment the eggs.
12. Discard the supernatant, and put the sediment with eggs in sterile tube with 1 mL of bleach solution or another surface sterilization solution in aseptic condition.
13. Shake delicately the tubes for 4 min, and then centrifuge for 2 min at 2000 rpm (total 6 min surface sterilization time).
14. Discard the supernatant, and add sterile yeast salt (YS) broth medium to fill the tubes.
15. Centrifuge at 2000 rpm for 2 min and repeat centrifugation once.
16. Put the eggs in WOUTS agar plates and incubate for 24 h at 28 °C (*see* **Note 14**).
17. Inoculate the pre-cultured bacteria in the plates that contain the eggs.
18. In the meantime, keep the bacteria freshly propagated in YS broth medium at 25 °C.
19. Inoculate the YS culture of the bacteria into the IJs developed at the WOUTS agar plates, and incubate at 25 °C until harvest (13–17 days post-inoculation).
20. Obtain only mobile IJs via suspending the nematodes in 30 μm sieve in water, and gather IJs that could pass through the sieve after 3–5 h in sterile Ringer's solution.
21. Keep IJs at 4 °C for use.

3.11.4 Solid Medium Method

1. Grind in a blender chunks of the dog food.
2. Blend 100 g of the grinded food with 500 mL of dissolved 1% agar for 3 min to obtain a homogenous mixture.

3. Autoclave (sterilize) for 15 min at 120 °C.
4. Cool down the mixture after autoclaving at 60 °C.
5. Add 25 mL of the medium solution into sterile plastic Petri dishes, and after cooling inoculate with 100–200 IJs on the medium surface.
6. Seal the dishes and incubate at 28 °C for 22–33 days.
7. Open the dish cover, and put its bottom containing the IJ-infested medium into a large pan or dish (e.g., 150 × 25 mm), with sterile distilled water for the IJ migration.
8. Reap the IJs in the pan (dish) every other day for about 10–20 days, and wash thrice in distilled water.
9. Count the collected IJs and store them for use.

3.12 Virulence Bioassays

3.12.1 Petri Plate Bioassay

1. Put a wetted Whatman No. 1 filter paper lined within sterile 9 cm diameter Petri dish (*see* **Note 15**).
2. Set the IJs of the intended species at the proposed concentrations (e.g., 5, 15, 25, and 30 IJs/larva).
3. Put 1 mL of IJ suspension of each concentration on the above-mentioned filter paper.
4. Use healthy and equally sized larvae of test insect species. Otherwise, any specific insect stage can be tested for this purpose.
5. Add a fixed number of the test insect species (targeted stage) to each dish.
6. Use proper number (3–10) of replicates per each treatment (concentration and/or insect stage).
7. Add untreated check (distilled water with test insect but no EPN-IJ).
8. Add natural food to the insects, if necessary, and cover the dishes.
9. Incubate the inoculated plates at 25 °C.
10. Check the insect lethality at a fixed time daily after inoculation.
11. Use Abott's formula (*see* **Note 12**) to correct the mortality data.
12. Use analysis of variance and mean separation statistical technique to find the significant differences among the corrected percent mortality data.

3.12.2 Sand Column Bioassay

1. Replace the wetted Whatman No. 1 filter paper in the above Petri plate bioassay by 1 kg sterilized moist soil to be put in proper sterilized glass/plastic jar for the test.

2. Repeat the same steps in the above Petri plate bioassay with the necessary modifications (i.e., cover jars with muslin cloth, and rub out the soil gently to search for and record the dead/live insects after 1, 2, 3, and 4 days).
3. Apply also the corrected percent mortality data via completely randomized design.

3.12.3 Leaf Disk Bioassay

1. Rinse the plant leaves, favorable for feed by the test insect host, with SDW.
2. Cut these leaves to form 9 cm diameter disks.
3. Add each disk of the leaves into the Petri dishes to offer food for the test insects.
4. Add 2 mL (200 IJ/ml) of IJ suspension on the leaf disk; use also untreated check (without IJs).
5. Add proper number (to comply with insect size and disk area) of healthy and target stage of the intended insect species (e.g., 8–12 insects) on the leaf disk.
6. Put the dishes at room temperature, and also check daily for about 1–4 days (depending on the percent mortality) as in the previous Petri plate bioassay.
7. Compute LC_{50} and LT_{50} values by Probit analysis.
8. Definite IJ concentrations are used in the study relying on experience with species/strain virulence.

3.13 Isolation of the Symbionts (Bacterial Partners) and Their Phase Variants

1. Surface sterilize small IJ suspension (fresh/alive 40 IJs/mL) via immersing IJs in a solution of 0.1% merthiolate (w/v) with streptomycin (5000 units/mL) for 2–3 h (make sure that the solution covers all surfaces of the nematodes).
2. Triple rinse IJs in sterile distilled water (SDW) via aseptically pipetting the IJs from one container to the surface of the next container, and allow them to settle (*see* **Note 16**).
3. Examine under a microscope to assure that the (sterilized) cuticle surface is free of bacteria.
4. Suspend these IJs in YS, and then homogenate the suspension in a sterile tissue homogenizer, and aseptically transfer to test tubes.
5. Let the bacteria multiply in YS, preferably on a shaker for 1–2 days.
6. Streak a loopful of the bacterial suspension to the indicator NBTA media plates.
7. Keep the plates at 28 °C for 1–2 days in dark condition.
8. Repeat streaking of the intended individual colonies onto fresh solid media to purify them (*see* **Note 16**).

9. Use either MacConkey (MAC) agar or NBTA media to ratify phase variation.
10. Phase I cells have brick/dark red colonies on NBTA, but red colonies on MAC agar. Phase II cells have yellow colonies on NBTA and transparent colonies on MAC agar (*see* **Note 16**).

3.14 Preparation of Whole Cell Suspension and Cell-Free Supernatant

1. Streak a stock of the intended symbiotic bacterial species on NBTA media plates.
2. Keep the plates at dark (28 °C for 4 days).
3. Relocate the characterized (intended) individual colony from plates to grow in 25 mL Luria–Bertani broth.
4. Put this broth on shaker (180 rpm for 24 h at ambient temperature).
5. Set the concentration of this whole suspension to 10^7–10^8 CFU/ml via phosphate buffer saline solution.
6. Take 10 mL of this cell suspension to be centrifuged (12,000–20,000 rpm for 5–15 min).
7. Collect the supernatant and filter it via a 0.22 μm filter.
8. The cell-free supernatant is the flow-through.

4 Notes

1. Host insects should be available as basic materials for EPN culturing/bioassays; *G. mellonella* is the most commonly used host insect due to its high susceptibility to most EPN species/strains, ease of rearing, wide accessibility, and large yields of EPN-IJs. Yet, others as *Tenebrio molitor* and *Spodoptera litura* may be utilized with different diets [11], whenever appropriate or to fit the goal of experimentation. During culturing process, hands should be disinfected by alcohol to avert contamination. For the same reason, the outer area of the trays should be wiped by ethanol; e.g., *Galleria* individuals are highly susceptible to fungal and bacterial infection. Get rid of any dead larvae using sterilized forceps, and avoid overcrowd of larvae which leads to insect death and contamination. Covering the trays with muslin cloth/mesh screen is recommended to enable air exchange and preclude overcrowding.
2. Sampling includes three main methods. Every member of the EPN population has an equal chance of being detected in random sampling method, but taking samples in a zigzag pattern across the field is recommended as it increases this chance. Stratified random sampling method divides the sampled area into strata where each stratum has specific trait. Systematic sampling method is chosen at regular intervals as

with equally spaced trees. It may be adopted in order to facilitate applying *X* and *Y* coordinates for spatial analysis by distance indices (SADIE). Although determining the coordinates exactly is mostly costly, SADIE is, unlike other indices, a realistic measure in case the goal is to study nematode spatial distribution [16]. Number of samples should be adequate to the sampling goal and sampled area [17]. Sampling may use fine (sample points 5 cm apart) [10] or large-scale/governmental [18] scale examination of EPN distributions. Sampling accuracy/precision should be considered [17]. Global Positioning System (GPS) might be used only for regional surveys as it has location accuracy from 100 to 10 m for common equipment. Also, using sentinel insect larva, where cages are buried at about 25–30 cm deep in soil and recovered after almost 7 days to check the larvae for EPN infection, can detect EPN population density in situ (*see* Fig. 3). To accurately characterize nematode abundance and their community structure, using morphological or molecular methods, a sample size of 200 g soil is suggested [19]. Also, a rational sampling scheme has recently been proposed to detect more EPNs with an accurate distribution patterns.

3. While Baermann funnel technique relies on nematode mobility, insect-baiting techniques depend on nematode infectivity to the bait (insect). Another direct extraction method of EPNs is the flotation technique [11] which may extract dead and live nematodes together. The baiting (trap insects) method is mostly used due to its merits; i.e., direct extraction methods have three flaws (inefficient EPN extraction, possible handling damage of EPNs, and time-consuming process to sort through all the material extracted) [20]. Identification of positive samples (EPN-infected cadavers) is based on the characteristics of the dead larvae. *Steinernema*-infected *G. mellonella* larvae are limp and show yellowish/whitish color, but *Heterorhabditis*-infected (limp) larvae show glowing reddish color. On the contrary, insects dying of other causes smell putrid and are usually blackish [20]. Koch's postulates should always confirm pathogenicity of the extracted nematodes. Other modifications of EPN extraction are in progress. For example, multiple *Galleria*-baiting cycles could detect more EPNs than single cycle [16]; each plastic cup is examined twice weekly for the first 4 weeks and once weekly thereafter in consecutive cycles. Thus, a cycle consists of about 3–4 days in the first 4 weeks or 7 days thereafter. A total of 11 cycles in 7 weeks are conducted in this multiple extraction technique. Also, the quantitative polymerase chain reaction (qPCR) assay, though not common, was found to be more efficient than the *Galleria* baiting method for detecting the EPN species composition in population

Fig. 3 Sentinel insect larva in cages buried in soil at a depth appropriate to the presence of the insect host in the soil. Cages are taken out of soil after about 1 week to check the insects for EPN infection [21]

mixtures [21]. A main limitation of qPCR assay, however, is its failure to detect EPN species matching the primer/probes that have not been developed or used yet. Therefore, metagenomic methods are emerging as alternative tools to detect any EPN species regardless of its presence in databases (e.g., Genbank). These metagenomics tools can detect, identify, and, to some extent, quantify EPN across various habitats [22].

4. The White trap is intended, well suited, for laboratory bench production of inoculum for experiments. Its scale-up for EPN production is generally difficult. Because nematode lateral migration from the host cadaver to the water reservoir is required, increasing dish diameter beyond 150 mm quickly becomes counterproductive. Early emerging IJs that migrate into the water in the first days may be collected separately and considered as the virulent juveniles. For *Steinernema glaseri*, "pre-IJs" usually emerge from the cadaver and cannot successfully complete development to IJs if these juveniles migrate to an entirely liquid medium. Therefore, the watch glass and filter paper are replaced by filling Petri dish by a thin layer (2 mm) of plaster of Paris, allowing the plaster to dry and then re-wetting the plaster just prior to use. The plaster should be wet, but there should be no standing water. Monitor the moisture level throughout the harvesting period, and add a few drops of water as necessary. *Steinernema glaseri*-infected insects are put on the wet plaster so that pre-IJs finish development on the plaster

substrate before migrating within 2–3 days into the water or formalin in the larger Petri dish or jar to be harvested [20].

5. Both EPN storage and formulation are intended to achieve maximum shelf-life period and maximum maintenance of their infectivity against the target pests. Small EPN samples can be stored within tubes in liquid nitrogen [23]. A good storage (formulation) method for large quantities of IJs is to coat moist autoclaved polyether polyethane foam (sponge) with 10 times its weight of a well-rinsed semisolid paste of the target nematode IJs before putting the foam in a sterilized container and forcing air via bacterial filters (0.45 microns). The aseptic conditions should be secured to avoid contamination of such a common IJ formulation method. The nematodes can be stored for 2–12 months depending on the used species/strain. At 4–15 °C *Heterorhabditis* spp. can live for 3–4 months, and *Steinernema* can live for 6–9 months [11, 20]. They may be safely transported on the sponge. It is less expensive for larger quantities of nematodes. When needed for use, IJs are extracted by squeezing the foam in water within a container until all the tan "nematode" coloration on the foam is gone. Other methods for EPN formulations are published [24] with details of definite ones as calcium alginate beads [25, 26].

6. To avert post-fixation deterioration or denaturation, EPNs are preferably relocated to glycerin using a dehydration technique [11]. This will render the EPN specimens more transparent and optimize their study with high-powered light microscopes while no stains are used. Clearly, outer structure of EPN body and its morphogenesis can be better described via using scanning electron microscopy (SEM), while transmission electron microscopy (TEM) can substantially improve studies of ultrastructural characteristics of EPN–host interactions, e.g., based on EPN responses to abiotic and biotic factors. So, SEM and TEM can provide valuable information adjunct to taxonomic/morphological information. Techniques for preparing nematodes in SEM and TEM studies [27, 28] are important tools too. Reliable EPN preparation for SEM comprises several steps: killing, fixing, and drying as the most impactful. While quick fixation reflects the need to rapidly arrest cellular activity and consequently avoids degradation of EPN tissues, sequential fixation reflects the impact of the concentration of the fixative on tonicity of the buffer [11]. Combining molecular characterization with morphological study and morphometrics is becoming a must for sound identification and classification of EPNs via several impactful tools [10].

7. Checks (controls) comprise crosses of the similar isolates. The test of mating should last for 10 days with proper repetitions to authenticate the results. If the mated pair gave offspring, then

the pair is of the same species (positive result). Absence of progeny/offspring indicates that the mated pair is different species (negative result). If hybridization assays of the target species with morphologically similar/closely associated species show negative results, then the target (identified) species/isolate is considered as a new species. Although the hemolymph hanging drop technique is the common method, other methods achieving the same goal such as lipid agar plate method [11] may be used.

8. Gel agarose casting and running the related electrophoresis are very useful in molecular characterization techniques as they can separate large macromolecules (e.g., RNA, DNA, and large proteins). For example, the restriction fragment length polymorphism (RFLP) technique of polymerase chain reaction (PCR), which can amplify product from specific regions of the genome, is used for the documentation of single nematodes [11]. Main regions of taxonomic value include the internal transcribed spacer (ITS) of the ribosomal DNA (rDNA) repeat unit and the region of the mitochondrial genome that divides the cytochrome oxidase subunit II (COII) and the 16S genes of EPNs [29].

9. RAPD is utilized for phylogenetic studies of diverse organisms and for resolving taxonomic problems based on the genetic variability and using nucleotide sequences of RAPD primers found in tables [11]. It reveals polymorphisms with a single primer of arbitrary nucleotide sequence. Its merit is that no previous information of the genome is required for reliable application. RAPD technique is a comparatively direct method to use, and the number of loci that can be tested is unlimited. RAPD data analysis [11] is usually used to define the magnitude of genetic variation among EPN strains.

10. MALIGN program [12] is routinely used for a phylogenetic analysis to construct the related tree as an important component of the comparative approaches [11]. The DNA sequences of the ITS1 region of the rRNA tandem repeat could supply the heritable features needed for a precise phylogenetic analysis [29, 30]. The oligonucleotides with the same sequences are utilized to sequence the purified *gyrB* gene PCR products from both ends [31]. Recently, whole genome for phylogenetic reconstruction is used. It revealed that *Xenorhabdus lircayensis* exhibited a unique clade in the genus. The 16S rRNA gene analysis is used to establish phylogenetic tree for *X. lircayensis* within the genus *Xenorhabdus*, but a phylogenetic reconstruction is also set to comprise *X. lircayensis* based on core genome sequences [32].

11. An evolutionary model like MEGA4 [33] may be used. Carry out minimum energy calculations via GeneBee-NET program [34, 35]. The Fitch package of PHYLIP [13] does not presume a molecular clock to analyze the generated genetic distance table. Apply same program to choose the best tree from 100 runs.

12. As EPNs act differently under a range of ecological conditions due to their various biological attributes, these ecological tests are essential to characterize each EPN species/strain. Such characterization is necessary to predict their potential biocontrol under certain conditions. Sterile sandy soil is a good alternative to filter paper bioassay in such procedures, IJs emerging from cadavers are collected via White traps [36] in both tests and stored as aforementioned. Usually, the IJs in water suspensions are placed at room temperature (about 23 °C) for 12 h to acclimatize them prior to application. Abbott's formula has become a standard in such bioassay evaluations to correct the obtained data for control (untreated check) response (i.e., data are corrected in order to compensate for natural mortality). Accordingly, % corrected mortality = $[(X\text{-}Y)/X]$ *100 where X is the percent living in the check and Y is the percent living in the treatment [37]. A proper number of well-plate replicates should be used ($\geq$ 10 wells as replicates and repeated at least once to get accurate results).

13. The original prototype of LOTEK system [38] provides process technology for low-cost, high-efficiency mass production. The concentrated suspension is in a 10-cm-deep container and should be agitated between dips. LOTEK has a related conditioning apparatus for incubating trays of host cadavers at a high relative humidity without free water to optimize nematode development within the used insect hosts.

14. Sound sterilization with access to monoxenic culture IJs on an artificial media is the key to various methods for in vitro mass production of EPN-IJs. Useful recipes and commercial products for laboratory tests of EPN and their symbionts are found elsewhere [14, 20, 36].

15. Virulence bioassays are usually laid out in a completely randomized design with at least three replicates and may preferably be repeated at least once. Use proper arena to avoid crowd of the insects. The untreated checks are maintained by adding 0.5 mL distilled water on leaf disks to avoid drying. Other IJ concentrations can be determined according to experience with the used nematode species/strain. Software is available to calculate LC_{50} and LT_{50} values (e.g., https://www.aatbio.com/tools/lc50-calculator).

16. The IJs may be surface sterilized in 1% sodium hypochlorite for 10 min, and then rinse IJs thrice in SDW. The surface-sterilized IJs should still contain their symbiotic bacteria internally; IJs may be placed on nutrient agar for 2–3 days to verify that only the symbiont colonies are growing via definite characteristics of the intended bacterial species/strains and their colonies [11, 20, 36]. Distinguishing primary and secondary forms of the symbiotic bacteria may not be 100% accurate because of natural variation among the bacterial strains.

Acknowledgments

This work was supported in part by the US–Egypt Project cycle 17 (no. 172), and the NRC help offered via "In-house project" headed by Mahfouz Abd-Elgawad in research plan No. 13 Project No. 13050112.

References

1. Abd-Elgawad MMM, Askary TH, Coupland J (eds) (2017) Biocontrol agents: entomopathogenic and slug parasitic nematodes. CAB International, Wallingford
2. Abd-Elgawad MMM (2021) *Photorhabdus* spp.: an overview of the beneficial aspects of mutualistic bacteria of insecticidal nematodes. Plan Theory 10:1660. https://doi.org/10.3390/plants10081660
3. Abd-Elgawad MMM (2022) *Xenorhabdus* spp.: an overview of the useful facets of mutualistic bacteria of entomopathogenic nematodes. Life 12:1360. https://doi.org/10.3390/life12091360
4. Kaya HK, Gaugler R (1993) Entomopathogenic nematodes. Ann Rev Entomol 38:181–206
5. Lacey LA, Georgis R (2012) Entomopathogenic nematodes for control of insect pests above and below ground with comments on commercial production. J Nematol 44:218–225
6. Koppenhöfer AM, Shapiro-Ilan DI, Hiltpold I (2020) Entomopathogenic nematodes in sustainable food production. Front Sustain Food Syst 4:125. https://doi.org/10.3389/fsufs.2020.00125
7. Hussaini SS (2017) Entomopathogenic nematodes: Ecology, diversity and geographical distribution. In: Abd-Elgawad MMM, Askary TH, Coupland J (eds) Biocontrol Agents: Entomopathogenic and Slug Parasitic Nematodes. CAB International, Wallingford, pp 88–142
8. White GF (1929) A method for obtaining infective nematode larvae from cultures. Science 66:302–303
9. Nguyen KB, Hunt DJ (eds) (2016) Advances in Entomopathogenic Nematode Taxonomy and Phylogeny. Nematology monographs and perspectives, 12. Brill, Leiden, p 438
10. Spiridonov SE (2017) Entomopathogenic nematodes of the families Steinernematidae and Heterorhabditidae: Morphology and taxonomy. In: Abd-Elgawad MMM, Askary TH, Coupland J (eds) Biocontrol Agents: Entomopathogenic and Slug Parasitic Nematodes. CAB International, Wallingford, pp 45–62
11. Sivaramakrishnan S, Razia M (2021) Entomopathogenic nematodes and their symbiotic bacteria, springer protocols handbooks. Springer Science+Business Media, LLC, Berlin. https://doi.org/10.1007/978-1-0716-1445-7_2
12. Wheeler WC, Gladstein DS (1994) MALIGN: a multiple sequence alignment program. J Hered 85:417–418
13. Felsenstein J (1993) Phylogeny inference package. Department of Genetics, University of Washington, Seattle
14. Wouts WM (1981) Mass production of the entomogenous nematode Heterorhabditis heliothidis (Nematode: Heterorhabditidae) on artificial media. J Nematol 13:467–469
15. Dye DW (1968) A taxonomic study of the genus Erwinia I. The 'amylovora' group. N Z J Sci 11:590–607

16. Abd-Elgawad MMM (2020) Can rational sampling maximise isolation and fix distribution measure of entomopathogenic nematodes? Nematology 22(8):907–916. https://doi.org/10.1163/15685411000003350
17. Abd-Elgawad MMM (2021) Optimizing sampling and extraction methods for plant-parasitic and entomopathogenic nematodes. Plan Theory 10(4):629. https://doi.org/10.3390/plants10040629
18. Abd-Elgawad MMM (2017) Status of entomopathogenic nematodes in integrated pest management strategies in Egypt. In: Abd-Elgawad MMM, Askary TH, Coupland J (eds) Biocontrol Agents: Entomopathogenic and Slug Parasitic Nematodes. CAB International, Wallingford, pp 473–501
19. Wiesel L, Daniell TJ, King D, Neilson R (2015) Determination of the optimal soil sample size to accurately characterize nematode communities in soil. Soil Biol Biochem 80: 89–91. https://doi.org/10.1016/j.soilbio.2014.09.026
20. Woodring JL, Kaya HK (eds) (1988) Steinernematid and heterorhabditid nematodes: a handbook of techniques. Southern Cooperative Series Bulletin No. 331, Arkansas Agricultural Experiment Station, Fayetteville, Arkansas
21. Campos-Herrera R, Johnson EG, El-Borai FE, Stuart RJ, Graham JH, Duncan LW (2011) Long-term stability of entomopathogenic nematode spatial patterns measured by sentinel insects and real-time PCR assays. Ann Appl Biol 158:55–68
22. Dritsoulas A, El-Borai FE, Shehata IE, Hammam MM, El-Ashry RM, Mohamed MM, Abd-Elgawad MM, Duncan LW (2021) Reclaimed desert habitats favor entomopathogenic nematode and microarthropod abundance compared to ancient farmlands in the Nile Basin. J Nematol 53:1–13. https://doi.org/10.21307/jofnem-2021-047
23. Curran J, Gilbert C, Butler K (1992) Routine cryopreservation of Steinernema and Heterorhabditis species. J Nematol 24:38–45
24. Grewal P (2002) Formulation and application technology. In: Gaugler R (ed) Entomopathogenic nematology. CAB International, Wallingford, pp 265–288
25. Kaya HK, Nelsen CE (1985) Encapsulation of steinernematid and heterorhabditid nematodes with calcium alginate: A new approach for insect control and other applications. Environ Entomol 14:572–574. https://doi.org/10.1093/ee/14.5.572
26. Kim J, Jaffuel G, Turlings TJ (2015) Enhanced alginate capsule properties as a formulation of entomopathogenic nematodes. BioControl 60: 1–9. https://doi.org/10.1007/s10526-014-9638-z
27. Eisenback JD (1986) A comparison of techniques useful for preparing nematodes for scanning electron microscopy. J Nematol 18:479–487
28. Yushin VV et al (2021) Electron microscopy techniques. CABI Books. CABI International. https://doi.org/10.1079/9781786391759.0008
29. De Luca F, Abd-Elgawad MMM (2017) Molecular systematics and phylogenetic reconstruction of *Steinernema* and *Heterorhabditis*. In: Abd-Elgawad MMM, Askary TH, Coupland J (eds) Biocontrol agents: entomopathogenic and slug parasitic nematodes. CAB International, Wallingford, pp 143–153
30. Joyce CM, Steitz TA (1994) Function and structure relationships in DNA polymerases. Annu Rev Biochem 63:777–822
31. Yamamoto S, Harayama S (1995) PCR amplification and direct sequencing of gyrB genes with universal primers and their application to the detection and taxonomic analysis of *Pseudomonas putida* strains. Appl Environ Microbiol 61:1104–1109
32. Castaneda-Alvarez C, Prodan S, Zamorano A, San-Blas E, Aballay E (2021) *Xenorhabdus lircayensis* sp. nov., the symbiotic bacterium associated with the entomopathogenic nematode *Steinernema unicornum*. Int J Syst Evol Microbiol 71:005151. https://doi.org/10.1099/ijsem.0.005151
33. Tamura K (2007) MEGA4: molecular evolutionary genetics analysis (MEGA) software version 4.0. Mol Biol Evol 24:1596–1599
34. Brodsky LI (1992) GeneBee: the program package for biopolymer structure analysis. Dimacs 8:127–139
35. Brodsky LI (1995) GeneBee-NET: an internet-based server for analyzing biopolymers structure. Biochemist 60:1221–1123
36. Kaya HK, Stock SP (1997) Techniques in insect nematology. In: Lacey LA (ed) Manual of techniques in insect pathology. Biological Techniques Series, Academic Press, San Diego, pp 281–324
37. Abbott WS (1925) A method of computing the effectiveness of an insecticide. J Econ Entomol 18:265–267
38. Gaugler R, Brown I, Shapiro-Ilan D, Atwa A (2002) Automated technology for in vivo mass production of entomopathogenic nematodes. Biol Control 24:199–206

Index

Sergio Molinari (ed.), *Plant-Nematode Interactions*, Methods in Molecular Biology, vol. 2756,
https://doi.org/10.1007/978-1-0716-3638-1,

O

P

Q

R

S

T

V

X

MIX
Papier aus verantwortungsvollen Quellen
Paper from responsible sources
FSC® C105338

If you have any concerns about our products,
you can contact us on
ProductSafety@springernature.com

In case Publisher is established outside the EU,
the EU authorized representative is:
Springer Nature Customer Service Center GmbH
Europaplatz 3, 69115 Heidelberg, Germany

Printed by Libri Plureos GmbH
in Hamburg, Germany